W9-AFP-514

Antisemitism
in America

DS
146
.46
.D555

MAY 1 7 1995

488186

Antisemitism
in America

LEONARD DINNERSTEIN

New York Oxford
Oxford University Press
1994

Oxford University Press

Oxford New York Toronto
Delhi Bombay Calcutta Madras Karachi
Kuala Lumpur Singapore Hong Kong Tokyo
Nairobi Dar es Salaam Cape Town
Melbourne Auckland Madrid

and associated companies in
Berlin Ibadan

Copyright © 1994 by Oxford University Press, Inc.

Published by Oxford University Press, Inc.
200 Madison Avenue, New York, New York 10016

Oxford is a registered trademark of Oxford University Press

All rights reserved. No part of this publication may be reproduced,
stored in a retrieval system, or transmitted in any form or by any means,
electronic, mechanical, photocopying, recording, or otherwise,
without the prior permission of Oxford University Press.

Library of Congress Cataloging-in-Publication Data
Dinnerstein, Leonard.
Antisemitism in America /
Leonard Dinnerstein.
p. cm.
Includes bibliographical references and index.
ISBN 0-19-503780-4
1. Antisemitism—United States—History.
2. United States—Ethnic relations.
I. Title.
DS146.U6D555 1994
305.892'4'0973—dc20 93-31187

2 4 6 8 9 7 5 3 1

Printed in the United States of America
on acid-free paper

For Myra

Preface

This is the first comprehensive scholarly survey of antisemitism in the United States. There are several reasons for this omission from the historical literature. Ethnic history emerged as a significant subject of inquiry for American historians during the past generation. To be sure, monographs appeared intermittently on ethnic groups but by and large not until the 1950s, and especially the 1970s, did the field begin to attract serious scholarly attention. Antisemitism, an aspect of American Jewish history, falls within that area. Although several American Jews have written about the "contributions" that Jews made to the development of the United States, these tracts were more concerned with showing why Jews should be accepted by other Americans. They made little effort to analyze carefully the complexities of the Jewish experience in America.

In 1950 Bertram Korn published an analysis of American Jewry during the Civil War and included chapters on antisemitism in the North and South during the conflagration. This was the first serious attempt by a professionally trained American historian to investigate the experiences of American Jews between 1861 and 1865. Also during the 1950s three other inquiries into American antisemitism, by Oscar Handlin, Richard Hofstadter, and John Higham, focused on events in the late nineteenth century as being crucial to the development of American antisemitism (Higham did note that economic and religious stereotypes of Jews were apparent as early as the Jacksonian era). The works of these last three scholars spawned several additional studies focusing on whether Populist ideology and rhetoric, which had specifically been singled out by Hofstadter, were in fact the sources of twentieth-century animosity toward Jews. Although newer inquiries tended to defend Populists from the charge, Irwin Unger's *The Greenback Era*

(1964), which exposed the antisemitic tendencies of some silverites and agrarians during the Gilded Age, and Michael Dobkowski's *The Tarnished Dream* (1978), which conclusively proved that the depth and breadth of American antisemitism from the 1870s through the 1920s was much greater than any American historian had heretofore suggested, indicated that further historical assessments of American attitudes toward Jews were necessary. Major efforts to meet this need began in the late 1960s.

The Civil Rights movement of the 1960s prodded scholars to examine histories of various minorities in the United States. I began my own inquiries during that decade with a study of Leo Frank, a Jew lynched in Georgia in 1915 for a crime he did not commit. Building on my study of the Frank case, I then investigated the history of other Jews in the South, explored the circumstances of a variety of other American ethnic groups, and then returned to American Jewish experiences when I wrote about the American reaction to helping survivors of the Holocaust. What I discovered during the course of that work was that American Jewish history is difficult to interpret and understand without recognizing how antisemitism affected the course of Jewish development.

Antisemitism has existed throughout American history. However, it has manifested itself in a variety of ways, waxing and waning according to other factors in society at any given time. Only after World War II did it emerge as a subject that most American Jews were willing to have explored. Many Jews, in fact, sensitive to the reaction of other Americans toward them, feared that a discussion of antisemitism would exacerbate its effects rather than lead to an understanding of how serious and humiliating bigotry actually was. Then came a stream of scholarly analyses of the American Jewish experience that were neither apologetic nor filiopietistic but brought a new scholarly perspective to hitherto overlooked or underdeveloped themes in American Jewish history. A list of the most salient of these works must include Ronald Bayor's *Neighbors in Conflict* (1978), Jerold Auerbach's *Unequal Justice* (1976), Deborah Dash Moore's *At Home in America* (1981), Morton Borden's *Jews, Turks, and Infidels* (1984), Jonathan Sarna's *Jacksonian Jew* (1981), and Naomi W. Cohen's *Encounter with Emancipation* (1984), although there are several others that are also significant contributions. Louise Mayo specifically focused on nineteenth-century antisemitism with *The Ambivalent Image* (1988) while Judith Endelman's *The Jews of Indianapolis* (1984) is among the best local studies providing insights into the American Jewish experience in places far from major ethnic concentrations. David Gerber put together a collection of original essays on the topic of *Anti-Semitism in American History* in 1986 and John Higham rethought, refined, and republished his observations on the subject in his collected articles under the title *Send These to Me* (1975, 1984).

By the 1980s enough literature was available to begin synthesizing various facets of the American Jewish past, and I embarked on my own overview of antisemitism from the colonial era to the present. While this work is not intended to be definitive, and does not explore in depth various religious, literary, psychological, sociological, or philosophical aspects of antisemitism, it does attempt to provide a comprehensive narrative of the subject. The focus is on mainstream, rather than extremist, antisemitism. What most responsible Americans thought about Jews was much more important in developing national attitudes than what members of the "lunatic fringe" believed or preached. With a topic as daunting as antisemitism, no one book could possibly hope to explore all its many facets.

To clarify readers' understanding of my use of terms, let me state that I use the word antisemitism in this narrative to denote hostile expressions toward, or negative behavior against, individuals or groups because of their Jewish faith or heritage. Prejudice reflects antagonistic thoughts but when those ideas are put into actions that restrict or condemn Jews they become forces of discrimination. Sometimes antisemitism has been blatant and unadulterated; on other occasions it has been part of a broader nativist wave that targets many outgroups.

Simply put, Christian viewpoints underlie all American antisemitism. No matter what other factors or forces may have been in play at any given time the basis for prejudice toward Jews in the United States, and in the colonial era before it, must be Christian teachings. The Christian heritage has been a powerful one, and, as the Prologue details, goes back for almost two millennia. No explanation of antisemitism in the United States can ignore this salient fact. Moreover, Christian culture so permeated American society that even the unchurched or those with the most tenuous ties to a religious organization still picked up popular attitudes.

The United States may be said to be a derivative of England, and to a great extent it is, but it has always been a polyglot nation as well. Scots-Irish, Germans, Swedes, Scots, Huguenots, and, of course, American Indians and African blacks populated colonial America. This nation's tradition glories in its multi-ethnic background but at almost every point in American history Caucasian newcomers were encouraged to shed their Old World characteristics and adopt those of the dominant culture. And most European Christians and Jews did so. Sometimes it took three or four generations to accomplish that goal but even in our own day there are pockets of Americans who nurture aspects of an Old World civilization out of harmony with mainstream American values. In Fourth of July speeches we honor these people but most of their neighbors resent existing differences. For all of American rhetoric glorifying multi-ethnic backgrounds, those who refuse to relinquish a different life style are not easily accepted in this country. Immediately the Amish of Pennsylvania and the Hutterites of North and

South Dakota and Montana come to mind but throughout the country there are small groups that do not quite fit in. The problems faced by American Indians, African Americans, and Asians, who are perceived as racially different, have been much more severe than those of Caucasians.

Until after World War II, most Americans thought of their country as a Protestant one and Christian generally meant Protestantism not Catholicism when applied to the United States. Therefore individuals and groups of Catholic and Jewish backgrounds were often victims of antagonism from others even though legally, especially since the nineteenth century, they had equal rights and opportunities. This created a paradox. While by an absence of restrictive legislation all white men had equal opportunities in the United States, most Protestants regarded Catholics and Jews as inferior and adherents of these faiths often perceived themselves as outsiders in a Protestant nation, at best tolerated but not embraced. Some individuals of these faiths always found a place for themselves among the Protestant elite. Many even prospered economically. As the numbers of Catholics and Jews increased in this country, however, and there were always at least ten times as many of the former as there were of the latter even in colonial America, until the twentieth century Catholics were more intensely abhorred than Jews were.

But the Catholic psyche was different from that of the Jews. In many European countries Catholics constituted the majority and their Pope, who reigned over Catholics throughout the world, was regarded as a head of state by most European governments. Jews, on the other hand, were always despised minorities in Europe who suffered legal disabilities in one country after another even when allowed to dwell there and even when individual Jews achieved outstanding social or economic recognition. Thus Jews, no matter how successful, were always wary that some crisis or other in a given nation would lead to contraction of rights, severe restrictions, or even expulsion. This was the Jewish experience in Europe since the end of the eleventh century.

Nevertheless, the United States offered opportunities that Jews had never before experienced in a Christian nation: a constitution that provided full legal equality for men and prevented the national government from showing favor or discrimination based on one's religion. Of course in actuality that equality rarely exhibited itself in the United States until after World War II when antisemitic manifestations, except from African Americans, followed a downward course. Even today some, although few enough to be generally overlooked, social disabilities still exist. On the other hand, legal equality went against the grain of Christian teachings and centuries-old folklore that perpetuated and elaborated on Christian attitudes toward Jews. American Christians periodically shunned or attempted to convert Jews and sought changes in the American constitution that would have recognized the divinity of Jesus, the appropriateness of maintaining a Christian sabbath by law, and allowing passages from a Christian Bible to be read in public schools.

At the same time, however, countervailing American traditions dictated acceptance of white people and thus two distinct conventions developed nationally and locally. Jews would be allowed to live in American communities, go their own ways, and frequently be accepted as worthy members of a community. On the other hand, as Jews they would always be a people apart and in times of economic crises, personal anxiety, or societal stress would often be regarded as causes of whatever unrest existed. The beliefs and behavior of community leaders usually set the tone for what others believed and how they behaved even though there was not always a direct correlation between the one and the other. Thus it was the pronouncements of the religious hierarchy in Europe that over the centuries affected the feelings of nobles and peasants toward Jews while in the United States spokesmen for various religious dominations prescribed the actions of other Americans toward Jews. Again, even when there was no obvious or direct relationship between pronouncement or edict, some Jews sensed the impact of church teachings in Christian perspectives and demeanor toward them. And it was often true that the thoughts and values of the elites, whether religious, economic, or social, found expression, sometimes in a twisted fashion, in the conduct of the middle and lower classes.

Thus Jews as a group, despite their opportunities in the United States, never quite relaxed and always kept a watchful eye open for Christian bias. Such prejudice was not uniformly exhibited and it often depended on historical circumstances and the strengths or trials of distinct Christian groups at different times in history as to how the beliefs would be exercised. Sometimes numbers made a difference; when Jews were strong in number they often felt more secure and comfortable. Other times local values dictated their reception and demeanor. In the American South after the Civil War, for example, where fundamentalist Protestantism held greater sway than in any other area in America, regional mores mandated surface cordiality and extreme conformity in deportment, Jews ostensibly got on well with their neighbors but lived with a sense of discomfort and unease not felt by their coreligionists elsewhere in the country. In ordinary times, when significant strains did not develop, one saw a cordiality and a distant, but not social, acceptance equal to, or better than, that witnessed in other parts of the country. But when southerners felt aggrieved because of economic crises, violations of social mores, or the civil rights struggles in the middle of the twentieth century, Jewish anxieties reached new heights as they witnessed the savage excesses that some of their fundamentalist neighbors engaged in.

Thus writing about antisemitism in the United States is not an easy matter nor can facile generalizations be made because Christian thought and Christian actions did not always jibe, historical circumstances altered the relationship between one and the other, and a legal tradition, which eventually developed into a stronger societal edict, preached tolerance and acceptance. Moreover, at different times and in different parts of the United States American attitudes not only toward religious minorities but toward racial and ethnic minorities as well modified and complicated responses to Jews.

After World War II a concerted effort was made in the United States to eliminate prejudice and discrimination. Ethnic defense agencies, primarily the Jewish ones, endeavored to have state legislatures outlaw discriminatory policies while they propagandized that bigotry subverted democracy. President Harry S Truman appointed major commissions to investigate higher education and civil rights in the United States and the reports of those commissions pointed out how discrimination curbed opportunities for minorities. Truman lent his weight to proposed changes and the U.S. Supreme Court also came forth with a number of decisions protecting minority rights. A culture began to develop in the United States that this nation, in thought and deed and by legal means where necessary, should promote tolerance. Gradual changes then began to occur, and, since the 1960s, massive exertions have been made in both the public and private sectors of the nation to achieve this goal. These endeavors proved successful. Not only has prejudice toward all groups declined since 1945, but Jews, more than any other identifiable group, have been the major beneficiaries as educational, employment, housing, resort, and recreational opportunities opened up for them. Paradoxically, as antisemitism declined among white Americans it increased in intensity among African Americans. Chapter 10 analyzes the basis for this discrepancy.

There is also one aspect of white antisemitism that, although diminished, seems most difficult to eradicate. And that is the general American belief that the Christian faith is superior to all others, that Jews stubbornly refuse to accept the truthfulness of Christianity, and that until they do—whether in Europe or in the United States—as a group they can never be given the respect that Christians receive. No matter how it is expressed on either continent, Jews know this, Christians know this, and Christian behavior toward Jews is often predicated upon this understanding.

In the past, religious teachings scorned Jews, folklore in Christian societies perpetuated this contempt, and children imbibed this lesson early and well. Living in a Christian society, whether one was or was not religious, meant absorbing antisemitic attitudes. And although expressions of such thoughts have been minimized in recent decades (except in Europe where they appear to be resurfacing) and Christian theologians have called for reexamination of church teachings and their impact on Christian attitudes toward Jews, both Jews and Christians have not been totally oblivious of such perspectives.

It cannot be emphasized too strongly that all aspects of American antisemitism are built on this foundation of Christian hostility toward Jews. To argue otherwise is to misread history. I have found no other explanation for the causes of antisemitism satisfactory. Economic and status anxieties and competition, hatred of the city, and maladjusted personalities have been some of the other reasons that have been given in the past to explain the development of antisemitism. But there are well-adjusted individuals whose economic security and status in society have remained unchallenged, and whose affection for city life is not tarnished, who have been antisemites. Antisemitic feelings and behavior can be exacerbated by a variety of influences; but without the underlying base of deeply ingrained and culturally accepted Christian teachings to build on, none of the other factors would explain a prejudice that has lasted for almost two thousand years.

And, in fact, the Americans who have stood apart as they expressed their venom toward Jews since the 1960s are the African Americans, perhaps the most church oriented and religiously influenced of all ethnic groups in the United States. No other American people have as prominent religious spokesmen who also serve as community and political leaders. To be sure, the greatest of all black leaders, Martin Luther King, Jr., was a force for unity and understanding among peoples, and one of the most articulate of the black antisemites, Louis Farrakhan, is not even a Christian although he speaks in the Christian idiom; nonetheless, the early influence of church teachings and the devotions of so many African Americans to Protestant theology must be acknowledged as the bedrock that has caused antagonism toward Jews. And just as with other American Christian groups, all other factors, real or alleged, conscious or unconscious, have been built on this base.

A work such as this cannot be done alone. During the course of my research and writing I have received assistance and encouragement from a wide variety of people. It is a pleasure to acknowledge them at this time. Dave Reimers and Bob Schulzinger read all the chapter drafts, provided powerful critiques and suggestions, and kept my spirits up when they flagged. George Lankevich did a superb job of editing a revised version. And my editor, Nancy Lane, gave much greater assistance than any author has the right to expect. Other friends and/or scholars who read one or more chapter drafts and gave me the benefits of their insights include Alan Bernstein, Seymour Drescher, Larry Glasco, Cheryl Greenberg, Mark I. Greenberg, William B. Helmreich, John Higham, Fred Jaher, Brian McKnight, Deborah Dash Moore, Roger L. Nichols, Susan Reverby, Milton Schain, Michael Schaller, and

Harvard Sitkoff. Alison Futrell shared her knowledge of the Roman Empire with me. Marilyn Heins and Milton Lipson carefully went over my paragraphs on Grosse Pointe and discussed their recollections of that community; Susan Roth recalled her experiences of growing up Jewish in North Carolina; Roger Daniels and Walter Ehrlich shared the fruits of their research; Eric Rothschild enlightened me about Scarsdale; Fred Byrne allowed me to read his essay on restrictive housing in Bronxville, and Donald Yacavone periodically sent references about antisemitism in nineteenth-century America and facilitated my use of the Massachusetts Historical Society. Margie Fenton of the Jewish Federation of Southern Arizona periodically sent me materials of interest. Naomi W. Cohen and James F. Moore also provided helpful nuggets; in numerous conversations about the manuscript Kate Wittenberg gave warm support.

During the course of my research I used several archives and libraries and the staffs of the 42nd Street Library in New York, especially the Jewish division, Columbia University Libraries, and the Library of Congress were always helpful. In New York I also had the use of the American Jewish Committee's Blaustein Library, which is a joy to work in. Not only does is it provide a gold mine of information on the American Jewish experience in the twentieth century, but its head, Cyma Horowitz, and her associate, Michele Anish, are extremely knowledgeable and go to extraordinary lengths to make scholars welcome. One of the other delights of my many visits to the American Jewish Committee was afternoon tea with archivists Helen Ritter and Ruth Rauch. Professionals at the University of Arizona Library have, of course, been put to the greatest test by my many demands and they have come through with flying colors. Ruth Dickstein, in particular, went out of her way to be helpful and saved me weeks of time; Andrew Makuch has always moved with dispatch whenever I requested additional library purchases.

I have also been helped by generous financial support from the National Endowment for the Humanities, the University of Arizona's Social and Behavioral Sciences Research Institute, the Franklin and Eleanor Roosevelt Institute, and the American Jewish Archives where I was the recipient of a Bernard and Audre Rapoport fellowship in 1989–1990. While I was working at the American Jewish Archives Jacob R. Marcus was extremely cordial and Abe Peck and Kevin Proffitt extended every courtesy. Shelly Newman, whom I met at a conference in Jerusalem, graciously purchased some needed research volumes.

The staff at the University of Arizona History Department has also been particularly helpful. Pat Foreman skillfully formatted the final manuscript while Mary Sue Passe and Jim Lombardo worked on earlier versions. A myriad number of secretarial services were also cheerfully provided by Monica Theis and Donna Watson. Graduate research assistants who helped ferret out material included Walter Rusinek, Virginia Scharff, Angel Cal, Melissa McKinnon, and Susan Hill.

My loving family, including my wife Myra, my son Andrew, and my daughter Julie, provided warm emotional support and allowed me to discuss various facets of the work with them. This book is dedicated, of course, to the woman who shares my life and makes everything worthwhile.

Tucson   L.D.
August 1993

Contents

Prologue
The Christian Heritage

Antisemitism is a real and ignoble part of America's cultural heritage. It was brought to the New World by the first settlers, instilled by Christian teachings, and continually reinforced by successive waves of Protestants and Catholics who populated American shores. Like a genetic disease, it has been transmitted from one generation to the next but, like a folk tale, it has been added to, transformed, and adapted to particular times, places, and circumstances. It has been present in Christian societies for almost two millennia but its manifestations have varied according to historic circumstances. One or more of the following symptoms, however, is used as an excuse whenever the disease occurs: the Jew is the Christ-killer, the economic exploiter, the eternal alien, the subversive element within Christian civilization, or the embodiment of evil.

Although hostility toward Jews predates the Christian era, it is the Christian image of the Jew that has dominated the consciousness of the western world. That image took over a thousand years to draw, and another three or four centuries to complete the most extraordinary details. By the middle of the fourteenth century the perception of the "perfidious" Jew had been so thoroughly embedded in the mind of Christian Europe that seven hundred years later its essential ingredients still remain in place. And across two millennia animosity toward Jews was not only encouraged but also lauded as a religious virtue. Indeed, the "alien race's" continuing resistance to Christian missionary activity was a major factor fanning hatred of Jews. Their refusal to acknowledge that Judaism had been "fulfilled" in the person of Christ, and hence made obsolete, stood as a constant rebuke to Christians everywhere.[1]

Conflicts between Christians and Jews evolved in the first century between those who believed Christ and those who did not. In the Acts of Apostles, Jews are portrayed as vicious murderers and on numerous occasions in the New Testament Jews are mentioned as continually plotting the deaths of Christians. While writings by Jews about Christians showed marked indifference to the new cultists, Christian writers were full of malice toward Jews for their alleged persecution of Christians. Accusations of mistreatment of Christians by Jews had no basis in fact but the opposite is not true.[2] Nonetheless, Christian writers continually attacked the Jews.

The fourth century proved crucial in determining the fate of Jews. In 337 Emperor Constantine converted to Christianity on his deathbed, thereby opening the door for the eventual establishment of that faith as the religion of the empire. Christianity spread throughout Europe during the next several hundred years. England became Christian in the seventh century, Saxony and Bohemia in the ninth, Scandinavia and Poland in the tenth, and by the eleventh century the Christianization of Western Europe was largely complete. As Christianity successfully spread, and allegedly indicated the will of God, the antisemitic tone of tracts and sermons increased. The question most frequently posed was why were Jews so reluctant to admit the verdict of history and accept the truthfulness of Christian teaching? A Christian approach developed that branded Jews as stubborn enemies of the human race whose only hope for salvation lay in conversion. In the meantime, God would punish them for their obstinacy, force them to wander the earth, and make them bear witness to the triumph of Christianity.[3]

In the fourth century Jews were depicted by writers as monsters with superhuman cunning and malice, and in league with the devil.[4] Greek theologian St. John Chrysostom (354–407), who until the advent of Martin Luther in the sixteenth century stood "without peer or parallel in the entire literature *Adversus Judaeus*,"[5] wrote that the Jews "have all the vices of beasts and are good for nothing but slaughter. . . . They behave no better than pigs in their lewd vulgarities."[6] St. Ambrose thought the burning of a synagogue would be "pleasing to God,"[7] while St. Augustine, perhaps the most influential theologian of his time, argued that the Jews had always been a wicked people and could never have been God's favorites.[8] Augustine thought Jews should be allowed to live in Christian societies so that their degradation would serve as a reminder to others of the consequences of "sinful" behavior. "Slay them not," he argued, "lest my people [Christians] forget; make them wander to and fro."[9]

Augustine did not inaugurate a new message. By the fourth century the populace had already accepted the idea of the reprobate status of the Jews. It was also universally assumed in the Christian world that

Jewish contemporaries of Jesus had murdered Him and that for all time thereafter their descendants would be held collectively responsible for His death.[10] This idea may have come directly from Matthew who wrote about how Pilate had given them the choice of releasing Jesus or the notorious criminal, Barab'bas, from prison and the Jews had demanded freedom for the latter. "Then what shall I do with Jesus who is called Christ?" Pilate asked. "Let him be crucified," Jews allegedly responded. Then, according to Matthew, Pilate replied:

> "I am innocent of this man's blood; see to it yourselves." And all the people answered, "His blood be on us and on our children!"[11]

No idea in Christian teaching has been more solidly implanted among adherents of the faith, and more devastating to Christian-Jewish relations, than the accusation that the Jews had killed their Savior.[12] For this alleged crime the Jews have supposedly been rejected by God and doomed to eternal punishment. Characterized in church teachings as a "demonic" and "accursed" people, only by embracing Jesus would there be any chance for acceptance and forgiveness.[13]

Thus, by the sixth century, the basis for Christian hatred of Jews had been firmly implanted. Nonetheless, Pope Gregory I, who reigned from 590 through 604, forbade persecution of the Jews and during the next five hundred or so years life for European Jews was not particularly uncomfortable except in Visigothic Spain where royal edict mandated forced conversion of Jews in 613.[14] After taking the reins of power in England in 1066, William the Conqueror imported Jews to the realm to serve, along with others, in setting up a new regime.[15] Despite numerous humiliations, restrictions, and second-class citizenship, they remained there until King Edward I banished them from the kingdom in 1290. The expulsion of Jews from medieval England reflected the triumph of the antisemitism that was always latent in the Christian world.

The almost three-hundred-year era of savaging Jews began at the end of November 1095 in Clermont, France, when Pope Urban II called for a Christian crusade to recover the Holy Land from the infidels. The first crusaders embarked in the summer of 1096, as Christian soldiers headed for the Holy Land. Along the way they zealously vented their wrath on the Jews in Europe. During the eleventh century the Jews had become increasingly unpopular as larger numbers of people borrowed money from them and found difficulty repaying the loans. The Crusades added to this resentment. It was costly for knights to outfit themselves for the march to Jerusalem and they again had to borrow from Jews. Therefore, there was little opposition to slaughtering those who not only aroused resentment because of their failure to embrace Christianity but who were also seen as the cause for trying economic circumstances. The first massacre of Jews by the Crusaders occurred in Rouen, France, and as the Christian armies proceeded through Europe they decimated Jewish communities in their path. Massacres of nonbelievers would be the rule, rather than the exception, so long as Christ's soldiers strove to implant Christianity on the known world.[16]

For the next two to three hundred years periodic crusaders marched through Europe on their way to "liberate" Palestine for the glory of God. The effects proved devastating. As Leon Poliakov, the French historian of antisemitism, has written:

> each time medieval Europe was swept by a great movement of faith, each time the Christians set out to face the unknown in the name and love of God, hatred of the Jews was fanned into flame virtually everywhere. And the more the pious impulses of the heart sought satisfaction in action, the worse became the Jews' lot. Virtually every time a Crusade was preached, the same consequences could be anticipated.[17]

The rabble and peasants who preceded or accompanied the Crusaders were the ones who actually annihilated the nonbelievers, Jews in their midst. Theologian Rosemary Ruether has written that it seemed preposterous to the lower classes to embark on a crusade to kill God's enemies far away when His greatest foes dwelled among them. Leaders of the Church could not understand, she continued, "that the mob merely acted out, in practice, a hatred which the Church taught in theory and enforced in social degradation wherever possible."[18]

One of the most bizarre accusations against Jews that took hold during the years of the Crusades was that they sacrificed Christian children for religious purposes. This belief had its historic origins in fifth-century tales that a group of Jews had tortured and murdered a Christian child to mock Christ.[19] Not until the twelfth century, however, did that story reappear. Then, in 1144, William of Norwich died of a cataleptic fit. Theobald of Canterbury, a Jew who had converted to Christianity, spread the word that Jews must have been responsible for the deed since the Jewish faith required the sacrifice of a Christian child annually "in scorn and contempt of Christ, so that they might avenge their sufferings on Him."[20] At the time of William's death few people accepted this ludicrous and inaccurate explanation. Nonetheless, the libel circulated and as the decades passed more credence was given to it. In England, ritual murder accusations were aimed at the Jews of Gloucester in 1168, of Bury St. Edmunds in 1181, and of Bristol in 1183. The same accusations were hurled at French Jews in Blois in 1171, at Jews in Erfurt, Saxony, in 1191, and later on in Paris, Weissenburg, Salzburg, Bern, and Seville. All told there were at least six such accusations in the twelfth century, fifteen in the thirteenth, ten in the fourteenth, sixteen in the fifteenth, thirteen in the sixteenth, eight in the seventeenth, fifteen in the eighteenth, and thirty-nine in the nineteenth centuries. Similar charges erupted in both Europe and the United States in the twentieth century.[21]

The most notorious of all ritual murder allegations occurred in

England following the death of Hugh of Lincoln in 1255.[22] After a woman found the body of her eight-year-old child in a well a Jew who lived nearby was arrested and tortured into confessing his culpability. On the basis of this "confession," he and ninety-two other Jews were imprisoned and had their possessions confiscated. Eighteen of them were hanged. Throughout the village of Lincoln people first whispered that Hugh had been tortured and crucified but as the story passed from one lip to another it was further embellished. The boy had supposedly been stolen away, his body fattened on white bread and milk for ten days, and then slaughtered so that his blood could by used by Jews for ritual purposes. The alleged crime had occurred near Passover and Easter, and over the years rumor had it either that during Passover week the Jews crucified Christian children to reenact the execution of Jesus, or that the child was killed so that Jews might use his blood in the food for their Passover service. "It never seemed to occur to the accusers," Father Edward Flannery has written, "that the doctrine of the transubstantiation was a dogma completely alien to Jewish monotheism."[23] Other contemporary explanations for the alleged ritual murders concluded that they occurred because Jewish men menstruated and needed to refill their blood, or because Jewish boys lost blood while being circumcised and as men they needed to replenish it.[24] The blood libel/ritual murder accusation has always been condemned as fantasy by church theologians and scholars, and by Popes Gregory X, Innocent IV, Clement IV, Innocent V, Mark V, Nicholas V, Paul III, Benedict XIV, Clement XIII, Clement XIV, and most recently by Pope John XXIII. Yet the idea maintained a life of its own and many Christians clung to its authenticity as justification for their most unchristian acts.[25]

Popes, like other authority figures, learned that no contrary explanations can shake something from peoples' minds if they want to hold on to it.[26] The ritual murder tale satisfied some deeply felt popular need and it spread from community to community, from century to century, modified, distorted, and horrifying. It received its most polished presentation in Chaucer's "Prioress's Tale," and has come down through the years in folk ballads known variously as "Sir Hugh," "The Jew's Daughter," or "It Rained a Mist."[27] The New York *Tribune* reported in 1902 that students of old English ballads assumed that the theme of ritual murder dated back almost two thousand years and was a "relic of a cruel and wicked superstition. A superstition that seems to be ineradicable."[28] In the numerous stories, songs, and poems that perpetuate this fable, the Jew, or his daughter, always has a knife, and circumcises, castrates, or murders the Christian child, who is unwittingly enticed before being slaughtered. A verse in one of these variations runs:

> She laid him on a dressing table
> And stichit him like a swine.[29]

Folklorist Alan Dundes has written that although "there isn't a shred of evidence whatsoever to indicate that Jews ever killed Christian children to obtain blood for sacrificial or ritual purposes," the legend developed out of the "Christian need for a scapegoat." Dundes explained that Christians blamed Jews for something that they psychologically needed to have happened. Christians decided that someone had to be guilty of killing Christ. The Jews became the culprits and have been scapegoats for Christian society ever since.[30]

Another seemingly ineradicable image dating back to medieval times is that of the Jew as economic exploiter. As part of the economic growth that made the crusades possible, money became an essential ingredient in the new European economy. Full-time bankers were necessary to lend and provide funds to kings and princes, merchants and peasants, yet the Church forbade Christians to engage in that "accursed" business; it was left to Jews to fill the need. Those already "accursed" for the rejection of Christ became even more despicable for being engaged in a sinful and disreputable occupation.[31] Although the vocation of banking was necessary, and many Christians practiced it, the activity aroused the ire of almost the entire populace. Borrowers saw lenders not as valued members of society but as usurers who charged too much interest.[32] Since usurers charged rates that borrowers often found difficult to repay, lenders incurred the rancor of both debt-ridden peasants and impoverished noblemen. The image of the Jew as usurer, therefore, combined with the metaphor of "bloodsucking" and coalesced with the blood-libel parable to complete the picture of Jews as parasites on society, performing no useful function but draining the blood out of the hardworking Christian peasantry.[33] As Joshua Trachtenburg has indicated, the usurer became "one of the most thoroughly despised and hated members of the medieval community."[34] The popular attitudes toward the detested money lender were symbolized by Shakespeare's depiction of the infamous Jew "Shylock," a man totally consumed with receiving what was owed to him even if it be a pound of Christian flesh.[35] Since the Middle Ages, the Jew has borne the image of economic exploiter grinding down the penitent Christian.

Also during the Middle Ages a strengthened Papacy institutionalized what would later be considered appropriate behavior toward those who lived among them. Pope Innocent III, the most powerful of the medieval Popes, presided over the fourth Lateran Council in 1215 and had the bishops decree that non-Christians must wear distinctive garments. Jews specifically were prohibited from appearing in public during Holy Week because their usual dress was interpreted as mocking the Christians clad in mourning. Decrees of the Council reemphasized the separateness of the Jews; their lower status in society allowed humiliating regulations to be imposed on them. Jews possessed some rights, albeit of a subordinate nature, but Innocent III stated that they were only for those "who have not presumed to plot against the Christian faith."[36] Innocent opposed the killing of Jews "against whom the blood of Jesus Christ calls out," and, echoing St. Augustine, stated that "as wanderers must they remain upon the earth, until their countenance be filled with shame and they seek the name of Jesus Christ, the Lord."[37] In 1239 Pope Gregory IX concluded that following Talmudic forms was the chief cause of the Jews' "perfidiousness," and within three years Parisians began burning Talmuds.[38] The cumulative effect of these allegations was that Europeans increasingly saw Jews as subversive elements in their midst, as agents of Satan intent on destroying Christian civilization.

The onset of that catastrophic European event in 1347, the plague, or "Black Death" as it was known, sealed the fate of the Jews in Christian minds. Within three years the "Black Death" combined with widespread famine and ultimately killed about one-third of western Europe's population, unhinged people's minds, aroused expectations of the apocalypse, and fostered popular obsession with the devil and his alleged servants, the Jews. Obviously some rationale had to be found to comprehend this horrible and apparently unending punishment from God. No phenomenon of such magnitude could be explained away without someone being responsible and Jews, believed to have rejected God and acting for the Devil, were the answer. Stories circulated that a Jewish conspiracy aimed at Christians existed and that Jews, either alone or in connivance with lepers, poisoned the wells in an effort to exterminate Christians. Ignoring the numbers of Jews who fell victim to the plague, peasants asked why else would Jews live and Christians die? Having answered their own question, Christians embarked on pogroms that led to the arrest, torture, and killing of Jews throughout Europe. In 1348 virtually every European Jewish community, save those in Vienna and Regensburg, was attacked. Every Jew in one French town was burned at the stake and over 200 Jewish communities in Europe were totally destroyed.[39]

Despite the belligerence that Jews encountered everywhere in medieval Europe, they always found a place to dwell because after the "Black Plague" one town or another needed the economic skills that many Jews allegedly possessed and rulers recalled them. Never much liked, they remained under the protection of the sovereign until, having served their purpose, they were expelled and forced to find residence elsewhere. In the German parts of the old Holy Roman Empire ghettos were established for Jews. The gates were locked at night but the residents were permitted to walk on Christian streets during the daytime.[40] On the other hand, as the mercantile age in Europe unfolded, especially between 1570 and 1713, the fortunes of many Jews in western Europe improved. During the Puritan interlude in England, from 1649 to 1661, Oliver Cromwell invited Jews to return to that country; the seventeenth century also found them in Holland. The perceived commercial skills of Jews made them valuable residents during the years of European expansion.[41]

The Protestant Reformation of the sixteenth century revolutionized religious practice in northern and western Europe but did little to alter popular attitudes toward Jews. Martin Luther at first tried to befriend Jews, but after they spurned his offer of love and conversion he, like both predecessors and successors, savagely turned against them. His venom then knew no bounds. Jews, he alleged, are foreigners, they do not work, yet they keep our money and have become our masters.[42] "Know my dear Christians," he wrote in 1542, "and do not doubt that next to the devil you have no enemy more cruel, more venomous and virulent, than a true Jew." Luther's heartless words have been studied continually since the sixteenth century and, as historian Heiko Oberman has pointed out, it is important to note that they were similar in tone and content to the writings of his contemporaries.[43] Calvin also thought Judaism a creed riddled with errors, but acknowledged that this faith prepared the world for Christianity.[44] In both Catholic and Protestant Europe, Jews found few friends—they remained always the alien, the other.

The views long enunciated by their leaders affected Christians everywhere. There was no need for independent thought when everyone and every institution agreed on what Jews were like. "If it is the part of a good Christian to detest the Jews," Erasmus, the Biblical scholar who lived in the Renaissance era, observed, "then we are all good Christians."[45] To function like a Jew had a negative connotation; to act in a Christian manner had a positive association. Thus the word "Jew" was always used to denigrate an individual or an activity, and continued to convey that meaning well into the twentieth century. Except when used by Jews themselves, no virtuous deed or act was associated with that term.[46]

Many of the most destructive modern images of, and visceral reactions to, Jews developed between the eleventh and fourteenth centuries. Virtually every affliction of those troubled centuries was attributed to Jews. During that period no crime was too fanciful to ascribe to them as they were accused of witchcraft, devil worship, and magic. They allegedly used sorcery, poisoned wells, tortured Christian children, and sacrificed unbaptized infants to Satan. In tapestries, drawings, and other forms of art Jews were pictured as having horns and sometimes a tail; Jewish males sported the beard of a goat, a particularly lecherous beast. They were made to seem inhuman because that is how many Christians viewed them.[47] Peter the Venerable of Cluny, who lived in the early decades of the eleventh century, doubted "whether a Jew can be human, for he will neither yield to human reasoning, nor find satisfaction in authoritative utterances, alike divine and Jewish."[48] People in the medieval world also believed that if the Jews were allowed to do so they would destroy Christian civilization. This theme was never far from the surface in subsequent centuries and in the twentieth *The Protocols of the Elders of Zion* spelled out and exaggerated the fears that millions of people in both the Christian and Moslem worlds had always harbored.[49]

On the stage, in literature, and in philosophical discourse Jews were thought of as strange and almost barbarous. They were portrayed with deformed bodies and misshapen character. Words like crafty, cunning, and sinful were associated with them. In medieval religious dramas Jews were false, felonious, wicked, perverse, disloyal, villainous, or traitorous. They were the enemies of Christ, either diabolical tempters or sorcerers trying to undermine Christians. Depictions showed them as physical cowards, materialists, concerned with money, clannish, secretive, and servile but always hostile to Christians. The Jew had a big nose, a bad odor, and a speech impediment.[50] In her study of Jews as they are delineated in drama, Ellen Schiff wrote:

> the Jew is cast everywhere along the gamut of malevolence, from the simple foil who provokes laughter by ludicrous contrast with the faithful, to the wicked perpetrator of cruelty conceivable only by a mind possessed of the devil.

The Jew, she added, had become "an increasingly spectacular personification of evil."[51] The allegedly demonic Jews also found themselves the subject of "learned" discussions as to whether cohabitation of Christians and Jews constituted sodomy since intercourse between men and women of these different groups "is precisely the same as if a man should copulate with a dog."[52]

Through the ages folklore has also contributed to the hostility with which Christians regarded Jews. "Individuals acquire stereotypes from folklore," Alan Dundes has written.

> Most of our conceptions of the . . . Jew come not from extended personal acquaintance or contact with representatives of these groups but rather from the proverbs, songs, jokes, and other forms of folklore we have heard all our lives.

Stereotypes may or may not be accurate, he continued, but they exist and "countless people" make judgments on the basis of them."[53] Other twentieth-century studies confirm Dundes' analysis.[54]

Defamation is one of the most common themes in folklore, a trait scholars have associated with superstition, legend, and myth. In these folk tales fabled qualities are implanted in real individuals and social groups to create powerful fantasy material, a common baggage with enormous potential to sustain evil thoughts and deeds. Thus the image of "Christ-killer" perpetuated in Christian societies became so engrafted into people's psyches that efforts to dispel it were useless.

Similarly, the images of the wandering Jew, the shrewd money-lending pawnbroker, and the ritual murderer persist in Christian mythology.[55]

Europeans who came to the New World in the sixteenth century and afterwards carried Christian thoughts and attitudes as parts of their cultural and intellectual baggage. And just as Jews had served as the eternal aliens and scapegoats to whom every ailment in European society could be attributed, so, to a certain extent, did they fill that function in America. On both continents they were to be victimized and blamed for major economic crises, ostracized in times of great national fervor, and condemned for whatever evils plagued a community at any given moment. Centuries of denigration had been so thoroughly absorbed by the populace that animosity toward Jews—for whatever reasons—seemed instinctive and appropriate.

In the New World, however, especially in those colonies that would later join together to form the United States, such rancor would lessen, become enmeshed with other characteristics and attributes developed in the fledgling states, and emerge in a weakened and less vitriolic form—but it would always be there. Once separated from the mother country, the United States never had an official church and the federal government never sanctioned antisemitic policies. Nor would there ever be any government inaugurated or supported pogroms. Rarely would an elected political official publicly state or endorse critical remarks about Jews and on several occasions Presidents and other governmental officers have specifically spoken out against flaying the Jews.

For reasons unique to the history of the United States, like the written compact, the alleged tyrannical rule of the British after 1763, the impact of the ideas of the enlightenment, the outstanding intellectual insights and abilities of the framers of the Constitution, and the polyglot nature of American society, this nation evolved differently. Policies of legal equality, individual liberty and civil rights for white men, religious freedom for white men and women, and all of these ultimately for African-Americans, American Indians, and Asians as well, developed. Unfortunately, however, the ancestral European Christian obsessions with Jews and their alleged attributes have also become an irrevocable part of the American heritage and periodically Jews have been accused of stubbornly refusing to embrace Jesus, plotting to undermine the American government, exploiting the economically less fortunate, being more concerned with their international brethren than with their American compatriots, and withholding patriotic fervor during the nation's battles with its enemies. The interplay of these various and mixed factors have contributed to Americans' perceptions of Jews and have determined as well the experiences and opportunities that Jews have had in the United States.

Antisemitism
in America

1

Colonial Beginnings
(1607–1790)

The founders of the American colonies, the Dutch, English, Scots-Irish, Germans, French, and other European settlers, brought with them to a New World the seeds of antisemitism.[1] In the colonies, however, new traditions developed that combined the heritage of Europe with the needs of people trying to establish a civilization in the wilderness. Thus, compared with the ghetto conditions that existed for Jews on the continent, the colonies offered vastly superior possibilities for Jews and the few who came received opportunities generally unavailable to their European brethren. Moreover, colonial interactions between Christians and Jews lacked the venomous tone so common in the Old World.[2]

The Europeans who came to the New World in the early seventeenth century established colonies that endured in Virginia, Massachusetts Bay, and New Netherland (later New York), developments that proved successful models for the others to follow. By the time of the American Revolution in 1776 thirteen colonies, which would later provide the nucleus for the United States of America, revolted from Great Britain and ultimately, in 1783, won recognition of their independence from the mother country. At first they formed a loose confederation but after only a few years' experience they recognized the need for a stronger union. It was a little more than 180 years from the founding of Jamestown until George Washington took the oath as the first President of the United States in 1789, but that time established in the national consensus most of the values and attitudes that Americans still hold dear. To be sure, significant changes later occurred and modifications of clearly enunciated policies have been made by history, but the essential thought and attitude of colonial Americans persists today albeit in muted form.

Perhaps the most outstanding characteristic of our current polyglot nation is its origin in an English Protestant heritage. Descendants of those whose forebearers belonged to the Church of England (the Anglican church in the colonies), the Presbyterian church in Scotland, or the Congregational churches that dotted the New England landscape have always been most easily accepted and identified as Americans. Protestant dissidents such as the Quakers, Baptists, and French Huguenots fared less well but were nonetheless vital to the success of the colonies. Catholics and Jews, however, and especially the former, were excoriated and considered outcasts in society, although colonies like Pennsylvania and Maryland allowed Catholics to settle freely; individuals of this faith were rarely found elsewhere in the original thirteen colonies. All told there were perhaps 25,000 Catholics among the 2,000,000 or more colonists who revolted in the 1770s but the "deep-seated prejudice with which they were everywhere regarded," scholar W. W. Sweet observed, "almost beggars description."[3] The anti-Papist strain in the American mind was strong and violent well into the twentieth century. Jews, too, were looked down on in colonial America but they were so small in numbers—perhaps 1,000 at the time of the Revolution—that they rarely encountered the same kind of vitriol colonists displayed toward Catholics. African Americans and American Indians in colonial America, as in our own day, occupied the lowest rungs of society.

It is hardly surprising that this religious and racial classification was a fact in the New World. Europeans who set the tone for colonial development brought with them the thoughts and prejudices of their native countries, including Protestant hatred of Papists. Contempt for Jews existed generally among Protestants and Catholics alike.[4] Historian Carl Wittke, writing in 1932, described the fate of the Jewish immigrant during colonial times as "one of persecution and discrimination, and of the long and eventually successful struggle for the removal of civil disabilities."[5] Jacob Marcus later elaborated:

> The first Jews to set foot on American soil found themselves objects of prejudice. The very word Jew evoked something negative in the minds of many, if not indeed most of their neighbors. The earliest American colonists were Europeans after all; they had brought with them their bigotries which Europe nourished almost as a matter of course, and among the emotional impediments they carried was a contempt for Jews.[6]

The first Jews recorded in what is now the United States reached New Netherland in 1654. Peter Stuyvesant, Governor of New Amsterdam (now New York City), was equally contemptuous of Quakers, Lutherans, and Jews and so tried to bar the 23 Portuguese Jews from Brazil from remaining in his colony. But the Dutch West India Company instructed the Governor to permit them to live and trade in New Netherland. The directors of the company sought to promote overseas trade, and encouraged settlement by any group or individuals who might help advance their commercial ends. But desirability for purposes of trade hardly equalled acceptance as individuals. Settlers in New Netherland did not want Jews in their midst and were refused by law permission to worship in public, vote, hold public office, purchase land, work as craftsmen, engage in retail operations, or trade with Indians. Moreover, although they were refused the right to stand guard with other community dwellers, a special tax was levied on them because they did not assume this task.[7] Expressing what was probably the dominant sense of his peers, one merchant castigated Jews for "having no other God than the Mammon of unrighteousness, and no other aim than to get possession of Christian property, and to overcome all other merchants by drawing all trade toward themselves."[8]

The attitudes observed in the residents of New Amsterdam paralleled those of most other colonists. Fully five years before the first Jews arrived in their colony, the Catholics of Maryland, fearful that a future Protestant majority might sever their own rights, passed an Act of Toleration prescribing death and loss of all property for those who denied the Trinity. Their original purpose was subverted, however, when Protestants regained control and repealed the act in 1654. When Maryland became a crown colony in 1692, the Anglican church was formally established, and citizenship was limited to church members.[9] Thus Catholics as well as Jews saw their rights curtailed.

Yet in surveying colonial legislation and practices over a century, it becomes obvious that restrictions on members of other religious groups were carelessly applied. Most of the colonies placed limits on the rights and duties of citizenship, including the opportunity to vote, but such policies were not uniformly implemented. The need for white people who would help build struggling communities and add to colonial wealth usually overrode stringent administration of prejudicial laws. Although both Jews and Catholics were subject to special taxes and limited opportunities for public worship, most of these burdens were later lifted or rarely enforced. Despite the existence of formal barriers, Jews and Christians often intermingled, especially in business affairs, and generalizations about their relationships with the dominant Protestant majority must be specifically defined by time and place. Just as in Europe over the centuries, the perceived talents and capital of Jews earned them a degree of tolerance in places where they appeared to be helping the community as a whole.[10]

Essentially there were five areas where Jewish settlements could be found in pre-Revolutionary America: New York, Philadelphia, Newport, Charleston, and Savannah. Although individual Jews might be located in any of the colonies, these five cities were the only ones to have a substantial Jewish presence even though the groups in Newport and Savannah disbanded and came together again at different times. Jews in New Netherland remained there after the Duke of York conquered the area for the English in 1664. The Duke's general concern for increased mercantile growth resulted in a lessening of restrictions on non-Protestants and a gradual tolerance for practices hitherto forbidden. By the end of the seventeenth century, Jews in New York City were able to purchase land, engage in retail trade, stand guard, and worship in public. The Mill Street Synagogue, functioning by the late 1690s, became the focal point for Jewish community life in the city. Called Congregation Shearith Israel, it was the first synagogue established in what is now the United States.[11]

Although individual Jews in other parts of seventeenth-century America are identifiable, New York City remained the only area of the mainland with a visible Jewish community until the late 1720s. Within the next decade new ones developed in both Philadelphia and Savannah, Georgia. In Philadelphia, Jews had earlier engaged in trade and commerce but no distinctive community was identified. Pennsylvanians allowed them most of the rights granted to citizens but they could neither vote nor hold office. And in Georgia, Jewish experiences in Savannah were reminiscent of the arrival of Jews in New Netherland. The colony was begun as a philanthropic experiment in 1732 but the arrival of a boatload of 40 Jews upset the trustees who hoped not "to make a Jew's colony of Georgia." Nonetheless, the newcomers remained for several years before moving on to South Carolina in the early 1740s. Periodically individual Jews returned to Georgia—and left again—before establishing a viable Jewish community around 1790.[12]

The colony that proved most hospitable to Jews was South Carolina where several families relocated after leaving Savannah. Charleston had perhaps the most vibrant mercantile economy in America by 1776, and treated Jews almost as equals. South Carolinians, in fact, elected Francis Salvador, Jr., to the First Provisional Congress in 1774 and so was the first of the colonies to allow a Jew to represent their interests. After the Revolution, South Carolina became one of the first American states where Jews could both vote and hold elected office.[13]

The only other pre-Revolutionary area where numbers of Jews could be found was Newport, Rhode Island, where about a dozen families dwelled in the 1670s and 1680s. For some inexplicable reason all relocated to other colonies. Only around 1750 did another group of Jews move to Newport, remaining until the Revolution when once again Jews dispersed to more hospitable climes. The 1760 Jewish community in Newport has attracted considerable attention from historians because of the noted Touro Synagogue and because of the economic opportunities Jews enjoyed there before the coming of the British during the American Revolution made life precarious again.[14]

The common denominators for the five areas where small Jewish enclaves could be found in colonial America was that all allowed freedom of worship—first in private, later in public— while providing economic opportunity to the enterprising. While Jews are often associated with the occupations of traders and merchants they were also butchers, silversmiths, soapmakers, distillers, and craftsmen. They were generally barred from practicing law; several individuals became physicians. But whatever their status, through trade and family connections Jews maintained a network among themselves throughout the colonies. Their European experiences made them wary of interaction with Christians and the restrictions in colonial America, though far fewer in number than those of Europe, kept them alert to both subtle and overt prejudices.[15]

Despite the sometimes significant gains, the overriding factor determining Jewish-Gentile attitudes toward one another was, as Jacob R. Marcus has observed, that "the good Christian of seventeenth and eighteenth century America was committed to a theology which represented the Jew as the villain in the drama and mystery of salvation."[16] Most colonists, especially in the early decades of the eighteenth century, did not give much thought to the philosophical nature of Judaism or to the beliefs of Jews as a group; however, rejection of Jews because they did not acknowledge Jesus Christ as their savior, and the belief that they could not be "forgiven" until they accepted Jesus as the Messiah, were equally common ideas among those who gave the subject any consideration at all.[17] All good Christians considered it their duty to help Jews "see the error of their ways and accept the divinity of Jesus," and Jews themselves were regarded as stubborn for clinging to their traditions.[18] William Penn, long considered a model of tolerance, implored Jews to accept the New Testament, repent their sins, and accept Jesus so that they might "enjoy the *Blessed Benefits* that accrue by *Him,* to all those that embrace Him."[19] "If you will but turn in your Minds," Penn continued, "and consider that *His Divine Light Shines in you,* and shows you the Error and Vanity of your Minds and Affections, and if you would but give up your selves to obey the same, through the Course of your Lives, you would become a *Tender People.*"[20] Penn's harsh evaluation perhaps reflects the deepest of his thoughts but his actions in Pennsylvania belied his beliefs. Jews were welcomed as colonists even if they did not share the religious views of the colony's founder. Lutherans and Anglicans, in Pennsylvania and elsewhere, seemed hostile to Jews and at times denounced the stiff-necked people who refused to accept the Protestant faith. Generally, however, Jews did not occupy the minds of most colonists and, even when they did, the expressions and demeanor of members of the majority group differed from the well-defined parameters of their thoughts.[21] In the colonies, and later in the United States, it must be continually emphasized, the harsh and rigid attitudes about Jews that most Christians embraced were not enacted in their day-to-day behavior. A stressful or exacerbating situation usually had to occur before the views that most Christians harbored toward Jews were articulated or acted on in public.

New England always had fewer Jews than Pennsylvania because the social and religious climate was less favorable. Congregationalists resented the Jews' refusal to accept Christianity as the only true faith and Yankee towns and communities remained hostile well into the twentieth century.[22] Religious leaders as varied in theology as Jonathan Edwards, Increase Mather, and Samuel Wilard all denounced Jews, though Edwards admitted that he knew of them only by reputation. Nevertheless, he stated definitively that "he would refute them should he ever encounter them."[23] Mather wrote that Jews existed in "fearful" and "horrid" sin because not only did they "blaspheme the tabernacle of God" but "they have been wont once a year to steal Christian children, and to put them to death by crucifying out of scorn and hatred against Christianity." He was also

> certain, that the most prodigious Murther that ever the Sun beheld (yea such Murther as the Sun durst not behold) hath been committed by the *Jews*, and that the guilt thereof lyeth upon the Jewish Nation to this day, even the guilt of the bloud of the Savior of the world.[24]

Nonetheless, he prayed for them to be saved and converted. In 1700, Mather's soon-to-be successor as administrator of Harvard, Samuel Willard, preached a sermon entitled "The Fountain Opened" in which he castigated the Jews for "the horrible contempt which they cast upon Christ and his gospel." As a result, they " 'were made a scorn and reproach to the world,' and their condition would not improve 'till the happy day of their conversion cometh.' "[25]

It is hardly surprising that the views of the intellectual and religious luminaries of the colonies were mirrored among most of the other colonists. The term "Jew" was everywhere used to express contempt and disdain; sometimes it meant sharp business practices, other times blasphemers. English folklore dating back to the thirteenth century portrayed the Jew as cunning and vicious, and in colonial America theatregoers enjoyed *The Merchant of Venice,* Marlow's *The Jew of Malta,* Dryden's *Love Triumphant,* and other plays depicting Jews as scoundrels, low comics, or fools. At no time before the twentieth century, in fact, did the Jew ever appear in a favorable light in the English-speaking theatre.[26]

Despite prevailing religious and cultural stereotypes of the Jew, in actual practice Jews and Gentiles had many pleasant and neutral interactions; the images both groups carried were rarely a hindrance to personal encounters among individuals. To be sure, most Christians and

Jews lived apart from one another and had little social intercourse. Yet there were many exceptions as in Newport, where the Lopez, Rivera, Polock, and Hart families achieved a certain prominence, as did the Seixas, Hart, and Hendricks clans in New York and the Gratz and Nones families in Philadelphia. And if such breakthroughs were possible in the cities, the relaxed and newer folkways and mores on the outskirts of civilization, and the greater willingness of Christians in America to convert Jews, led to widespread intermarriage. Every Jew who settled and remained in colonial Connecticut before the Revolution married a Christian. Similarly, large numbers of Jews in New York and Philadelphia married, raised their children in the community's dominant faith, and sometimes became Christians themselves.[27]

Ironically, more intermarriages occurred between descendants of Sephardic Jews—who arrived in the colonies before the 1740s—and Christians, than took place among Sephardim and Ashkenazim, those Jews who started coming from central Europe in the mid-eighteenth century. By the time of the American Revolution, the Ashkenazim in America far outnumbered Jews descended from the Iberian peninsula, and they too found it easy to intermarry. By the end of the eighteenth century only a tiny Jewish community survived in New York City, while slightly larger ones remained in Philadelphia and Charleston. Intermarriage reflected the social reality of colonial America. Fewer Jews were ostracized in the New World than in the Old because societal demarcations were still fluid. Economic prominence in a community led to greater social interaction, white people of whatever lineage were needed to propel growth, and, importantly, the small numbers of Jewish people in every community limited the options of eligible maidens and bachelors. Finally, many Jews wanted to avoid the degradation associated with their heritage and welcomed the opportunity to escape from a group constantly reminded of its alleged inferiority and merge with a member of the dominant society.[28]

Not unimportant, however, were the influences of the European Enlightenment and legislative changes in England in 1740 and 1753. The Enlightenment encouraged greater tolerance of divergent thought and adopted a more critical approach to traditional Christian teachings. And in England the Parliament in 1740 granted Jews in the colonies the right to become citizens after seven years' residence, a privilege also embracing Protestant sects but not Roman Catholics. To a certain extent this act reflected a greater tolerance for diversity within the British empire, but primarily it indicated the British government's desire to people the colonies with all settlers who would contribute to mercantile growth. Not until 1753 did a similar law permit nationalization of Jews within England, and opposition from powerful groups in the Church resulted in its repeal the following year.[29] Although the 8,000 Jews in England constituted less than one-tenth of 1 percent of the population[30] the "indignant fanaticism"[31] of the protesters revealed an

enmity that echoed the intensity of medieval feelings. The opponents
of the 1753 naturalization bill argued

> that the peculiar rites of the Jews were formidable obstacles to their incor-
> poration with other nations; and that if they were admitted to rank of
> citizens, they would engross the whole kingdom, gain possession of the
> landed estates, and dispossess the Christian owners. They also asserted
> that it was impious to gather a people whose dispersion was foretold in
> the sacred Scriptures, and who, according to the prophecies, were to
> remain without country or habitation until they should be converted and
> collected together in the land of their ancestors; and that an attempt to
> incorporate them, previous to this renouncing their religious tenets,
> directly opposed the will of heaven; by endeavoring to procure for them
> a civil condition while Jews, which, it is predicted, they should not enjoy
> till they become Christians.[32]

Similar attitudes were carried by English settlers who came to the
colonies but the ramifications of these thoughts were somewhat muted
in the New World where the need for colonists was paramount. Moreover, the thirteen British colonies also contained almost a half a million
other peoples including Germans, Scots-Irish, Huguenots, Dutch, and
Scandinavians who held European traditions, yet who had also learned
to adapt to interchanges with strangers. Both tolerance and intolerance
of Jews existed in colonial America; the same may also be said about
Quakers, Baptists, and Scots-Irish. Roman Catholics were treated more
harshly than any of these groups, a condition directly traceable to the
animosity that European Protestants displayed toward Rome. Awareness that England favored more white people—even Jews—in the colonies in the 1740s and 1750s fostered a greater Jewish exodus from
continental Europe to America. A change in the background of all Jews
in the colonies became apparent during the eighteenth century since
those who migrated from the German states in central Europe—the
Ashkenazim—far outnumbered those who had already arrived from
Brazil and the Iberian peninsula—the Sephardim—decades earlier.[33]

The main impetus for the Jewish movement from Europe to the
British colonies was the Naturalization Act of 1740. At a time when
trade restrictions, exclusion from guilds, and general medieval persecution still reigned on the continent, England had already passed legislation providing for greater opportunities in its overseas possessions.
Thus the 1750s and 1760s mark the real beginnings of substantial Jewish
enclaves in Philadelphia, Charleston, and Newport. The meager population figures that we have—intelligent estimates rather than accurate
numbers—suggest that in the 1770s there were 1,000 or fewer Jews in
all the British colonies on the mainland out of a total population of
about 2,000,000 individuals. There may have been 30 to 50 Jewish families in New York City and a somewhat smaller number in Philadelphia. Newport had perhaps 150 to 175 Jews while the estimate for

Charleston, where the largest number of them dwelled, is anywhere from 150 to 700 people, figures that highlight the imprecision of scholars' knowledge of colonial Jewry.[34]

The population figures indicate there were too few Jews in the British colonies to warrant either concern or attention by the dominant community, and usually Jews were ignored. The more pressing needs of fledgling developments in the wilderness and the presence of so many other non-English and non-Anglican groups made the Jewish presence somewhat of an afterthought. Only occasionally did active and conscious antisemitism erupt but it festered beneath the surface. European and American Christians carried centuries of hostility toward Jews and periodic rancor appeared. New York led the way when in 1737 a dispute arose over whether Jews should be allowed to vote;[35] six years later a mob assaulted a Jewish funeral procession in the city.[36] Some colonies considered Jews about as socially desirable as Indians and during the Revolutionary War many were denounced as Whigs by the Tories or as Tories by the Whigs.[37] Writing in 1837, a Pennsylvanian looked back and observed that during the Revolutionary era, "the Jews were yet a hated and a despised race."[38]

John Quincy Adams probably reflected prevailing colonial sentiments when in Amsterdam, Holland, in 1780 he visited a synagogue there and then confided to his diary that the Jews in Amsterdam "are all wretched creatures for I think I never saw in my life such a set of miserable looking people, and they would steal your eyes out of your head if they possibly could."[39] And in 1786, France's minister to the United States, writing to the Minister of Foreign Affairs in Paris, observed that the prejudices in the United States "are still too strong to enable the Jews to enjoy the privileges accorded to all their fellow citizens."[40]

Nevertheless, as the colonial era ended Jews found themselves with more rights and opportunities in America than they had ever enjoyed in Europe.[41] Yet the mixed condition of tolerance, ambivalence, and rejection Jews experienced in the colonies was to persist throughout American history. European attitudes toward Jews and Catholics arrived in the American colonies as part of immigrants' cultural baggage and the impact was lasting. Colonial conditions and revolutionary enthusiasms about the possibilities of a more democratic society would modify some legal restrictions and mitigate the harshest attitudes toward Jews. But, as we will soon discover, strong religious prejudices were more difficult to discard. Moreover, in times of crisis, many Americans displayed virulent animosity toward Jews and other outgroups. All periods of national stress would witness increased violence toward Jews while there would be continual revivals of Christian proselytizers trying to induce Jews to accept Jesus as their savior.

In 1776 the patriot Sam Adams called on Americans to inaugurate "a reign of political Protestantism," and for the next two centuries many United States citizens believed that Protestantism, liberty, and Americanism were equal parts of the democratic trinity. Protestantism was seen by Adams as "the religion most conducive to the Arts, Science, Freedom, and [the] consequent temporal Happiness of Mankind." Moreover, " 'it was the religion of the greatest, wisest, and best men this world has ever seen.' Government must 'honor' Protestantism by affording it 'every possible distinguishing mark of preeminence.' "[42] Adams spoke for millions of Americans, then and later, who shared his outlook on the world. While the creed Adams expressed was not specifically antisemitic, it left little room to respect and understand any other religious belief or cultural value system. The seeds of intolerance in the United States, therefore, were not only planted by the Europeans who settled the New World but nurtured as well by the patriots who equated Protestant thought with American virtue. Such ideas would never be far from the surface in the United States.

2

Developing Patterns
(1790s–1865)

The creation of the United States did not significantly alter either American Christian perceptions of Jews or Jewish attitudes toward themselves and Gentiles. The small number of American Jews in the United States remained in the few urban centers where they had been in previous decades, and folk wisdom, literary sources, and varieties of Christian teachings rather than personal interactions continued to shape Americans' judgments of them. At odd times and places, and in a variety of circumstances and settings, antisemitic outbursts came to the fore; their roots, however, had already been firmly planted. Most citizens of the new nation had never seen a Jew but many of those who had warmly embraced the individual. Thus, paradoxically, one pattern that developed in America held the "mythical Jew" in contempt while praising and respecting the Jew who was known. Existing religious prejudice in the United States continued to be directed mainly against Catholics; animosity toward Jews was less frequently, but no less vitriolically, expressed. On the other hand, America's image of itself as a tolerant and accepting nation prevented the kind of physical attacks and segregation that Jews experienced in Europe. People in the United States welcomed Caucasian newcomers and thought that eventually these people would accept prevailing values.[1]

The belief that the United States was, and should be, a Christian-Protestant nation stood out as a dominant theme throughout most of

13

American history but also served as an ominous portent for interfaith friction.[2] Protestant missionaries strove to bring Jews and others inside the great family of those who accepted the teachings of Jesus Christ. Jews did not appreciate these efforts and, as a small and scattered group, managed, for the most part, to ignore sporadic efforts to convert them. Although the national Jewish population tripled from around 1,600 in 1800 to perhaps 4,500 thirty years later, and then mushroomed to about 150,000 by the time of the Civil War, Jews still remained a minuscule group almost impervious to Christian missionary activity. Not until the burst of Jewish immigration in the late 1830s did this situation begin to change. Then, however, the xenophobic outbursts that almost always occur concurrently with a massive influx of foreigners in the United States (a similar pattern exists in other nations) resulted in vociferous anti-Catholic and antisemitic diatribes. After the middle of the nineteenth century American racism dictated attitudes, assaults, and policy changes toward Asians but European stereotypes permeated the reactions to Catholics and Jews.

From 1789 through 1839 the population of the United States more than doubled from 3.9 million to 9.6 million people. In that same period the country's small Jewish population multiplied tenfold from around 1,350 persons in 1790 to perhaps 2,700 in 1820, 4,500 in 1830, and 15,000 in 1840. All told Jews constituted less than 1 percent of the nation's inhabitants and outside of a few cities—New York, Philadelphia, Charleston, Baltimore, Richmond, and Savannah—they were only dots in a Protestant landscape. Estimates vary as to the size of the largest population centers but New York, with 600 to 1,000 Jews in 1825, and Charleston, South Carolina, with about 700 at that time, held the largest Jewish concentrations. By 1830 the 35 or so Jewish households in New Orleans constituted the largest Jewish center outside the East Coast.[3]

The early immigrants to the new nation met religious bigotry head-on. Except in New York, new state constitutions after the Revolution continued the colonial practice of restricting voting and office holding to Protestants, Christians, or specific denominations. In 1790 five states—Massachusetts, Connecticut, New Hampshire, Maryland, and South Carolina—still had established churches. New Jersey limited office holding to Protestants; Connecticut discriminated against atheists, Catholics, Jews, and non-Congregationalists; North Carolina required state officials to accept the divinity of the New Testament and prohibited Jews, Quakers, atheists, and sometimes Catholics from holding office. Americans generally were not concerned with religious freedom and they did not want the federal government to legislate or make pronouncements in this area. Furthermore, they wanted Protestants governing in the state and national governments and non-Protestants on the sidelines. Sometimes existing laws, as in colonial times, were not obeyed and occasionally individuals were elected to, and/or held positions of authority in, the government despite state constitutional restrictions. In 1809, for example, a colleague in North Carolina's House of Representatives challenged Jacob Henry's right to a seat in the chamber on the ground that he did not accept the truthfulness of the New Testament. Henry successfully defended his right to remain. But in 1835 the state amended its constitution to prohibit anyone who denied "the truth of the Christian Religion, or the divine authority of the Old or New Testament" from holding "any office or place of trust or profit in the civil department within this State." As late as 1858 a legislative committee rejected an attempt to permit Jews to hold office in North Carolina.[4]

Between 1789 and 1792 Delaware, Pennsylvania, South Carolina, and Georgia, under the influence of the European Enlightenment and post-revolutionary ideas about reform and the possibilities for a great new nation, lifted religious requirements that would have barred Jews from voting, but they did not extend full equality of rights. In Georgia, David Emanuel, thought to be a Jew, was elected Governor but only judges and Christian ministers could legally perform marriages in the state until the late 1840s. In the early nineteenth century Jews could practice law only in Pennsylvania, Virginia, South Carolina, and New York. After 1818, when Connecticut disestablished the Congregational church, Jews could hold office there but the new state constitution granted full equality of citizenship only to members of Christian denominations. Voting restrictions against Jews remained on the books in Rhode Island until 1842, in North Carolina until 1868, and in New Hampshire until 1877.[5]

The most celebrated crusade to win the vote for Jews occurred in Maryland. As early as 1797 Jews petitioned various legislators for the right to vote, hold office, and be able to engage in the practice of law, but not until 1826 did Maryland pass the "Jew Bill." The fight to achieve this reform lasted three bitter decades because significant opposition regularly prevented its passage. Rural residents and Federalists, in particular, resisted the change and in 1823 Christian opponents of the "Jew ticket" successfully defeated 16 Christian incumbents who wanted to broaden the franchise. Finally, however, on January 25, 1826, the Maryland "Jew Bill" won enough adherents to become law, most probably because of pressure from affluent Baltimore Jews. The state needed merchants and traders, and restricting the vote to Christians not only tended to place limits on those who might contribute to the state's wealth but also provided a reason for some Jews to seek other abodes. Legal discrimination continued in Maryland, however, and in 1830 the state refused to charter a Hebrew congregation solely on the grounds that it was Jewish. Maryland's fight over giving voting privileges to Jews was hardly antisemitic in nature since most Americans believed that the franchise should have been restricted to practicing

Christians. But once the issue arose, latent negative attitudes quickly surfaced. Opponents of expanded voting rights neither liked Jews nor regarded them as equals in society.[6]

In the late eighteenth and early nineteenth centuries, Christian Americans commonly characterized Jews in European Christian stereotypes. They saw Jews as cheats and blasphemers, and condemned them for avoiding agricultural labor.[7] Little, if any, change in these sentiments occurred over time as many individuals made no secret of their visceral hatred of Jews. Rebecca Samuel, writing to her parents from Virginia in 1791, acknowledged that she and her family were left to live and work unmolested but "the German Gentiles [in our midst] cannot forsake their anti-Jewish prejudices."[8] New England poet James Russell Lowell thought that Jews were plotting to take over the world, while William Ellery Channing, pastor of Boston's Federal Street Church from 1803–1842, saw Jews as "closed, circumscribed by the arrogance of self-interest and the idiocy of 'closeness' "; Ralph Waldo Emerson regarded Jews as usurers.[9]

In politics, as in other walks of life, religious beliefs underscored anti-Jewish sentiments. Northern Federalists espoused privileges for Christians and favored an established church for the nation. They equated morality with Christianity and dismissed non-Christians as *ipso facto* immoral. A New York Federalist denounced the local Democratic-Republicans because its vice president was "of the tribe of Shylock." No safety existed in the other party, however, because Democratic-Republicans in New York and Philadelphia in the 1790s also condemned Jews because of their faith. In the 1820s Stiles Ely of Philadelphia opposed any candidates for public office who were "infidels, Socinians, or Jews."[10] Such outspokenly negative opinions reiterated conventional stereotypes about Jews.

The best known Jewish politician in the first half of the nineteenth century, Mordecai Manuel Noah, a minor luminary in New York City, was taunted as "the Jew" and reviled as "Shylock" during much of his adult life. Even though both his father and grandfather had been in the American Revolutionary army, and that on his mother's side he was fifth-generation American, Noah, who grew up in Philadelphia and spent most of his adult life in New York City, was always the "marginal man," never fully accepted in society. He served as American consul in Tunis, sheriff of New York City, newspaper editor, playwright, and, during the Jacksonian era, self-appointed leader of the American Jewish community. His opponents, however, never let him forget his religious affiliation. During election campaigns they excoriated him as a Jew and in later life, when he defended the institution of slavery, abolitionists denounced him and other Jews as well.[11] The most famous of them, William Lloyd Garrison, attacked Noah as "the miscreant Jew" and "a Jewish unbeliever, the enemy of Christ and Liberty." On another occasion he referred to Noah as "the lineal descendant of the monsters who nailed Jesus to the cross. . . . Shylock will have his flesh at any cost."[12]

The views of Uriah P. Levy, another Jew whose public career spanned half of the nineteenth century, on the subject of slavery are not known but he also encountered public opprobrium because of his faith. Levy, who may have been the first Jewish officer in the United States Navy, served from 1812 until his death in 1862. An irascible man, Levy felt victimized by antisemitism in the service. He endured several decades of "stormy weather," including half a dozen court martials, often for minor infractions of the rules but mostly because he was Jewish. Frequently referred to as "the damned Jew," treated disrespectfully by peers and subordinates alike, and scorned, condemned, or snubbed by fellow officers—some of whom absolutely refused to speak to him because he was a Jew—Levy lamented late in life:

> I have been marked out to common contempt as a Jew until the slow unmoving finger of scorn has drawn a circle round me that includes all friendships and companions and attachments and all the blandishments of life and leaves me isolated and alone in the very midst of society.[13]

Levy's experiences in the military reflected the traditional and almost universally accepted values of American society. Christian religious teachings continually emphasized the shortcomings of Jews, therefore mocking or mistreating a Jew seemed natural to many servicemen. They had grown up in communities with other Christians, had listened to sermons about Jews as "rebels against God's purpose" and as people who had rejected and crucified Jesus.[14] *Niles' Register*, the closest thing the United States had to a national newspaper in antebellum America, informed its readers that Jews were "everywhere despised and maltreated."[15]

This point was further emphasized by Christian crusaders who sought to embrace non-Christians and bring them into the fold. After the pro-Christian religious revivals that occurred between 1798 and 1800, fundamentalists appeared obsessed with the idea of Christianizing everyone in the nation and making Christianity the state religion. Strong attempts to evangelize the Jews began in 1812 and gained support as the century progressed. Many Americans believed they had a special duty to convert Jews. Not only Jews but American Indians, Africans, and Hawaiians were among the targets of those desirous of bringing to Christ all who had escaped His influence.[16]

Throughout the nineteenth century, and especially from the 1830s through the 1860s, many Christians joined missionary societies aimed at "enlightening" Jews and others. Groups like the Female Society of Boston and Vicinity for Promoting Christianity and the American Society for Evangelizing the Jews, which in 1820 became the American Society for Meliorating the Conditions of the Jews (ASMCJ), worked tirelessly to achieve their goal. From 1823 to 1826 the ASMCJ published a periodical, *Israel's Advocate,* exhorting Christians to make every effort to convert Jews. Some of the most distinguished Americans, including John Quincy Adams and the Presidents of Yale, Princeton, and Rutgers Universities, supported them in this endeavor. As a result of the missionaries' efforts the first Jewish periodical in America, *The Jew,* which Solomon Jackson edited during its two-year existence (1823–1825), appeared to fend off those who offended Jewish sensibilities by trying to undermine their faith and convert its adherents. But Christian zealots would not stop. "Confident that they were on the side of truth," historian Robert T. Handy observed, "Protestants generally believed that in the end they would prevail and that good persons would be persuaded to the truth of their claims."[17]

Dedicated Christians instructed their children in traditional values and also reached out to the unconverted. Throughout the nineteenth century more than 150 million copies of the McGuffey readers, first published in 1836, also informed American students that the United States was a Christian country. Schoolbooks throughout America reiterated Protestant homilies while portraying Jews as crafty, greedy, dishonest, sly, selfish, unkind, unethical, disobedient, and wicked. Biblical accusations were so continually repeated in these works that hatred of the Jews was almost universally instilled. Even the unchurched, who probably constituted a majority of Americans in the mid-nineteenth century, absorbed some of the Christian values and attitudes peddled in the culture and in the schoolroom. Children learned how the Jews "were the bitter enemies of the early Christians," that they rejected and killed the Savior, and that even in the nineteenth century they still denied their responsibility for the crucifixion of Jesus.[18] Therefore, although their punishment for these faults was heavy, it was nonetheless just.[19]

From their earliest years children imbibed society's image of the Jew. One of the many verses in the *Mother Goose* rhymes, which began publishing in England in the seventeenth century before making their way to the colonies and then to the United States, emphasized the allegedly disreputable features of the Jew:

> Jack sold his egg
> To a rogue of a Jew
> Who cheated him out
> Of half of his due.[20]

As early as 1813 Boston's Elizabeth Peabody had prepared a slim volume of *Sabbath Lessons.* Lesson 10 read:

Q. What is recorded in the twenty-fifth chapter of Matthew?
A. The conspiracy of the Jewish rulers against Jesus Christ. . . .[21]

Lesson 22 discussed the General Epistle of James:

Q. What does this Epistle contain?

A. It principally contains severe reproofs to the Jews, for pretending to faith without works; for indulging themselves in instability, partiality, reviling, covetousness, oppression, vain swearing, etc.[22]

A book intended for children, *The Young Jew*, began:

> The Jews are to be seen in every land. What is the reason for this? Why do they not live in a country of their own as we do? The reason is that they were disobedient to God.[23]

A youth reading this book then learned that wherever Jews go they are "as a nation, despised and ill-treated. . . . But the worst part of the Jews' punishment is, that their hearts are hardened, so that they still will not believe that Jesus is the Son of God."[24] In another children's story one of the young characters states, "I have heard that hatred of a Jew has been reckoned a virtue."[25] On the other hand, it was cause for rejoicing when disbelieving Jewish children accepted Jesus and converted to some Protestant denomination. The same characterizations that appeared in children's literature made their way into plays, songs, and books for more mature persons. Hannah Adams's 1818 history of the Jews informed readers that the "Jewish nation" had put Christ to death and "horrendous calamities" befell Jews for perpetrating "this horrid crime."[26]

State and local governmental polices and edicts, along with judicial decisions, continually reinforced cultural prejudices and popular beliefs that the United States was a Christian country. Thanksgiving Day proclamations, like that issued by Governor James H. Hammon of South Carolina in 1844 calling for "citizens of all denominations . . . to give thanks to God . . . and . . . Jesus Christ" aroused Jewish ire and protests because they felt left out.[27] In 1843 when Jews in New York City complained about the use of passages from the New Testament in the public schools a Board of Education committee dismissed their objections. This was a Christian country, the Board replied, and indicated that Jews should conform to the established customs of the nation. Catholics, on the other hand, who opposed using a Protestant Bible sought state assistance to open their own schools.[28]

Economic folklore reinforced views about Jews being outside the American fold. In Europe Christians denigrated alleged Jewish attitudes toward money, economic transactions, and lack of commitment to ethical intercourse in business endeavors. A composite portrait evolved of the Jew as a vastly powerful, manipulative, corrupt, devious, cunning, greedy, tricky, materialistic, dishonest, shrewd, grasping, and close-fisted man who would do anything to acquire and hold onto gold.[29] Some people even thought that Jews would sell their own children for money.[30] The composite "Shylock," developed centuries earlier, incorporated all or most of these negative characteristics attributed to Jews long before Shakespeare created the figure. Since the sixteenth

century, however, the impression of the mercantile Jew as "Shylock"
has been as deeply imbedded in the Christian culture as the conception
of the Jew as "Christ-killer." Moreover, Jews, who had been barred
from agriculture in medieval Europe but allowed to work as bankers,
also acquired the image of non-producers who preyed on unwary Gentiles. As *Niles' Register* observed, Jews

> do not appear identified with those of the communities in which they
> live, though there are some honorable exceptions to this remark. But they
> will not sit down and labor like other people—they create nothing, and
> are mere consumers. They will not cultivate the earth, nor work at
> mechanical trades, preferring to live by their wit in dealing and acting
> as if they had a home no where. It is to this cause, no doubt, that an
> hostility to them exists so extensively; and that hostility is again, perhaps,
> a cause why they do not think and act like other people, and assume the
> character and feelings of the nations to do with their rights as men.

Thus, as the outsider who rejected Christ, and who lusted after gold,
the Jew could not be trusted—especially in any dealings involving
money.[31]

Well-established European and American stereotypes made it difficult for Jewish businessmen from the Jacksonian era through much
of the Gilded Age to obtain credit from non-Jews. Investigators from
credit-rating bureaus in cities as diverse as Buffalo, Indianapolis, Cincinnati, Columbus and Cleveland, Ohio, New York, New Orleans, and
San Francisco identified Jewish applicants by their faith or heritage and
used almost identical language when describing their credit-worthi-
ness. These men, following age-old traditions that obviously continued
into the nineteenth century, did not pause to think or examine the justification for their views—they simply assumed Jewish dishonesty.
Excerpts from individual credit reports indicated: "We should deem
him safe but he is not a *white* man. He is a Jew, and that you can take
into account." "Responsible now," read another, "but he is a Jew; there
is no telling how long he will remain so." Others noted, "Jew ... be
careful," "is considered a sharp and shrewd Jew," "Jew, not to be
trusted," "are Jews, and therefore cannot be well-estimated." In
describing one Cincinnati Jewish firm, the credit investigator reported:
"They are Jews and little reliance can be placed on their representa-
tions.... Creditors had better send their claims at once as delay is
always dangerous with Jews."[32]

Another myth about Jews, their wealth, and their alleged power
throughout the world revolved about the stories told of the Rothschilds, a major European banking family who had branches in Frankfurt, London, Paris, Vienna, and Berlin. Some of the feelings Americans
held about the Rothschilds no doubt came from their own family her-
itages where Jews, as the alleged murderers of the son of God, were
always regarded with loathing. Newspaper accounts of the legendary

Rothschilds, whose name suggested "visions of untold wealth and unrivaled power,"[33] enhanced popular impressions. *Niles' Register* informed its readers in 1829 that the Rothschilds were "wealthy beyond desire, perhaps even of avarice"[34] and had supposedly purchased the city of Jerusalem from the Turkish sultan. Six years later the paper added that the family, who sprang from a "mysterious race," were the "wonders of modern banking." Allegedly, they

> govern[ed] a Christian world. Not a cabinet moves without their advice. They stretch their hand, with equal ease, from Petersburgh to Vienna, from Vienna to Paris, from Paris to London, from London to Washington. Baron Rothschild, the head of the house, is the true king of Judah, the prince of the captivity, the messiah so long looked for by this extraordinary people. He holds the keys of peace or war, blessing or cursing.

The Rothschilds were, the newspaper added, "the brokers and counselors of the kings of Europe, and of the republican chiefs of America."[35]

Throughout the rest of the century the impression of the Rothschilds controlling enormous sums of money and governments throughout the world held sway among the untutored and even among many educated Americans.[36] For example, fully twelve years after the erroneous tale appeared concerning the Rothschilds' purchase of the city of Jerusalem, and six years after the account of the family's alleged political influence, New England abolitionist and reformer Lydia Maria Child wrote:

> The sovereigns of Europe and Asia, and the republics of America, are their debtors, to an immense amount. The Rothschilds are Jews; and they have wealth enough to purchase all Palestine if they choose; a large part of Jerusalem is in fact mortgaged to them.[37]

In 1858 the editor of *Harper's Magazine* reemphasized the belief in the extraordinary influence of the Rothschild family. He wrote that "at this moment, as power is estimated, a Jew is the most powerful man in the world. Rothschild holds in his pocket the peace of the world."[38]

In addition to firmly held religious and economic stereotypes of Jews, a third and extremely important aspect of American attitudes toward them can be seen in how the term "Jew" evolved in American language. "Jew" always had a negative connotation and was often used as a reproach. ("Israelite" was the favorable or respectful terminology in mid-nineteenth-century America.) Dubbing a political candidate a "Jew" could mean that he was vile, greedy, sneaky, untrustworthy, or despicable in myriad ways. Thus, from the 1830s through the 1850s, some Kentuckians who clashed with Senator Henry Clay accused him of consorting with Jews; in other states people labeled President John Tyler "an accursed Jew," and some Americans in different parts of the country erroneously dismissed the first Republican Party presidential nominee, General John C. Fremont, as "a Jew." African American Christians shared the prejudices of white Christians (this theme will be expanded in Chapter 10); when black abolitionists attacked proslavery white churches they characterized them as "synagogues of Satan" and stigmatized the parishioners of those churches as " 'worthy' successors to the Jewish 'swine of the Scriptures' who refused to accept the liberating word of Christ."[39]

The term "Jew" also substituted for a verb as in "jew down," "jewed," or "jewing," which meant to haggle, to bargain shrewdly but not quite honestly, or to cheat in economic endeavors.[40] The words "Jew" and "Jewish" were also used as synonyms for "rogue," "swindler," "selfish," "mean," or "cold-hearted." Finally, as early as the 1830s, before the major migrations of the central European Jews, caricatures of Jews appeared in posters, cartoons, and theatres in the United States mocking alleged Jewish speech patterns and substituting v for w, t and d for th, and exaggerating the sibilant sh for the s sound. As historian Hasia Diner noted,

> American popular and elite literature, theatre, cartoons, advertisements, and jokes portrayed the Jews in a set of flat images that, whether positive or negative, had little to do with Jews as real people.[41]

Yet, despite these prejudices, Jews were generally left alone. They were not as victimized and as exploited as Irish Catholics; they were not pushed out of society as the Indians were; and they were not enslaved like the Africans. Thus while they were objects of prejudice in people's minds they were not thwarted in the United States in anywhere near the same fashion as members of other marginal groups. And while Jews were allowed to follow their own pursuits and receive assistance from one another, they often moved in elite circles and found friends and companions among respected Christians within both local and national communities. Sometimes their individual talents or accomplishments were recognized; other times their assets—such as retail businesses—provided needed services in communities; and sometimes the wealth that a few acquired provided access to places and persons that impoverished Christians or coreligionists would never have considered within the realm of possibilities. This was especially true in locales where there were no significant crises or disturbances. Members of minority groups of color in the United States rarely, if ever, enjoyed those opportunities.

Extremely important contrasts between Jews and other minorities—African Americans, American Indians, and later Asians—are that Jews occupied positions of political and economic power in some communities, intermingled socially with prestigious individuals in the dominant culture, and intermarried with white Christians when they chose to do so. There are no statistics on how many intermarriages occurred in much of the nineteenth century but in several American cities it happened as a matter of course. Thus Jews who married within their own group as well as those who chose Christian partners could be found among the elites in New York, Philadelphia, and Charleston. Although their numbers were small, they included the Hendricks, Gomez, Seixas, and Lopez families in New York, members of the Gratz family in Philadelphia, and the Moise and Mordecai families in Charleston. Moreover, in the early decades of the nineteenth century Jews in Charleston functioned as professionals, political officials, and state employees at the highest levels of government. Elsewhere in America, the heads of prominent Jewish families usually engaged in trade and commerce, moved easily in cultural and philanthropic circles, and showed no indication that the antagonism many Americans exhibited toward their coreligionists significantly affected their lives. And, just as in a later period established German Jews would avoid social contact with the newly arrived east European Jews, the elite Jews in Jacksonian America, mostly of English and Dutch backgrounds, stood apart from the growing influx of German and other central European Jews whose appearance began attracting attention as the nation moved into its second half century.[42]

Immigration of Jews to the United States in any appreciable numbers began in the 1840s but thousands trickled in earlier and made a niche for themselves in several areas. By the 1830s communities had been planted along the East Coast as well as in the hinterlands. In addition to the cities where Jews had established themselves earlier, one found enclaves of them in places like Albany, New York; Cincinnati and Cleveland, Ohio; and in outposts far from New York City such as St. Louis and New Orleans. Major migrations of Protestant and Catholic German immigrants began in the late 1830s and Jews from various parts of central Europe emigrated at that time as well. About 200,000 Jews, mostly from the German states of Europe but also from France and areas of what is now Poland, arrived in this country between 1840 and 1880 and contributed to the growth of the Jewish settlements already in place. Thus, not only did small groups of Jews increase their numbers in the Jacksonian era and in the pre–Civil War period but they mushroomed in successive decades. No sooner did the central and western European Jewish migrations begin tailing off in the 1880s than hundreds of thousands of east Europeans, including Jews, swelled the American population. Like the others, only more so, Jews preferred settling in areas near the water or along railroad routes. By the 1840s and 1850s, additional Jewish settlements appeared in Rochester, New York; Chicago, Milwaukee, Boston, New Haven, Cincinnati, San Francisco, Los Angeles, and Charlotte and Statesvile, North Carolina.[43]

In the 1840s and 1850s the European migration to the United States exceeded 4,000,000 people, the largest concentration of newcomers in a twenty-year period before the Civil War, and considerably more than the total colonial population of 2,500,000 on the eve of the American

Revolution. Significant changes occurred in America as a corollary of the influx from abroad but the results did not become obvious until well into the 1850s. Thousands, then hundreds of thousands, of people arrived from Ireland and the German states. The Irish went mainly to the large urban areas on the East Coast, to cities in the Midwest, and to a few places in the South. Germans avoided New England but settled almost everywhere else. They were in both urban and rural areas, along the mid-Atlantic, in the Midwest, and in the South. Among the Germans were tens of thousands of Jews.

The Jews who left Europe for America in the 1840s and 1850s did so for a number of reasons. European population growth in the early nineteenth century narrowed opportunities for young people. With the spread of industrialization the need for skilled artisans declined. Then the German states tightened conscription regulations, placed restrictions on occupations that Jews might enter, and limited the numbers of Jews permitted to marry. The revolutionary spirit exhibited in the uprisings of 1830 and 1848 promoted the Jewish exodus as young people wished to expand their horizons and secure their futures. Word from previous immigrants to America had already reached Europe that the United States government did not restrict access to occupations and possibilities for achievement were real. Thus what started as a trickle in the 1830s developed into a flood by the 1850s. The American Jewish population of about 4,500 in 1830 rose to approximately 40,000 in 1845 and leaped to 150,000 by the time of the Civil War.[44]

The new immigration, along with the political crises over the expansion of slavery, contributed to a wave of national hysteria and xenophobia, especially in the decade preceding the Civil War. Both the Irish, and the Germans to a lesser extent, were victimized by the prejudices and anxieties that erupted.[45] As usual in times of crises and xenophobia, the Jew also felt the sting. "Suspicion and contempt met him at every step," Gustave Gotheill wrote in 1878,

> and forced him not seldom, to hide his origin and to bury his faith in his bosom. Unless he did that, he could not ply his trade, nay, would be refused shelter and food. On this free soil he was often obliged to perform the rites of his religion and offer his prayers behind locked doors. It was not until personal contact had proved him to be a man, that he could safely avow himself a Jew.[46]

Most Americans experienced their first encounter with any Jew when the itinerant peddler arrived at their door. Most Jewish males who emigrated to the United States engaged in that occupation—some for a brief period, others for a few years. Thousands of them later established dry goods stores. Other occupations that attracted Jews included various facets of retail endeavors such as tailoring, business clerks, and jewelers. Only a handful of Jews in mid-nineteenth century America served as unskilled and manual laborers or engaged in professional or semi-professional pursuits. For example, of 2,000 Jews in New York City in 1838 there were only three physicians and three lawyers; of 15,000 Jews in Cincinnati in 1858, there were but three physicians, four attorneys, and ten teachers. Several of the peddlers-turned-merchants eventually developed larger retail emporiums; some moved into man-ufacturing, especially clothing; and a smaller number chose factoring, credit, and banking—ultimately a few, like the Guggenheim, Lehman, and Seligman brothers, became international financiers of wide renown. Centuries of rejection by, and discomfort in the presence of, Christians led most of the immigrant Jews to socialize primarily with one another wherever communities afforded that opportunity.[47]

The new migration included dynamic individuals who had strong views about how to shape the development of fledgling Jewish com-munities in the United States. Outstanding among them were Isaac Leeser who arrived in Philadelphia as an eighteen year old in 1824 and emerged as one of the prominent national Jewish leaders in the 1840s, and Isaac Mayer Wise, sometimes considered the father of Reform Judaism in America, who arrived in Albany, New York, in 1846 and moved to Cincinnati in 1854.[48] (In successive waves of Jewish migration in the nineteenth and later twentieth centuries leaders continually emerged who tried to move coreligionists in particular directions. Sometimes they succeeded, more often than not they met indifference from their intended followers. Not until the end of World War II, how-ever, when the most significant Jewish organizations coalesced around realistic programs that met the immediate needs of a majority of Jews in the United States, did the self-appointed leaders make a major dif-ference in the lives of American coreligionists.)

Before the 1840s Mordecai Noah stood out as the most prominent Jew in America but in that year Issac Leeser, by then an orthodox rabbi, emerged as leader of a group of American Jews protesting what was then known as the Damascus affair. The Damascus affair began with the murder of a monk and his servant in Damascus, Syria, in February 1840. Thirteen Jews were arrested, accused of having committed the crime, and then tortured into confessing that they had done so to use the blood of the Christians for Jewish ritual purposes. Jews throughout the world protested the barbarity of the Syrians and demanded the release of those imprisoned. In August 1840, after President Martin Van Buren had made efforts to free those who had been unjustly arrested, Jews held protest meetings in New York, Philadelphia, Cincinnati, Richmond, Savannah, and Charleston calling for American interven-tion on behalf of their coreligionists. Leeser spoke at the Philadelphia rally, and in his speech he proposed a national union of all American Jews. Such a union did not occur at that time but the idea would be revived, and to a limited extent accomplished, in subsequent decades and especially in the twentieth century. The significance of the Damas-cus affair, however, was that it was the first display of American Jewish unity. Jews felt secure enough in the United States to seek the federal government's intervention on behalf of coreligionists in another part of the world. This first attempt at exerting group influence would later become a regular occurrence.[49]

As a result of the Damascus affair Lesser emerged as the Jewish community's preeminent spokesman. In subsequent years he took forward positions on a number of other problems facing American Jewry. He objected fiercely to religious instruction, prayers, and Bible-reading in the public schools, argued for the right of a Jewish student to be excused without penalty from school on a Jewish holiday, and established a national Jewish publication in 1843, *The Occident,* to help counteract the efforts of zealous Christian missionaries who wanted to convert the Jews. At the same time he encouraged immigrant Jews to acculturate as quickly as possible, drop German and use English as their language of intercourse, and diversify their places of abode and occupational choices as a way of lessening American antisemitic tendencies. He believed that if Jews worked in agriculture or made handicrafts and scattered themselves throughout America prejudice against them would be reduced.[50] (Leeser would not be the only Jewish leader who believed that prejudice would decline if Jews were not so heavily concentrated in a few geographical and occupational areas.) At the same time, however, he advised parents to prepare their children for victimization. "The Jewish child," he wrote, soon observes that "his *religion* is ridiculed and heartily despised by the great majority around him."[51] Throughout his life Leeser tried to serve as a bridge between Jews and American Christians, to explain one group to another, and to lessen Gentile hostility. Despite his orthodox commitments, Leeser encouraged Jewish immigrants to Americanize the synagogue and observe greater decorum there, cut the length of religious services, and make traditional Judaism more acceptable to the Gentile majority.[52]

Another prominent Jewish immigrant, Isaac Mayer Wise who arrived at Albany, New York, in 1846, also urged Jews to acclimate themselves to the United States. Wise, much more accommodating than Leeser, sought to change both the form and substance of Judaism so that outwardly it could resemble a branch of Christianity. He praised the Jews of Charleston who had already placed an organ in their temple and switched to a more Americanized liturgy. Wise favored Reform Judaism with abridged services conducted decorously in the English language. He also wanted men to discard their hats and prayer shawls during services, favored families of men and women sitting together in the synagogue, and urged the abandonment of Jewish dietary laws. In other words, aside from accepting the basic tenets of Judaism—broadly interpreted—Wise was not averse to having synagogues look like churches and services that resembled liberal Protestant discourse.[53] "Whatever makes us ridiculous before the world as it now is," he wrote, "may safely be and should be abolished."[54]

Attempts of Jews in the United States to acculturate, however, had little impact on general American perceptions of them. As immigration from abroad increased there was a parallel growth in nativism throughout the country. Augmented numbers of Irish and German Catholics in the United States in the late 1840s and early 1850s led to the rise of the Know-Nothing Party throughout the country in the mid-1850s. Primarily anti-Catholic, the Know-Nothings appeared less concerned with Jews.[55] The Germans and Irish, like other Christians before and since, brought their cultural baggage with them and that included strong antisemitic feelings that had been nurtured by churches, rulers, and peers in most of Europe for centuries. Thus not only did Jews have to contend with the prejudices of those whose families had been in the United States for generations, but also with the living enmity that European immigrants possessed.[56] As a result of the rancor created among Americans by enlarged immigration—over three million newcomers arrived between 1845 and 1854, the largest total ever to enter the United States in one decade before the 1880s—there was also increased hostility to Jews from both Americans and immigrants. The number of people of varying stripes verbally, and sometimes physically, attacking Jews in their midst rose precipitously in the 1850s and positively exploded in both the North and the South during the Civil War.

Partially to protect Jewish interests, partially to defend Judaism, and partially to give American Jews a sense of belonging, Wise founded *The Israelite* when he moved from Albany to Cincinnati in 1854. *The Israelite*, as with Leeser's *Occident*, published stories, editorials, and the doings of American and world Jewry. Leeser and Wise were also swift to inform readers of attacks on Jews in this country and abroad, to condemn and warn against proselytizing movements, and to defend Jews from those who abhorred them. Most early Jewish newspapers, and this included the short-lived *Asmonean* in New York City (1849–1858), and the voice of Jewish orthodoxy in the city, *Jewish Messenger*, which began in 1857, functioned almost as defense agencies to protect the good name of Jews and quickly responded to slurs and assaults. The short-lived *Jewish South* also began in 1853 but suspended publication three years later before reappearing for another five years from 1897 through 1902. These publications constantly reminded readers that antisemitism stemmed from Christian beliefs about Jews rather than Jewish behavior in the presence of Christians.[57]

In the 1850s, a decade in which slavery, and whether a free society could justify its continuance, dominated political discussions in the North and South, nativism and antisemitism swept the country and particularly the Whig Party.[58] Then, as in earlier and later periods of national stress and contentiousness, increased animosity developed toward immigrants and minorities and the number of publicized incidents of violence against Jews increased. Wise wrote in his autobiography that in the mid-nineteenth century:

every pastor and every insignificant little preacher, every common jester, and every political rogue raised blows upon the Talmud and the Jews. A rascally Jew figured in every cheap novel, every newspaper printed some stale jokes about the Jews to fill up space, every backwoodsman had a few such jokes on hand for use in public addresses; and all this called forth not one word of protest from any source.[59]

Detroit newspapers in 1850 referred to Jews as "mysterious," "cursed," and "wanderers,"[60] while the April 6 issue of the New York *Herald* ran a front-page story accusing Jews of bleeding a Christian missionary to death in the Middle East, grinding up his bones, and mixing his blood with unleavened bread for the Passover feast. Then in September, on the eve of Yom Kippur, rumors spread in New York City that Jews had murdered a Gentile girl for the holiday. As a result of this fable, the next morning about 500 men, mostly of Irish descent and led by the police, invaded and ransacked a synagogue on the holiest day of the Jewish year.[61]

Attacks on Jews in the 1850s were not restricted to members of any particular geographical locale. Catholic Germans clashed with Jews in cities like Philadelphia, Cleveland, Syracuse, Detroit, and Baltimore, and sometimes beat them up. Several priests in a Cincinnati German Catholic Church in the 1850s told employed domestics that they would not be allowed to confess if they continued working for Jewish employers.[62] On the West Coast the *Oregonian* regularly assailed Jews and in a singularly inept recommendation in 1855 the Speaker of the California House of Representatives, William W. Stowe, proposed levying a high tax on Jews to keep them out of the state. This motion came during the legislative debate on Sunday-closing laws that were aimed specifically against Jews. Stowe, who had been angered by a single Jew and who had carried this animosity over to include all Jews, declared that Jews came to the state only to make money and leave, and not to invest their acquired wealth to help develop California. Stowe also added that "the Bible lay at the foundation of our institutions, and its ordinances ought to be covered and adhered to in legislating for the state."[63]

Stowe's remarks may have highlighted his own bigotry but California housed a myriad diverse individuals and groups whose attitudes could not be definitively categorized. In San Francisco and other parts of the state Jews, and especially those of French ancestry, were generally accepted into the elite social circles. Nativism prevailed in the mining communities but even there Jews were not the major target; Latinos, Chinese, and American Indians bore the brunt of West Coast bigotry.[64] However, where prejudice toward one outgroup exists manifestations toward all are noticed and Jews were the reason that an 1851 state law required peddlers to purchase expensive licenses.[65] As California historian Hubert H. Bancroft has written, the antipathy manifested toward the Jew was perpetual and unattended by violent demonstrations, while repugnance to the Chilean and China-man broke out into occasional bloody encounters. In this inspiring of dislike [Jews] excelled all other people.[66]

The disdain and petty violence that Jews encountered throughout the country annoyed but did not daunt them. It was, in fact, that pattern that had been established early in national history. Life in the United States afforded too many opportunities, especially in contrast to their experiences in the Old World, for them to cave in to an antisemitism that they expected. To be sure, American antisemitism was on a much lesser scale than Jews had known in Europe, and with no legislative restrictions or governmental support for opposition to them, Jews maneuvered around the bigotry. Most Jews preferred living in neigh-borhoods inhabited by coreligionists and interacted socially with one another whenever possible. Other methods of accommodation that one would see over and over again in the United States witnessed some Jews who kept their heritage secret, others who changed their names, and thousands, if not tens of thousands, of individuals who found ways of ingratiating themselves among the Gentiles within the communities where they dwelled.[67] As Harold Hyman noted of Harris Kempner who settled in Galveston: "Though perhaps no Jew could wholly escape bigotry[,] Kempner managed to accommodate his behavior to his neighbors' ways in order to minimize discriminations."[68] And in the face of pervasive antisemitism, several Protestant denominations and churches preached respect and tolerance for Jews while at the same time scorning their faith. It was a source of pride to many Christian Americans, historian Louise Mayo reminds us, "that these ancient peo-ple could worship with complete freedom only in America."[69]

The privileges guaranteed in the United States did not extend over-seas. In 1850 President Millard Fillmore's government negotiated a commercial treaty with Switzerland that granted full rights to Ameri-can Christians traveling or working in the country but permitted Swiss cantons to deny entry to American Jews. When American Jews pro-tested American government participation in a discriminatory treaty, Fillmore renegotiated the agreement. However, when it finally passed Congress during Franklin Pierce's presidency, in 1855, only the lan-guage of the treaty had been modified, not the provisions. Despite pro-tests to Pierce, and later James Buchanan who succeeded Pierce in the White House, no changes were effected. Not all Jews were aware of the treaty or responded in any way when they learned about it, but the fact that the American government accepted it indicated that Jews were not an important enough constituency to negate matters considered more significant in the overall picture of foreign policy.[70]

Another untoward European event in the 1850s of concern to Jewry received widespread attention in Protestant and Catholic circles as well. In June 1858 papal authorities forcefully removed seven-year-old Edgar Mortara from the home of his Jewish parents in Bologna. At the age of two, Edgar had been ill and supposedly on his deathbed when a Catholic servant secretly baptized the child. The boy lived and the servant later informed officials of her action. Papal authorities regarded the child as a Catholic and in need of a Catholic upbringing. They then seized him from his parents. Many people in the Protestant world were shocked by such inhumane behavior (Catholic spokesmen and periodicals defended the Church's action) and Jews and Protestants in the United States, as well as in England, called for the boy's return to his parents—but to no avail. American Jews petitioned Secretary of State Lewis Cass and President Buchanan for intervention but although these men sympathized with the request they believed it inappropriate for the United States to intervene in the domestic affairs of another government, in this case the Papacy, or even to make a humanitarian call for Edgar's return. That Van Buren had intervened in the Damascus affair in 1840 did not appear to influence Buchanan, probably because American conditions were different in 1858 than they had been eighteen years earlier. The debate on slavery expansion consumed much of his time in the White House and showed signs of weakening Democratic party unity. Jews were insufficiently important as a group for him to give much care to their concerns, and, as a Democrat, he did not want to split the party further by antagonizing northern Irish Catholic voters who supported the Pope's position.[71]

Jewish inability to influence the American government on behalf of Edgar Mortara convinced leaders that they needed a defense group "to secure and maintain Jewish civil and religious rights at home and abroad."[72] To this end in 1859 some prominent Jews founded the relatively weak, and short-lived, Board of Delegates of American Israelites. The Board had little power or influence, although it did protest incidents of discrimination against American Jews in the United States. For the most part it gathered statistics, fostered Jewish education, and, when asked, arbitrated congregational disputes. Occasionally, the Board represented "Jewish" positions and interests to political figures. Most American Jews failed to become involved in the Board's activities or concerns, and in 1878 the Union of American Hebrew Congregations, founded in 1873, absorbed the Board into its orbit.[73] Not until the formation of the American Jewish Committee (AJC) in 1906 would there be a sufficiently powerful and influential organization to speak for Jewish interests in the United States, and even the AJC did not represent most of American Jewry or reach its peak effectiveness until after World War II.

The Board of Delegates convened on the eve of the American Civil War, an event that coincided with the worst period of antisemitism in the United States to that date. Economic distress and political tensions ignited vicious reactions to Jews as Americans, in both North and

South, presumed them to be disloyal profiteers and blamed them as well for major societal problems. Jews were often held responsible for the frustrations, anger, disappointments, fear, insecurity, anxiety, shame, and jealousy that the war generated.[74] Issac Leeser attributed the increased hatred of Jews to the willingness of Americans to accept Christian prejudices. Clerical intolerance, he wrote, "is the seat of danger to liberty of conscience and perhaps the permanence of free institutions of all kinds."[75] The Civil War era was the first, but certainly not the last, time in American history that Jews were targeted during periods of anxiety, uncertainty, and crisis. In fact, attributing the nation's economic and political woes to Jews would reverberate among several disgruntled groups of Americans in succeeding decades.

In 1861 Congress, while authorizing the raising of troops, included a provision that allowed only Christian ministers to serve as chaplains in the army and navy. After protests from the Board of Delegates and others, the law was amended to allow Jews to serve in that capacity. In the North, Jews were presumed to be southern sympathizers of questionable loyalty to the Union and on one occasion the chief of the War Department's Detective Bureau arrested Simon Wolf, a member of the Jewish fraternal order B'nai B'rith, because the detective assumed that B'nai B'rith was some kind of disloyal organization assisting southern traitors.[76] Horace Seaver, editor of the *Boston Investigator,* denounced Jews as "about the worst people of whom we have any account" and a "troublesome people to live in proximity with."[77] Seaver's slanderous misrepresentations enraged Ernestine Rose, a Jewish abolitionist and champion of the rights of women and she replied to Seaver in no uncertain terms. In a series of about ten letters, their exchange of views appeared in several issues of the newspaper in the fall and winter of 1863–1864.[78]

Many northerners also imagined that Jewish financiers benefited at the expense of patriots and prominent northern newspapers attacked them. In 1863 the *Detroit Commercial Advertiser* referred to Jews as "the tribe of gold speculators" and "hooked nose wretches" who speculated "on disasters and a battle lost to our army is chuckled over by them, as it puts money in their purse."[79] In September 1864 the Chicago *Tribune* editorialized that the Rothschilds, their "Jewish" agent, August Belmont, "and the whole tribe of Jews," sympathized with the South. Although Belmont had no formal affiliation with Jewish synagogues or organizations, married a non-Jew, and reared his children as Christians, hate mongers labeled him a "Jew banker" surrounded by other "Jew bankers" and "Jew brokers" who allegedly favored the Confederate cause. When Pennsylvania tried to repay its war debt to the Rothschilds in depreciated currency, the Rothschilds, through Belmont, asked the state to hold onto the bonds until gold became available. The state treasurer responded to Belmont: "We are willing to give you the pound of flesh, but not one drop of Christian blood." Other attacks on

Belmont cited his service as chairman of the Democratic Party during the Civil War. In 1864 the *New York Times* noted that "the great Democratic Party has fallen so low that it has to seek a leader in the agent of foreign Jew bankers." The Rothschilds did not appreciate the unjustified denunciations of them either and asked the American consul in Frankfurt to inform the State Department that neither Baron Rothschild nor members of his family were aiding the Confederacy and that collectively they opposed slavery. The message was sent.[80]

The most publicized antisemitic incident during the conflagration occurred after General Ulysses S. Grant issued General Orders #11 that expelled all Jews from his military district in western Tennessee on December 17, 1862. Banishment followed a series of scandals during the previous year in which smuggling, thievery, chicanery, malfeasance, and irresponsibility were rampant inside of the army and in the surrounding area. Speculation and contraband abounded, cotton went to the highest bidders, trade permits were bought and bartered, and vagrants of all kinds were seen inside and around army camps. The army tried to establish guidelines for the exchange of goods but too many of the servicemen were either lax in carrying out directives or personally profiting from a chaotic situation. Legions of unsavory characters near army posts included Yankees, Confederates, treasury agents, army officers, and vagabonds; the motley crew seeking advantage included both southern and northern Jews. Although they were neither the most numerous nor the most iniquitous of the various characters causing problems, Jews were the most physically visible because their manners, accents, and surnames invited special attention. Existent American antisemitism made it inevitable that Jews would become a convenient scapegoat for the frustrations and annoyances experienced by the Union Army.[81] Isaac Leeser commented on this situation in the *Occident*: "It has been the fashion to call all who were engaged in smuggling or blockade running, as it was termed, Jews."[82] A scholarly assessment a century later noted: "General Orders No. 11 was a logical culmination of the history of anti-Semitism in Grant's army and his own intensifying bigotry."[83] American Jews, their spokesmen, and their newspapers loudly protested Grant's action and demanded its recision. Their complaints went directly to the White House where President Abraham Lincoln understood both the error of Grant's blanket condemnation and the justice of the Jews' demands. He rescinded the order,[84] although on January 21, 1863, General-in-Chief of the Army Henry W. Halleck informed Grant that "the President has no objection to you expelling traitors and Jew peddlers."[85]

Antisemitism also erupted in the South during the war. Antipathy to foreigners, which included Jews, was more severe in this region than in the North. Moreover, in ways that were easy to sense but more difficult to measure, observers noted that the region's intense commitment to fundamentalist Christianity encouraged greater disparagement of Jews. Larger numbers of people were wary of merchants and suspicious of their methods and acquisitions, and, since southerners suffered more atrocities and devastation than people in the North, they shared a heightened sense of frustration and despair. Jews were denounced for being insufficiently proslavery and disloyal to the Confederacy. It was commonly assumed that Jewish merchants hoarded merchandise and sold goods at extortionist prices, thereby intensifying the South's great deprivation. The diary of John Beauchamps Jones, a clerk in the Confederate War Department, reeked with antisemitic references; the citizens of Talbotton and Thomasville, Georgia, voted to expel Jews who lived in their towns; and members of the Confederate House of Representatives openly denounced Jews in their midst.[86]

For southerners, Judah P. Benjamin, who served successively as Attorney-General, Secretary of War, and Secretary of State in the Confederate government, stood out as the archetypal perfidious Jew. Benjamin had been born Jewish but, like August Belmont, married a non-Jew, raised his daughter as a Christian, and did not affiliate with any Jewish organizations. Southern antisemites resented him and although he suffered no undue attacks while an attorney in Louisiana, or as a United States Senator representing the state from 1853 through 1861, many confederates attributed military losses and diplomatic failures to his being Jewish.[87] A Methodist parson in Nashville, characterized Benjamin as "a little pilfering Jew . . . one of the tribe that murdered the Savior,"[88] stories circulated in the Confederate Army that he lived on "fine wines, fruits—the fat of the land,"[89] and one person even believed that the prayers of the Confederacy would have been more efficacious without a Jew in the Cabinet.[90]

Publicly articulated venom toward Jews in both the South and the North subsided considerably after the war ended, and the intensity of the enunciated prejudices abated. Basic sentiments, however, did not change. Thus as the nation prepared to enter the last third of the nineteenth century, clearly established views had been expressed about Jews. They were not Christians, although enormous efforts had been made to convert them to Christianity and Gentiles believed that Jews generally, although not necessarily individuals with whom Christians were personally acquainted, were crafty and dishonest in economic dealings. In times of crisis, moreover, Jews were considered outsiders rather than members of the community. All these ideas existed in Christian societies prior to the establishment of the British colonies and the United States but in the first six decades of the nineteenth century they solidified in the minds of the American people, natives and immigrants alike.

There were countervailing traditions, however, in the United States that allowed for divergence and difference, and many Americans, even some of the most prejudiced, engaged in self-congratulation as they pointed out that Jews and other "nonbelievers" had the right to live and prosper in this country even though they may have been misguided and wrongheaded. In addition, the skills that Jews possessed were often utilized in helping the nation develop and in small communities, on a one-to-one basis, and when crises were not in the offing, people's prejudices did not generally surface in public. And finally, as Jonathan Sarna has so insightfully observed, and as has been noted at the beginning of this chapter, it was the "mythical Jew" and not the Jew "next door" who was hated.[91]

Individual Jews either remained in their own enclaves, unknown to most other Americans, or else engaged in cordial relations with Gentiles. The "mythical Jews" possessed the enmity of the Christian world. During the last three decades of the nineteenth century the Jews became better known and had more frequent contacts with Gentiles. Nonetheless, this did not lead to altered perceptions; in times of crises, in fact, just as during the Civil War, Jews stood out as scapegoats. And as the newly rich of the Gilded Age assumed a pompous grandeur they especially wanted to separate themselves from Jews of equal wealth and accomplishment so that European nobility might give them the recognition they craved. Moreover, for Christians without established positions, associating with Jews never brought them status in society and could very well have endangered the mobility they sought.

3

The Emergence of an Antisemitic Society (1865–1900)

From the end of the Civil War until the beginning of the twentieth century, the United States witnessed the emergence of a full-fledged antisemitic society. Like the hysteria exhibited during the war, the institutionalized bigotry that developed afterwards reflected the biases of practically every stratum in society. As immigration figures soared, and as a significant Jewish presence emerged in the United States, people in every walk of life, from respectable working, middle, and upper classes to agrarian protesters, Protestant and Catholic spokesmen, and members of the lunatic fringe increasingly focused on the allegedly deleterious characteristics of Jews that they believed impinged on American lives. Having been thoroughly indoctrinated as children, and having absorbed conventional attitudes simply by living in the United States, Christians believed in the superiority of their faith and few Gentiles questioned the fact that the United States was, and of right ought to be, a Christian nation. They held on to traditional views of Jews as Christ-killers who remained obstinate in their determination not to accept the truthfulness of Christian teachings; as dishonest businessmen always out for material gain; and as strange, crass, and aloof individuals who insisted on standing apart from the community in which

35

they lived. Thus, despite their acceptance as citizens in the United States, the Jew was "everywhere an alien."[1] Stories and cartoons in popular periodicals of the era, such as *Puck, Judge,* and *Life,* reinforced these interpretations as they perpetuated the image of the Jew with an outlandish nose, hoarding his gold, and gesticulating wildly. Writers of American fiction also depicted the stereotypical evil Jew whom readers of Chaucer, Shakespeare, Marlowe, Scott, and Dickens knew so well.[2]

Offensive attitudes toward Jews were particularly apparent in the New York City area where a plurality of them dwelled after the Civil War. The lowest classes were often described as "Chatham Street Jews" because that avenue and its surrounding streets housed so many of their shops. Israelites were considered disagreeable foreigners whose presence was "a nuisance to any Christian neighborhood."[3] "Their social customs and habits, their pastimes, and the manner in which they spend the Sabbath," a contemporary observed, "are so unlike our own, that it is impossible to dwell with them with any comfort. When they get into a neighborhood, in any numbers, it is deserted by all others."[4] So pervasive was the image that one Jew lamented in 1872, "to a large share of America the Jew of Chatham Street is the typical Jew, and so the whole race suffers by being judged by its worst part."[5] Alternatively, Jewish efforts to Americanize also brought forth scornful observations. Charles Dana, editor of the New York *Sun,* wrote that

> Thousands of *skys* . . . were being made over into Gordons, without dropping their old *sky* characteristics or taking on manners and accents belonging to their new name.[6]

The dislike expressed by Dana quickly appeared in business circles. Financiers resented Jews who had profited greatly as clothing manufacturers and retail merchants during the war and who were now influential among the city's bankers. The generally negative impressions Americans cherished of Jews created a simmering antagonism that would soon erupt in public.[7]

American Jews suffered also from the suspicions of Gentile businessmen. After a series of fires occurred in New York City's clothing district in 1866, as well as in several Jewish owned properties in the South and the West, insurance companies assumed that the perpetrators were Jews destroying their own businesses to collect insurance money. Within months seven major fire insurance companies—Aetna, Manhattan, Niagara, Germania, Hanover, Phoenix, and the Republic—agreed not to insure Jewish businessmen in the future. The ban was not the first time that insurance companies had discriminated against Jews; a Georgia company had already classified them in 1852 as "people of doubtful reputation."[8]

The 1866 pact among insurance companies to cease dealing with Jews was supposedly a secret but word leaked out and unleashed Jewish fury. Jewish businessmen held protest meetings in Nashville, St. Louis, Richmond, Cleveland, and New York City. So many of them cancelled existing policies that the seven companies lost business and shortly thereafter began wooing Jews again. Not a single case of arson was ever proven against any Jewish merchant[9] yet "adjuster," a correspondent of the *Banking and Insurance Chronicle*, stated that "not one in twenty of these [fire] claims is honestly made up." The editors of the journal responded with skepticism:

> If "Adjuster" will furnish us the figures showing the over-proportion of losses on this property of Jews, of any class, we will not only publish them, but will call to them the special attention of every insurance paper in the country.[10]

Neither "adjuster" nor anyone else accepted the challenge but specific facts were less important than general impressions. No matter how Jews conducted their businesses, they met with suspicion and distrust from Gentiles.

The bigotry adult Jews experienced was visited on their children as well. Historian Jonathan Sarna tells us that "the most common anti-Jewish schoolchild ditty in the English speaking world of the nineteenth century" was a vicious three-line attack with the title, "The Pork on the Fork." Repeated in books and on the lips of children who may have liked the rhyme or who enjoyed taunting an odd Jew, it went:

> I had a piece of pork, I put it on a fork
> And gave it to the curly-headed Jew.
> Pork, Pork, Pork, Jew, Jew, Jew.[11]

Christian hostility toward Jews was especially harsh during the Easter season and Jewish children, in particular, noticed increased animosity from their fellows. One Jew who grew up in Syracuse, New York, reminisced how on Good Friday

> the readings always related to the crucifixion and the teachers seemed to have the habit of intoning their reading, and especially when the word 'Jew' was mentioned in such a manner as to convey the idea not only of contempt, but also of hatred. This was always followed during the recess and for several days after by the most hostile demeanor on the part of the Christian boys and girls of the school, some of whom resorted to physical violence and most of them to the calling of names and the making of scurrilous remarks.[12]

Jews who attended schools in other cities recalled similarly cruel treatment. Emanuel Haldeman remembered his Philadelphia schooldays as a period in which he "was continually persecuted by Irish-Catholic children who accused him of being a 'Christ-killer.'" Nina Morais, another Philadelphian, noted that at the schools that she attended, absences for religious purposes were punishable offenses. Morais, moreover, pointed out how one teacher stated with pity in her voice, "I am sorry for you, it is your misfortune, not your fault, that you are a Jew." In Cincinnati the most prestigious college preparatory schools for girls refused to accept Jewish students while in Nashville two Jewish girls withdrew from a private school in 1878 because Christian mothers threatened to pull their daughters out if the non-Christians remained.[13]

College students and young adults also fell victim to Christian prejudices. Oscar Straus, later Secretary of Commerce during Theodore Roosevelt's administration, entered college in 1867 but was barred from the undergraduate literary society because of his faith. Bernard Baruch, the financier and friend of many Democratic politicos in the twentieth century, was blackballed by fraternities at the City College of New York in the 1880s for the same reason; and the New York Bar refused to admit a candidate in 1877 simply because he was a Jew.[14]

Clearly the religious attitudes held by millions of Americans contributed to antisemitic attitudes. While the whole population may not have been devout churchgoers, few questioned the assumption that this was a Christian country. Despite the First Amendment, periodic revivals exhorted Christians to support political changes designed to institutionalize Christian beliefs and undermine individual freedoms. For decades, but especially in the 1860s and the early 1870s, a movement existed for a constitutional amendment acknowledging the authority of God, Jesus, and spiritual law in the United States. Protestant revivalists supported this goal and preached an evangelical Christianity as the finest base for American society. Many devout Christians also believed deeply that it was their "religious duty" to convert the Jews and save them from eternal damnation. The proposed amendment died in the House of Representatives in 1874 although the high point of the post Civil War revivalist crusade occurred in the winter of 1875–1876 when the Reverend Dwight L. Moody toured the country reminding his listeners that the Jews killed their Savior and enjoining audiences to convert the Jews to Christianity.[15]

Another facet of Protestant fundamentalism appeared in the Sabbath Crusade as devotees of the movement vigorously attempted to have their beliefs requiring a day of rest on Sundays written into state and national legislation. No other social issue save Prohibition motivated as many committed Christians to petition the government for redress as did the Sunday law movement. Most of the people involved in this endeavor believed that "God's law should be the basis of human legislation" and that whatever they considered sinful should be outlawed. Zealous in pursuit of their cause, which reached its high point during the years 1879–1892, they predicted devastation for the United States lest the government proclaim its loyalty to Jesus.[16]

Despite the farfetched goals of religious zealots in the Gilded Age, social discrimination loomed as a far more significant problem for Jews who had already acquired wealth in the United States. Social exclusivity, which had not been a major factor in ante-bellum America, emerged in full bloom during the last third of the century. American upper classes and parvenus embraced an idolization of mimicking everything they associated with the British artistocracy on their way to solidifying what they believed to be a superior enclave within a republican framework. In this niche they carved out for themselves they found no room for Jews and, in fact, castigated the wealthier among them for exhibiting characteristics that the parvenus so fulsomely possessed.

By being so enamorored of British gentry, America's aristocracy tried to outdo them in all of their habits. An 1887 article, for example, reported the following interchange:

> "What is an American?" an English traveler asked, some years ago; and answered in the same breath: "A more or less successfully disguised Englishman."
> "Very much disguised," I remarked.
> "I must disagree with you," said the Briton; "the less disguised he is—the nearer he comes to the English prototype—the better he is satisfied with himself."[17]

Nevertheless, the need for upper-class Americans to ape British behavior includes an ironic twist, especially in their ostracism of Jews. While the British may have been quite antisemitic, the Prince of Wales always included Jews within his own personal circles and other European Jews of high status moved more easily in society abroad than did their similarly situated coreligionists in the United States. Another irony is that while Jews were chastised for being clannish and remaining distant from Gentiles, the social exclusion began with those who wished to remain aloof from the Jews. Jews, moreover, had tried to emulate the most successful Christians in their deportment. Yet they were criticized for conduct that some of the wealthiest parvenus like Jim Fisk and "Diamond Jim" Brady displayed to a much greater extent. The rapid Americanization and financial success of German Jews began that basic contradiction that persisted well into the twentieth century. While urging Jews not to be so clannish, barriers were erected to prevent those who wished to interact with non-Jews from intermingling with their Christian counterparts. And to justify their exclusionary positions, members of the American gentry labeled wealthy Jews as crass and uncouth.[18]

The most singular incident occurred in 1877 when Judge Henry Hilton barred the German-Jewish banker, Joseph Seligman, from registering as a guest at the Grand Union Hotel in Saratoga Springs, New York. Although not a unique event, it was notable since Seligman was a banker of great prominence and a friend of the late Abraham Lincoln and Ulysses S. Grant. Therefore the affair made headlines in places as distant as New York City, Philadelphia, Cleveland, and Chicago.

Newspaper editorials differed as to whether Hilton was right or not in barring Jews but the distasteful episode indicated that no matter how well-to-do or refined Jews might be they were socially undesirable and some later chroniclers erroneously marked this incident as the beginning of antisemitism in America.[19]

Another well-publicized exclusion of Jews occurred two years later when Austin Corbin, president of the Manhattan Beach Corporation in Coney Island, announced that he would not allow them in his exclusive hotel. Corbin unabashedly stated:

> Personally I am opposed to Jews. They are a pretentious class who expect three times as much for their money as other people. They give us more trouble on our [rail]road and in our hotel than we can stand. Another thing is that they are driving away the class of people who are beginning to make Coney Island the most fashionable and magnificent watering place in the world.[20]

He also added that his hotel would not be attractive to the "highest social element" if he accepted Jews. "We do not like Jews as a class," he told a reporter for the *New York Herald*. "As a rule they make themselves offensive. They are a detestable and vulgar people."[21] Just as the Hilton-Seligman affair in 1877 aroused newspaper controversy so too did Corbin's stand on refusing to allow Jewish guests at his Coney Island hotel. A storm of protests resulted as newspapers received over 150 letters to the editor expressing different opinions of Corbin's position. The controversy, however, did not move the hotel magnate.

In June 1879 a meeting of the short-lived American Society for the Suppression of the Jews took place at the Grand Union Hotel in Saratoga. Judge Henry Hilton presided over a group of perhaps 100 people while Austin Corbin served as Secretary. Corbin thought that refined Christians ought to put themselves on record as opposed to the tolerance of Jews in the United States. "If this is a free country," he asked an enthusiastic audience, "why can't we be free of the Jews?" Listeners greeted this remark with loud applause. Those present then adopted resolutions not to elect Jews to public office, not to attend theatres where Jewish composers wrote the music or where Jewish actors performed, not to buy or read books by Jewish authors, not to ride on Jewish-owned railroads, and not to do business with Jewish-owned insurance companies. In conclusion, the gathering decided that "Jews must be excluded from all first class society."[22]

It was the wealthiest Americans who put that resolution into effect. America's upper classes were seeking ways to distinguish themselves and disassociating from Jews was one of the first items on their agenda. Within the next decade gentlemen's clubs, exclusive resorts, and private schools began to bar Jews. In 1879 the *Elite Directory of San Francisco* included both a Christian "Calling and Address List" and a "Jewish List." By the early 1880s social discrimination against Jews was obvious wherever prominent members of Gentile society gathered in cities like Los Angeles, New Orleans, Mobile, Portland, Oregon, Cincinnati, Columbus, Akron, Cleveland, Denver, Philadelphia, Rochester, N.Y., and, of course, Boston and New York.[23]

During the 1880s patricians finally erected significant dividers between themselves and others whom they considered less worthy members of society. In 1881 President Charles Eliot of Harvard built a summer home in Northeast Harbor, Maine, beginning the trend of the upper crust spending summers in cooler climates. Brookline, Massachusetts, elites founded "The Country Club" the following year and in 1883 the "Sons of the Revolution" inaugurated the movement for exclusive patriotic societies open only to those who could trace their lineage to colonial America. Shortly thereafter Endicott Peabody founded Groton, modeled on the tradition of the British boys' public schools, and soon other secondary institutions designed for the male scions of the nation's elite sprang up in the Northeast. By 1887, with the appearance of the first *Social Register,* people who counted in society had a book to inform them of who belonged to their most exclusive circles.[24]

The increasing distance that prosperous Gentiles put between themselves and Jews was especially hurtful to those Jews who had succeeded in various walks of American life. Some tried to make light of the situation. In a self-effacing comment in 1883 the editors of *The American Hebrew,* the publication of well-to-do, reform-minded Jews, wrote:

> If, in truth, New York knickerbockerism is going to avoid us, and ask, "Who's Who?" when a Jew is a candidate for social entree, we of older patricianship can well afford to smile at the question as we repeat it to ourselves with differing application. . . . We have our own social circle . . . we are not desirous of seeking other circles—certainly not when unwelcome there.[25]

Other Jews, however, were angry about the exclusions and plainly let the world know what they thought. Nina Morais, a Sephardic Jew from an old Philadelphia family, noted how "in the popular mind the Jew is never judged as an individual, but as a specimen of a whole race whose members are identically of the same kind." She described how non-Jews, in general, found the Jew

> an objectionable character, whose shrewdness and questionable dealings in trade enable him to wear large diamonds and flashy clothes. He raises his voice beyond the fashionable key, in a language execrable to the ears of the English-speaking people. For the proprieties and amenities of cultured life he has no regard. His conversation rings upon the key-note of the dollar; his literature is the quotations of the market. Mean in pence, he spends his pounds with an ostentation that shocks people. Of the higher sympathies he has none; the finer feelings he cannot appreciate. In a word, he is foreign—outlandish—a Jew.[26]

Thus the social exclusion that received widespread publicity in the late 1870s contributed to the blatant antisemitism that characterized the 1880s, a decade during which the image of the Jews deteriorated. Already thought of as the quintessential alien upstarts trying to wedge their way into restricted social circles, Jews then became increasingly criticized by groups outside of America's elites. Social discrimination, however, did not trickle down from one rung of society to another but appeared, almost simultaneously, in different social stratums. In 1887 a Methodist newspaper in North Carolina blamed the Jews for the fact that a Prohibition amendment to the state constitution failed to win enough votes for passage.[27] Two years later a Baptist publication noted that "the Hebrews are still as distinct a race among us as the Chinese" and indicated that "so long as the Hebrews remain a separate race—a caste, as it were—in American society, there will infallibly be a prejudice against them."[28]

During the 1890s Baptists officially opposed immigrants coming to urban America,[29] and their ministers expressed a strong distaste for newcomers. In an 1893 sermon in Baltimore the Rev. A. H. Tuttle preached that the Jew was

> greedy, merciless, tricky, vengeful—a veritable Shylock who loses every sentiment of humanity in his greed. Of all the creatures who have befouled the earth, the Jew is the slimiest.[30]

That same year a Detroit minister argued that if the Jew would identify in spirit with Protestant Americans, cease maintaining social relations exclusively with other Jews, and mix with other Americans "much of the prejudice that exists against him would soon pass away." Until then, he condemned the Jews with traditional and time-worn accusations that had little to do with anyone's religious heritage:

> certainly the craftiness of the Jew in trade, his success as a schemer, his ability to get hold of the purse strings of the nations while not amalgamating with them, his insatiable thirst to get the best end of the bargain, his unscrupulousness in many of his money transactions, has, far more than any religious prejudice that may exist against him, created the anti-Semitic feeling both on the continent of Europe and here.[31]

Another Baptist minister in Detroit feared that the new Jewish immigrants would destroy America's "grand Christian civilization" while a visiting minister from Scranton told parishoners of Detroit's Fort Street Presbyterian Church that immigrant Russian Jews constituted a greater threat to the United States than the "yellow peril."[32] In 1896 a Baptist minister in New York, repeating what others had already speculated on, reiterated worn-out shibboleths that the Jews controlled the largest commercial interests in New York, owned the best parcels of land, and the best palaces in Europe. "The Jew," he concluded, "is the financial master of the world."[33]

Nondenominational American weekly and monthly periodicals, which mushroomed in the 1880s, also contributed to the negative image of Jews. These magazines, so reflective of contemporary sentiments, portrayed Jews to readers as intelligent, cunning, clannish, cowardly, unpatriotic, averse to manual labor, and guilty of deicide. Jewish business acumen was seen less as a virtue than as a quality that might lead to Jewish control of the United States and even dominance of the world's economy.[34]

Jews were not the only quarry of American bigots in the 1880s.[35] Congress restricted further Chinese immigration into the United States in 1882, anarchists and foreigners in general suffered from the public's shock after an 1885 bomb-throwing incident at Haymarket Square in Chicago resulted in the killing of a policeman, and a movement to bar Catholics from American shores erupted with the formation of the American Protective Association (APA) in Clinton, Iowa, in 1887. Members of the APA feared Catholics might soon come to dominate the nation and reverse established policies and traditions. Anti-Catholicism could trace its roots back to the colonial era and to the fear of Papists in Protestant Europe before that time. Nonetheless its reappearance so strongly in the 1880s was just one more indication of the prevalent nativism in a decade of economic uncertainty, fundamentalist rigidity, and upper-class parochialism.

Therefore, although antisemitism was not the only or primary prejudice of other Americans in the 1880s, it continued growing. Expanded east European Jewish immigration threatened people who feared an erosion of American values. To some, the augmented numbers as the decade proceeded showed signs of becoming a "menace" to America as the percentage of east European Jews among all incoming peoples to the United States moved from 0.9 percent of the total annual influx in 1881 to 6.5 percent of the gross in 1887.[36]

Both Jews and Gentiles viewed the rise as an ominous sign. For different reasons members of labor unions, Baptists, and Catholics, among others, noticed the increased numbers and questioned whether the types of new immigrants entering the country were really in the nation's best interests. A Protestant minister, Josiah Strong, wrote in 1885 that the newcomers possessed an inferior culture and required the inculcation of "superior Anglo-Saxon ways."[37] Many Americanized Jews worried that coreligionists among the recent arrivals, orthodox Jews with traditional customs, would undermine what they foolishly believed was their acceptance by non-Jewish Americans. They feared that these east Europeans would cause new outbursts and an upsurge of American antisemitism. Future events justified such apprehensions.

Prejudice toward foreigners and Jews escalated in the 1890s as calls for immigration restriction arose. Across America Protestants became even more outspoken on this question even as Catholics and Jews refrained from public denunciations because it was their coreligionists who were the targets of the exclusionists. Typically, Boston led the assault and in 1894 young scions of patricians banded together to form the Immigration Restriction League (IRL). The nucleus of the new group consisted of idealistic recent Harvard graduates concerned about maintaining America's Anglo-Saxon heritage and traditions. Led by Prescott F. Hall, who had just embarked on a career as an attorney, the IRL held meetings for an intimate circle that rarely exceeded a dozen participants. These men believed the nation could not endure, nor could its values be sustained, without restricting "the scum of Europe" from American shores. Boston's latest immigrants consisted not of sturdy Anglo-Saxons with a proud heritage, but people "of objectionable races" from parts of Europe Americans had never visited. The habits, values, and behavior of these "undesirables" struck members of the IRL as anathema to everything that "real" Americans cherished. These patricians feared that eventually the foreigners would undermine American culture and superimpose their own throughout the land.[38]

The IRL cause attracted some of the most prestigious names in the Boston community including historian John Fiske, who served as president of the organization, A. Lawrence Lowell, who would assume the presidency of Harvard University in 1909, and Leverett Saltonstall, father of the future Massachusetts Governor and United States Senator. In its early years, the late historian Barbara Solomon told us, the Immigration Restriction League acted as "the mouthpiece of older New Englanders who had wearied of the burdens of humanitarianism." Within the next two decades the group gathered support from similar clubs in New York, Philadelphia, Albany, Chicago, San Francisco, and the states of Alabama and Montana. Prominent academicians like Richard Ely, John R. Commons, and E. A. Ross, all of whom would serve on the University of Wisconsin faculty during the Progressive era, also offered their names and talents to this cause.[39]

The specific means that IRL members embraced to achieve their goal was a literacy test. They believed that the least desirable foreign elements were illiterate and that a simple, nondiscriminatory, test would weed them out.[40] In Henry Cabot Lodge, the patrician United States Senator from Massachusetts, the IRL found its political spokesman in Washington. At one time a supporter of free and unlimited immigration, in the 1880s Lodge had come to see the "ineradicable differences between the races of men."[41] He now questioned whether those who came from countries unused to self-government could be counted on to uphold the American system. Spurred by the encouragement of his fellow Bostonians, and sensing qualms of other Americans about the immigrant inflow, Lodge proposed to amend the immigration and contract-labor acts of 1891 requiring all persons over 14 years of age to read and write "the language of their native country or some other language." His proposal exempted parents and grandparents of such persons and, as eventually amended, excluded females as well. Lodge argued that the suggested amendment would keep out "the totally ignorant" but acknowledged

> that the illiteracy test will bear most heavily upon the Italians, Russians, Poles, Hungarians, Greeks, and Asiatics, and very lightly, or not at all, upon English-speaking emigrants or Germans, Scandinavians, and French. In other words, the races most affected by the illiteracy test are those whose emigration to this country has begun within the last twenty years and swelled rapidly to enormous proportions, races with which the English-speaking people have never hitherto assimilated, and who are most alien to the great body of the people of the United States.

These were the people, Lodge went on, who dwelled "in congested masses in our great cities" and who furnished most of the urban criminals, paupers, and juvenile delinquents.[42]

Lodge's campaign reflected the bias of most Americans favoring restricted immigration but it came just a little too late to accomplish its goal. Submitted first in March 1896, his proposal did not win Congressional approval in an amended form until February 1897. President Grover Cleveland vetoed the bill just before leaving office in March. The intensity of America's anti-immigrant feeling declined somewhat as prosperity returned and foreign affairs came to dominate American thoughts after 1896, although the IRL continued its quest for a literacy bill.

Of greater concern to the Jews at the time, however, was the viciousness of Catholics toward them. Opposition to Jews from urban Catholics, whether they were Irish, Polish, or German in origin, grew more out of a religious heritage than from any contemporary frustrations. Historian Egal Feldman, who studied American Christian attitudes toward Jews, has written:

> it is difficult to measure precisely the intensity of anti-Semitism in the American Catholic community, although its presence was unmistakable. It varied from city to city, from editor to editor, even from week to week in the same periodical. Although the Church did not endorse the ancient aberration, there is little evidence that its leaders, not even the most popular and venerated prelate of his time, Archbishop John Ireland, tried to curb it.[43]

In the 1870s the *Irish World* printed a cartoon showing Baron Rothschild dressed as Shylock and wielding a knife; an Irish rug merchant in upstate New York advised others: "Don't sell to them thieven' Jews"; and Boston's Catholic schools mentioned Jews only in connection with Christ's death or when accusing them of being jealous of the Savior.[44]

Catholic journals in the United States were also harsh in the late nineteenth century as they reflected the antisemitic views of members of the American hierarchy. Baltimore's *Katholische Volkszeitung* reprinted antisemitic articles from Germany, and other Catholic newspapers like the *Freeman* and the *Catholic Mirror* contained materials hostile to Jews. In March 1889 the *Catholic World* included an essay by A. de Glequier of Vienna entitled "The Antisemitic Movement in Europe." It was perhaps the most malevolent antisemitic article ever published in a reputable journal in the United States to that date.[45] The piece began with the old canard that antisemitism originally arose because

> the Hebrews, the chosen people of old, possessed even in pre-Christian times an ineradicable propensity to worship Mammon. The selling of the birth-right was a bargain, and they have bargained ever since.[46]

Glequier recounted how the Jews had rejected Jesus, how they were strangers wherever they settled, and how their concern with "money, money, money" reverberated in every century. Wherever they went, he argued, Christians fell prey to the wiles of Jews who took everything of value. "The aggregate money power of the Jews all over the world," he continued,

> is something incredible. In all the leading banking institutes of the world the Jews hold the reins. The power of the Rothschilds, the Bleichroeders, and a host of others is so great that modern governments are practically dependent upon them in foreign policy.

Glequier noted that "if the Jews are dangerous corrosive elements of civilized society, they are so because of their religion."[47]

The dominant elements in the American Catholic church appeared to be comfortable with Glequier's positions. Moreover, in a series that also ran in the *Catholic World* from 1891 through 1893, Manuel Perez Villamil presented what one scholar later characterized as "an extremely anti-Semitic interpretation of the Jewish role in Spanish history."[48] Father Frank McGloin and layman Thomas P. Hart, editors of Catholic diocesan newspapers in New Orleans and Cincinnati, respectively, also had reputations for being quite antisemitic. So, too, did Father Michael Earls, S.J., a close associate of Archbishop, later Cardinal, William O'Connell in Boston. Earls' *Marie of the House of d'Anters* characterized Jews as villainous socialists likely to advocate free love.[49]

One 1992 scholarly analysis of editorials published in Baltimore's three Catholic newspapers from 1890 through 1924 found that all the editors harbored traditional antisemitic images, portrayed Jews as Christ-killers, and were hostile toward the Rothschilds. In the *Katholische Volkszeitung*, over 80 percent of all editorials on Jews during this 34-year period were virulently antisemitic and one of the essays in 1900 specifically criticized President William McKinley as a "puppet of international Jewry." At the inauguration of the victorious President and his running mate, Theodore Roosevelt, on March 4, 1901, the *Katholische Volkszeitung* wrote that the event had been attended by 40 people with the name of "Rosenfeld." The author of this Baltimore newspaper survey believed that sentiments among all German Catholics in the

United States were similar to those expressed in the Baltimore Catholic newspapers. "There was probably no way of being a member of the German-Catholic *milieu*," he concluded, "without being antisemitic."[50]

Other aspects of Catholic behavior also suggest that church teachings and values led to antisemitism among the faithful. One Irish woman who had arrived in the United States in the 1890s quit her job as a domestic two months after she had begun working. She told her employer that she had never seen a Jew before and had she known of the family's religion she certainly would not have taken the position. "I beg your pardon, ma'am," she recalled herself saying, "but I can't eat the bread of them as crucified the Savior."[51] The attitude of this maid was that of the North End of Boston, where a social worker noted "a special antipathy on the part of the Irish for the Jews."[52] A parish priest complained that Jewish landlords were "encroaching upon our domain"[53] while Irish policemen in Boston not only harassed Jewish peddlers and grocers but also disrupted Jewish weddings on the pretext that paid musicians violated city ordinances prohibiting work on Sundays. Similarly, in Brooklyn, a Catholic priest complained to the courts that the running of sewing machines in a building near his church disturbed congregants during the worship services.[54]

Events abroad also contributed to Catholic attitudes toward Jews in America. A major antisemitic strain was regaining strength in Europe. One saw the sentiment in the renewed emphasis on nationalism in Bismark's 1870s *kulturkampf,* in the pan-slavic movements in eastern Europe during the last decades of the nineteenth century, and, most dramatically, in the Dreyfus affair in France during the 1890s when a Jewish captain in the Army was unjustly charged with, and convicted of, espionage. There army officers, titled upper classes, antibourgeois, antirepublican forces and the Catholic church aligned themselves with those who thought Dreyfus guilty. The reasons for this association of peoples, in the words of one scholar, were that "the wealthy young Jews represented all that they detested. Anti-Semitism ran through the veins of these monarchists and Catholics."[55]

Not only in France but throughout Europe antisemitism permeated society from the highest to the lowest levels. On Christmas night in 1881 Christians attacked Jews on the streets of Warsaw in the worst European massacre of a minority since 1572 when the Huguenots were shot down on the streets of Paris. Within months of the Warsaw bloodbath the Russian Tsar promulgated his May Laws severely restricting the rights of Jews to live where they chose, own and manage real property, and even their ability to conduct business in general. During the same era German and Austrian universities reflected prevailing antisemitic sentiments while throughout Europe Jewish financiers were held responsible for every crash in banking endeavors.[56]

In the Hapsburg Empire Jews could not be received at Court[57] and for that reason the United States remained without a Minister to the

Austro-Hungarian monarchy for some time in the mid-1880s. President
Grover Cleveland's first nominee, Anthony M. Keiley of Virginia, had
a Jewish wife and the government in Vienna refused to accept Keiley
as a representative on that account. Cleveland's Secretary of States,
James Bayard, rebuked the Austrians for their stance with a particularly
sharp note that emphasized the United States' official commitment to
an open and nonsectarian society:

> It is not within the power of the President, nor the Congress, nor any
> Judicial tribunal of the United States to take or even hear testimony in
> any mode, or to inquire into or decide upon the religious belief of any
> official, and the proposition to allow this to be done by any foreign gov-
> ernment is necessarily inadmissible. . . . To suffer an infraction of this
> essential principle would lead to a disfranchisement of our citizens
> because of their religious belief, and thus impair or destroy the most
> important end which our constitution of government was intended to
> secure.[58]

Keiley, who appreciated the government's support, did not wish to
prolong the awkwardness of the situation and resigned his commission.[59]

During the 1890s, antisemitic sentiments visible earlier crystallized,
intensified, and evolved into more urgent apprehensions. This syn-
drome was shared by a larger group of Americans than had previously
been affected. One scholar asserted that "by the 1890's a doctrine of
anti-Semitism had penetrated virtually every creedal, social and economic segment of society." Moreover, European ideas about the racial
inferiority of Jews made some sense when economic and foreign affairs
took on greater import to the American people. Nationalistic ferment
swirled particularly around the cult of Anglo-Saxonism,[60] the belief that
Americans had a mission to remake the world in their own image, and
the conviction that "lesser" peoples and individuals ought to accept
American Christian precepts. America's ideological crusade, therefore,
became clearer and more pronounced just as southern and eastern Europeans gravitated in greater numbers to our shores, as the most mean-
ingful agrarian protest movement in the nation's history crested, and
as the fear of displacement surged through the bastions of patrician
conservatism in New England, the Middle Atlantic states, and the Old
Northwest.

No economic group in the United States was angrier or better articulated its goals in the early part of the decade than the agrarians. Since
1865 farmers had seen prices of their produce drop on the world markets; their costs for manufactured goods, credit, and transportation
soared; and they were saddled with debt. Then, in 1873, the federal
government discontinued the coinage of silver, causing its price to fall
while that of gold rose. An ensuing depression created a growing furor
among farmers. The interplay of products and prices on the world mar-

ket involved an intermixture of complexities far too difficult for them—or even some of the most sophisticated politicians—to grasp. Nonetheless, angry people believed that someone or some group was responsible for their losses and difficulties. A scapegoat was clearly needed.

Toward the end of the 1870s those suffering most from the agricultural depression increasingly began to identify Jews as sources of their woes.[61] Agrarian antisemitism focused on the difficulties that they perceived had resulted from the federal government's decision to cease coining silver in 1873. Farmers especially disliked Jews, the "detested middlemen" who did not work with their hands or till the soil, and whom they associated with wealthy bankers who had allegedly forced the demonetization of silver. Agrarians wanted silver coinage resumed, and supported the Greenback Party in the hope that the federal government would redeem in gold all the paper greenbacks issued during the Civil War.[62] Those who tilled the soil in the 1870s did not focus on Jews as the major source of their difficulties, but as the plight of the farmers worsened in the 1880s and 1890s antisemitic rhetoric increased because it had already existed as part of the general American culture and had been expressed earlier in agrarian circles. "At its most unbalanced," historian Irwin Unger tells us, "rural anti-bankism of the 1860's and 1870's was tinged with anti-Semitism."[63]

To protest conditions under which they had to live and work farmers' alliances grew up in the Midwest and the South during the 1880s. In 1890 the two major groups united into the People's or Populist Party. By 1892 the Populists had developed a major program for significant change in American society and had even won several important congressional and gubernatorial elections. Their future seemed pregnant with possibilities but then in 1893 a terrible depression swept the country. In its wake, conservatives retreated to the comforts of establishment politics and the gold standard, and voted overwhelmingly for the Republicans, while Populists banked everything on their campaign for free silver in the presidential election of 1896. After Republican governor William McKinley of Ohio, a mild supporter of the gold standard, emerged victorious in the presidential election that year, the Populists faded away; many of their ideas, however, re-emerged in legislation passed by Congress and the state legislatures during the next three decades.

Unfortunately, one Populist notion voiced often concerned the influence of Jews over the American economy and the American banking scene. Jews were not the major concern of Populists but the agrarian protesters frequently sprinkled their rhetorical utterances with references to "Rothschild" and "Shylock." Even more significant, Populists strengthened their cause by using religious metaphors to link money with a Jewish conspiracy. Thus, in 1896, Democratic presidential candidate William Jennings Bryan, speaking in an idiom Protestant fun-

damentalists were fully conversant with, could easily intersperse biblical imagery with economic necessity when he thundered, "You shall
not press down upon the brow of labor this crown of thorns, you shall
not crucify mankind upon a cross of gold." The antisemitism evoked
by the metaphor of the crucifixion was powerful and appealed to rural
Protestants who possessed a similar religious and cultural heritage
with other Americans in the South and the West. Such imagery and
figures of speech reflected the conventional wisdom in their own circles
rather than any particularly radical philosophy.[64]

An Associated Press reporter, who had already covered the Populist convention that met in St. Louis earlier that year to choose a presidential candidate in 1896, wrote at the time:

> One of the striking things about the Populist convention . . . is the extraor-
> dinary hatred of the Jewish race. It is not possible to go into any hotel in
> the city without hearing the most bitter denunciation of the Jewish race
> as a class and of particular Jews who happen to have prospered in the
> world.[65]

Historian Irwin Unger suggested one reason for such fervid antagonism toward Jews:

> To Populists, the Jew was a "non-producer," a mere manipulator of
> money, a parasite, and at the same time representative of the sinister and
> forbidding power of international finance. "In the evil conditions, made
> by bad laws, the Jews alone thrive," wrote Ignatius Donnelly. "The rea-
> son is they deal only in money; they have no belief in farming, manu-
> facturing, or any other industry; they are mere money-managers. As
> everything else goes down, money rises in value and those who control
> it become masters of the world."[66]

Racist ideas traveled easily in agrarian circles in the late nineteenth
century and served as simple explanations for complex economic
changes threatening most people. They existed before the rise of the
Populists, and would be resurrected in the twentieth century by disillusioned individuals like Henry Ford, demagogic leaders like Father
Charles Coughlin, and disgruntled farmers during the 1980s. One did
not have to be a Populist, however, to view Jews as exploitative capi-
talists. Antisemitism was "endemic throughout the rural West," Victor
Ferkiss has written, where "the prairie farmer associated the Jew with
the merchant, the financier, and the corrupt and domineering Eastern
city."[67]

Traversing the United States in the 1890s one found abundant evidence of open hostility to Jews in both rural and urban areas. In northeastern Iowa and northwestern Illinois Jews were not welcome in the
civil and general affairs of the communities while intermarried Jews
were scorned by their own coreligionists and barely tolerated by Christians. Edna Ferber, the Jewish author and playwright, grew up in the

Welsh mining town of Ottuma, Iowa, and later wrote about the community's unreasoning and widespread antisemitism. Ottuma's half-dozen Jewish families were continually ridiculed and subjected to anti-semitic verbal attacks. The bitterness of her seven years in Ottuma made such a deep impression that in her 1939 autobiography Ferber acknowledged how "the brutality and ignorance of that little town penetrated to my consciousness."[68]

Antisemitism also flourished in urban America. There was hardly a major city in the United States where it could not be found. In Cleveland Jews could no longer join the German *sangerbunds*, the Knights of Pythias, or the city's exclusive Union Club.[69] Detroit's "Protestant churches were social garrets in 1880," one scholar has written, and "remained so in 1940, in the suburbs and in the city center, in middle-class as in working class communities." Louis D. Brandeis, while living in Boston where the faith of his ancestors kept him a perpetual outsider, believed that antisemitism reached its pinnacle in Detroit.[70] The black-balling of Jewish businessman Herman Freund from the Detroit Ath-letic League in March 1893, however, engendered a fierce debate in the editorial pages of the Detroit *Evening News* on "Why Is the Jew Hated?"[71] "What is the point or characteristic in the Jew, which sets the Christians' teeth on edge all over the world?" one of the editorials asked.[72] The conclusion, another of the opinions decided, was that Jews had brought the world's wrath down on themselves because they refused to marry or integrate with Christians. By choosing to preserve themselves as a separate nation they had invited others to scorn them.[73]

The theme of Jewish separatism from the larger society was also commented on by a San Franciscan who complained that the Jew never assimilates, "he commingles but never becomes one with; he associates, but never in sufficiently intimate relations to fuse and amalgamate."[74] Yet San Francisco's Jewish population generally fared better in terms of acceptance than coreligionists elsewhere in the country. Of those who arrived in the city before 1852 there was a fairly high persistence rate—an indication that things had gone well for them—and by the end of the century, despite social discrimination at the highest levels, San Francisco's Jews felt comfortable with their positions.[75] Nonethe-less, a prominent San Franciscan objected when asked to write about Jewish contributions to the city's development. He claimed that a sim-ilar question had not been put to members of other groups. And he also complained about something that all minorities earlier, then, and later resented: that whenever a coreligionist or fellow ethnic was engaged in some negative or criminal activity newspapers always iden-tified the culprit by the individual's ethnic or religious background.[76] In the late nineteenth century, however, Californians seemed more tol-erant of Jews than were compatriots in the East.

Philadelphia and Boston were among the most difficult cities for Jews to find acceptance. In the city of brotherly love, a wide-ranging plan to improve the quality of life for slum dwellers foundered because it was formulated by a Jew. Several Presbyterian church members refused to participate in the project and when a wealthy Jewish woman wrote a generous check to support the tenement improvement scheme it was returned with the suggestion that she spend her money elsewhere. Christians killed the reform plan because they simply would not work with Jews no matter how worthy the cause.[77] And the prohibition on contact was even more absolute at the highest levels. A recent chronicler of the fortunes of the Guggenheim family observed that "society in Philadelphia still remained so exclusive that no newcomer, and certainly no Jew, could hope to get even close to it, no matter how wealthy he became."[78]

Antisemitism also prevailed among the old stock patrician families in Boston.[79] That Brahmin barriers remained almost insuperable for Jews to traverse no matter how accomplished the individual or how close the working relationship was amply proven by Louis D. Brandeis' experience. Brandeis was Sam Warren's law partner, and, as much as any successful Jew, aped the manners and mores of the city's upper classes. But he was not invited to Warren's 1891 wedding because the bride refused to entertain a Jew. (When Brandeis married, only a few of his prominent clients sent gifts or left calling cards.) Warren at first remained a loyal friend of Brandeis and when Warren helped found, and served as President of, the Dedham Polo Club Brandeis received an invitation to join. But once in the club he was allowed to "flock by himself" because most other members chose not to associate with a Jew.[80]

If the "better classes" ostracized even so successful a man as Brandeis, they hardly hid their antipathy to Jewish immigrants. In 1891 a patronizing author described how Jews had come to the city after the Civil War, settled in the North End, and purchased a number of the tenements and stores in the area. These Jews allegedly bore "the marks of the degradation of centuries upon them"[81] when they arrived, the writer observed, yet they subsequently earned respectable livings as peddlers and then as merchants. This pious Unitarian commentator knew that not all Jews were dishonest, but "when a Jew is dishonest, he can be meaner and smaller and sharper in his dishonesty than most others." The writer claimed that the majority of Jews did not "seem to be fond of soap and water" and so their "tenements rarely present the neat appearance of being scrubbed that one finds in the houses, however poor, of the sober Irish." Nonetheless he praised Jews for their philanthropy and their tendency to work hard. "The Jew is sober, industrious, and law-abiding; he minds his own business and cares for his own people. Taken all in all, his virtues outweigh his faults."[82]

Few non-Jewish Americans or immigrants agreed with that tepid evaluation. Poor urban Jews found themselves under continual physical assault in the northern and midwestern ghettos. Their daily lives often included assaults from street urchins and indifference from the policemen on the beat. Children pulled old men's whiskers, peddlers had to protect themselves from the taunts and stones of idle youth, and cops wielded night sticks indiscriminately. Jewish schoolboys were victimized by Irish, Polish, or German assailants whether they lived in Detroit, Chicago, Boston, New York, or Philadelphia. Complaints about hooligans were also made in Rochester, New York, Holyoke, Massachusetts, Baltimore and Brooklyn.[83] As far as most observers could tell no provocation was necessary. "No cause was required for hostilities to start," the historian of Brownsville, a section in Brooklyn, New York, wrote. "It was enough that a Jewish boy appeared on the street for gangs to set upon him."[84]

Because the urban onslaughts were frequent and violent Jews banded together to defend themselves in cities like Brooklyn, Cincinnati, Cleveland and Detroit. Police officers seemed to enjoy seeing Jews being beaten up and ignored demands for protection. As a result Cincinnati Jews created the Hebrew Protective Club with about 400 members, Cleveland Jewry organized a Jewish Protective Association, and Detroit established a Jewish Peddlers Protective Association. "There has been so much brutality shown to the poor Jews in Chicago, Brooklyn and other cities," *The American Hebrew* wrote in July 1899, "and their appeals to the police have been so utterly futile, that they have finally felt compelled to form societies to demand proper police protection."[85]

Escalating hostility in the country toward east European immigrants of the lower classes did not escape notice from German Jews who had arrived in the United States decades earlier and had achieved some social and economic successes despite existing bigotry. They recognized the threat that increased Jewish immigration posed to their own positions in the United States. They also knew that they were not well liked and were excluded from general social intercourse with Christians of the same socioeconomic status as their own. But they feared, correctly, that the presence of the east Europeans would create new tensions and that all Jews would be judged by the characteristics and deportment of the worst of them. On the other hand, they assumed a burden of extra responsibilities to help coreligionist immigrants acculturate. The behavior of established German American Jews, therefore, reflected a mixture of fear, disgust, and responsibility in promoting the welfare of their coreligionists. In New York many established upper-class Jews regarded the Russians as "cunning knaves," in need of moral and ethical uplifting, while in Detroit, Atlanta, Boston, and Richmond, Virginia, German Jews systematically refused to allow their east European brethren into their social circles.[86]

Yet throughout the country Jews who had arrived in the United States earlier devoted time to helping the newcomers adjust. Welfare agencies were opened, educational alliances formed, English classes were given to help people learn to read, write, and speak the language, and job bureaus were established to help place individuals in appropriate positions. In addition, settlement houses similar to the one Jane Addams devised at Chicago's Hull House, came into being. There, a variety of educational, recreational, and economic needs were addressed while the German Jews tried to hasten the Americanization of coreligionists whom they considered somewhat bizarre and even barbaric. The varied attempts to "uplift" these people showed how desperate the Americanized Jews were for acceptance and how deeply embarrassed they were by the habits and interests of the Russians, Poles, and Rumanians. Some of the American Jews mistakenly assumed that if the immigrants behaved properly, attended synagogue services regularly, and maintained decorum throughout the prayers, then Gentiles would respect them. A rabbi in Denver advised fellow Jews to refrain from encouraging coreligionists to emigrate from Europe while doing everything they could to Americanize those who had already arrived in the United States. Rabbi Joseph Krauskopf of Philadelphia, like many Jews in New York, sought desperately to get the newer Jewish immigrants out of the cities and onto farms throughout the nation. Krauskopf assumed that if Gentiles saw Jews working as farmers and producers assimilating the agrarian ethos, it would dispel antisemitic stereotypes. In Boston, still another rabbi surmised that if Jews would modernize their beliefs and religious behavior, avoid exclusiveness, and try to understand Christian neighbors, then barriers would be broken down and harmony would reign between the two groups.[87]

Such solutions sprang from the heads of desperate men. No behavioral modifications would lessen antisemitism nor could any particular acts significantly impress Christians who did not like Jews. But when people are frantic, insecure, and aware of their marginal position in society they often delude themselves into thinking that certain changes would enhance their image among those who they most want to impress. In truth, the only alteration that would have made any difference—and it is questionable about how many generations it would have taken to do even that—was conversion to Christianity. And most Jews refused to consider that route. Perhaps the wisest observation came from an editorial in St. Louis' *Jewish Voice* in 1898:

> When Christians will cease trying to convert Jews, and Jews will desist from their endeavor of pleasing Christians more than themselves, then the dawn of an era of peace and good will might be advanced more effectually than is done at present.[88]

Established American Jews, however, were too insecure and too anxious to prove their mettle to accept advice from the *Jewish Voice*. They desperately emulated the standards of the nation's elite and as Christian patricians sought to distance themselves from the new waves of immigrants and tie themselves to a glorious past, Jews did the same.

One of the ways in which American Christians tried to cement their status as "the true" Americans was to revive the memories of their ancestors who ostensibly fought to bring forth a new nation a century earlier. The first of these so-called patriotic groups, the Sons of the American Revolution, had been founded in 1883 for those who could trace their roots in the country for more than one hundred years. Other such organizations appeared within the next decade. The Scotch-Irish Society began in 1889, the Colonial Dames and the Daughters of the American Revolution organized the next year, the Pennsylvania-German Society followed suit in 1891, and 1892 saw the appearance of the Aryan Order of St. George of the Holy Roman Empire in the Colonies of America.[89] Inevitably, upper-class American Jews followed patrician examples. Several American Jews gathered in 1892 to form the American Jewish Historical Society, which would be used as a vehicle to prove that their ancestors had also been in the United States since its inception, had contributed to its development and its greatness, and therefore had earned Jews a place of honor and respect in society. For the next few decades members of the organization met regularly to present papers detailing Jewish activities in colonial and ante-bellum America. They were not trained historians, were not concerned with historical method, and did not particularly care about impartiality. Their role, as they saw it, was to show Jewish participation in American development and therefore impress other Americans with earlier Jewish accomplishments.[90] In that endeavor they failed; they convinced one another of their worth but most Gentiles did not even know that the American Jewish Historical Society existed.

Other motivating factors in the formation of the American Jewish Historical Society were the slurs cast on Jewish patriotism and the questioning of Jewish commitment to America. With every attack on them, their values, or their accomplishments, Jews felt the need to prove that the affronts were without foundation. Thus, the appearance of an article in the December 1891 issue of the *North American Review* by J. M. Rogers stating that the author, a Civil War veteran, had served eighteen months but had never met a Jewish soldier had to be contested. It led to the preparation of a book by Simon Wolf called *The American Jew as Patriot, Soldier and Citizen* (1895). Wolf poured through military and naval records of the Civil War culling so-called Jewish names. He then produced a volume that listed every Jewish sounding name to support his contention that Jews were patriotic. It hardly mattered that the list was inaccurate, that many names included were those of non-Jews, and that other Jews who had served were not mentioned. What was of significance was the Jewish need to prove themselves to people who could not care less what Jews said or did. Those who believed Jews to be patriotic did not need a list, and those who did not were not convinced by a book of names.[91]

In a further attempt to highlight their patriotism Jewish members of the Civil War veterans alliance, the Grand Army of the Republic, formed the Hebrew Union Veterans Association in 1896. Three years later Jews who had served in the Spanish American War would organize "The Hebrew Veterans of the War With Spain." These associations endeavored to combat antisemitism in the United States by identifying with those who had fought in the armed services, but their goal was not achieved.[92] The Central Conference of American Rabbis had also adopted a patriotic resolution expressing its deep attachment and loyalty to the United States during the Spanish-American War. In July 1898 the conference announced that it

> rejoices at the enthusiastic participation of American Jewish citizens in the present war, which again evidence the fact that the Jew, in equal degree with his fellow citizens, is always ready to sacrifice life and fortune in defence of the sacred standard beneath which his fathers fought in the War of the Revolution, the Mexican War, and the Civil War.[93]

In retrospect, it is embarrassing to see how desperately Jews tried to be accepted and how futile their efforts were. But their attempts to identify with the majority also highlight how far the majority distanced itself from the Jews. No matter what Jews did or said they were always regarded as a group apart.[94]

As the century drew to a close, Mark Twain summed up the conventional wisdom about the Jews. He had written a piece about Austria the previous year in which he indicated that while factions in that country differed on most issues, all agreed that they disliked Jews. Several readers wrote in commenting on this and one asked

> why the Jews have thus ever been, and are even now ... the butt of baseless, vicious animosities? ... Can American Jews do anything to correct it either in America or abroad? Will it ever come to an end? Will a Jew be permitted to live honestly, decently, and peaceably like the rest of mankind?

Twain responded in a separate article, "Concerning the Jew," in which he claimed that he bore no prejudice at all against Jews or other groups and therefore felt qualified to try to answer the question. He acknowledged that "in all countries, from the dawn of history, the Jew has been persistently and implacably hated, and with frequency persecuted" but he assumed that that had occurred primarily because of economic reasons. Europeans have had to exclude Jews from various occupations and universities because they were brighter and more talented than Christians and thus bested them in every endeavor where they competed. Nonetheless, Twain added, Jews always "found ways to make money, even ways to get rich."

Twain's comments about the Jews seem to reflect more the conventional wisdom of the time than a thoughtful assessment of the group. He believed that Jews had the "cunningest brains in the world," and that "the Jew is a money-getter; and in getting his money he is a very serious obstruction to less capable neighbors who are on the same quest." Twain had no hesitation in acknowledging that the Jew

> is quiet, peaceable, industrious, unaddicted to high crimes and brutal dispositions; that his family life is commendable; that he is not a burden upon public charities; that he is not a beggar; that in benevolence he is above the reach of competition.

Yet he also subscribed to the views that the Jew

> has a reputation for various small forms of cheating, and for practicing oppressive usury, and for burning himself out to get the insurance, and for arranging cunning contracts which leave him an exit but lock the other man in, and for smart evasions which find him safe and comfortable just within the strict letter of the law, when court and jury know very well that he has violated the spirit of it. He is a frequent and faithful and capable officer in the civil service, but he is charged with an unpatriotic disinclination to stand by the flag as a soldier—like the Christian Quaker.

Was there any hope for change, one of his Jewish correspondents inquired? Was there anything that Jews could do to change their image? Twain thought not:

> You will always be by ways and habit and predilections substantially strangers—foreigners—wherever you are, and that will probably keep the race prejudice against you alive.[95]

Regarded in that fashion, Jews could invariably be singled out as culprits responsible for some societal woe when realistic explanations either failed to illuminate phenomenon or proved too difficult for any group of people to comprehend. Then they could be pointed to as the reason for whatever mishap had either transpired or was about to occur. Evidences of this type of behavior had already been seen in the Civil War and during the agrarian crusade. It would be much more common in the twentieth century and the ripple effects would create much greater concentric circles in the next five decades. But all the major accusations save one—Jews and their various political philosophies—had already been clearly enunciated in an earlier period.

Although Twain thus offered no encouragement for Jews to think that things could get better for them, American Jews had difficulty in accepting this assessment. Yes, there was antisemitism in the land. Jews did not deny that. But they assumed that the future would be brighter and that the twentieth century would see their full integration into American life.

4

Racism and Antisemitism
in Progressive America
(1900–1919)

During the first two decades of the twentieth century racism became a central component in the elixir of American antisemitic sentiments. Indeed, the racial components of antisemitic thought in America, always inherent yet mostly hidden, became obvious in the period known as the Progressive era. The belief that Jews were a separate race with distinctive mental and physical characteristics that were genetically acquired was widespread and accepted by members of the government, much of the scientific community, the media, and the general public.[1] Two elements were primarily responsible for adding the additional ingredient of race. One stemmed from the concept of racial superiority that developed out of America's Christian heritage along with the idea that emphasized the responsibility of Anglo-Saxon Americans to effectuate God's will and remake the world in its image; the second evolved from the massive influx of southern and eastern Europeans to the United States that began at the end of the nineteenth century. From 1890 to 1914 16,516,081 immigrants arrived in the United States of whom 1,694,842, or slightly more than 10 percent, were Jewish. Since the total of all officially recorded immigration until 1890 came to only 15,436,042, mostly from northern, central, and western Europe, of whom fewer than 2 percent were Jewish, the enormous impact immigration had in those twenty-five years is obvious. It was the deluge of immigrants into the United States that led to a major rethinking about who had helped to build the nation and which groups would blend best with those already here. Many of the upper- and middle-class Americans regarded some of the new immigrants as barbarians. Influenced by the racist thoughts permeating the European intellectual scene,[2] scions of New York and New England patricians articulated their concerns about the future of the nation. And while they generalized about the negative worth of most of the newcomers from Europe and Asia, their most severe racial animus was directed toward the Jews.[3] That phenomenon created the harsh antisemitism that has characterized much of America's modern history.

The influx of east European Jews had an unsettling effect on ordinary Americans as the twentieth century began. No longer were Jews merely those who had failed to accept the Savior, savvy Shylocks undermining American entrepreneurs, or uncouth parvenus, but additionally the newest arrivals from eastern Europe came to be regarded as "moral cripples" with warped souls[4] who emanated from an inferior racial stock.[5] A Unitarian minister in Denver articulated the new sentiment on December 4, 1903: "It is the Jewish race, not the Jewish church that is disliked."[6]

Most important, this view was not restricted to a handful of patricians. One need only look at the vast array of articles, including some by the nation's most reputable journalists and scholars, that derided the Jews during the Progressive era. From these essays readers were taught that the Jew was everywhere an alien, at best to be tolerated. Jews pushed themselves forward where they were not wanted, those with money were ostentatious, and those without it would seek any means to gain it. In the face of strong evidence that the Jewish crime rate throughout the world was low, perusers of numerous contemporary articles would discover "that criminality was an inherent Jewish trait."[7] Moreover, not only did Jews allegedly have bad manners, but they supposedly lacked a sense of fair play. "The Jew is winning everywhere. By fair means or by foul means he wins. . . . [Jews] beat us in the schools, in the colleges, in business—everywhere, and we are not used to being beaten." [8] Even a cursory reading of journals from the Progressive era shows that antisemitic feelings were widespread, and growing to embrace racial, religious, economic, and social concerns; the clear foundation of the prejudice was the fact that Jews were not Anglo-Saxons and consequently thought to be unassimilable.[9]

Not before or since in American history have periodicals been as influential as they were in the Progressive era in helping to shape political solutions to contemporary problems. Hardly a week passed that some topic was not unearthed indicating a sorry state of public affairs. Before the turn of the century American periodicals were rather staid in their coverage and rarely, if ever, published stories unfit for the eyes of the most innocent family members. Beginning in 1902, however, a new breed of journalism, perhaps best epitomized by S. S. McClure and R. S. Baker, strove to uncover and publicize aspects of corruption and malfeasance in American society. Editors and reporters dug up the dirt that the most unscrupulous elements in society wanted withheld from the general public and President Theodore Roosevelt dubbed these journalists "muckrakers." Since they covered popular topics, and because the essays were generally accurate exposes of areas demanding reform, the journals prospered. As readers reacted with alarm, a new breed of politicians came to the fore pledging themselves to end corruption in politics, make business more responsible, and uplift the morality of society in general.

There is little question that the problems considered by muckraking journals needed disclosure. Criminal politicians, shady dealings in the oil industry, tampering with food, abuse of children, tenements in decay, Senators more responsive to business interests than to the needs of their constituents, insurance fraud, and prostitution, all drew muckraker attention. For practically every problem raised in the journals, politicians attempted to find solutions and many bills dedicated to the elimination of a variety of injustices passed city, state, and federal legislatures. It was unfortunate, therefore, that one problem to which journals devoted considerable attention was the composition of the immigrants entering the United States. Most Progressives, no matter what their particular issues and no matter how they may have differed with one another on the problems of the day, as well as most Protestant Americans, considered the hordes of lower class foreigners from southern and eastern Europe detrimental to the welfare of the nation. America's latest newcomers were not perceived to be of as high a caliber as previous immigrant groups had been, their physical and intellectual traits supposedly paled in comparison with those of previous generations of newcomers, and it was thought that they simply would not be able to assimilate into American society without mongrelizing the race. Even the existing character and culture of the "best" Americans might be threatened.

Despite the venom and hostility displayed toward a variety of European immigrants, especially the Italians, no new group received as much attention and scorn as the east European Jews.[10] Americans had historically displayed sufficient antisemitic feeling before the arrival of more than one million Jews from Austrian Poland, Russia, and Rumania. Despite the American tradition of acceptance of healthy Europeans regardless of religion, political philosophy, or ancestry, the onslaught of so many lower-class people from nontraditional sources of immigration alarmed Protestants throughout the nation. And of particular concern was the fact that the influx was overwhelmingly of people who were Roman Catholic, Jewish, and Greek Orthodox.

Given the character and concerns of the era, therefore, it is not surprising that articles with the following titles appeared in the most frequently read periodicals: "Secretary Hay's Note and the Jewish Question"; "Is a Dreyfus Case Possible in America?"; "Race Prejudice Against Jews"; "Jewish Criminality"; "Prejudice Against Jews in the United States"; "Because You're a Jew"; "The Jews in the United States"; "White Slave Traffic and Jews"; "Jews in the White Slave Traffic"; "Is the Jew Getting a Square Deal?"; "Will the Jews Ever Lose Their Racial Identity?"; "Are the Jews an Inferior Race?"; "The Hebrews of Eastern Europe in America"; and "The Conquest of America."[11]

The central issue discussed in these essays was whether Jews would or could adapt in the United States and what kinds of citizens they would make. None of the articles, however, was as devastating or as detailed in analysis as were those by patrician Burton J. Hendrick. In two pieces written for *McClure's Magazine*, "The Great Jewish Invasion" and "The Jewish Invasion of America,"[12] Hendricks expressed what much of America was thinking. Hendricks was openly uncomfortable with both Jews and their purported character. His phrasing turned Jewish virtues into vices, and characteristics for which other Americans have been praised—steadfastness, self-reliance, and the ability to succeed within the capitalistic system—appeared somehow sinister when applied to Jews. "No people have had a more inadequate preparation, educational and economic, for American citizenship," Hendrick wrote, than the Russian Jews yet their "economic improvement is paralleled by that of no other immigrating race."[13] The Jew was not praised for working hard and being thrifty in his expenditures but disdained for being "a remorseless pace-maker. He allows himself no rest nor recreation, and works all hours of the day and night. He saves every penny, will constantly deny himself and his family nutritious food, and until he has made his mark will live in the most loathsome surroundings."[14] This last observation contrasted sharply with the assessments of community workers who noted that "the physical health of Jews was remarkably good in comparison to that of both non-Jewish immigrants and native-born Yankees." Even the poorest of Jewish immigrants cleaned their homes on Friday, and on Saturday the orthodox took a day of rest. Jews also had the lowest infant mortality rate of any group. Contemporary medical and social investigators attributed the comparative well-being of the group to rare alcoholism, the dictates of Jewish religious laws that restricted the kinds of foods they might eat, and the religious mandate that there be a thorough house cleaning before the Sabbath.[15] Hendricks, needless to say, did not incorporate such contradictory evidence in his writings.

Typically, the words Hendrick chose to describe the entry of Jews into the garment industry smacked of subversion. "Precisely as they supplanted the Irish and Germans in their homes, have they taken their places in these trades. Fifty years ago all our tailors were native-born Americans . . . now they are Jews." Jews were intensely ambitious and individualistic, he noted, and then, in the broadest of stereotypes, concluded that "whenever you see a Russian Jew, however insignificant his station, you see a prospective landlord." Hendrick was as absolutely wrong on that score as he was when he wrote that "trade unionism is antagonistic to the Jewish character."[16] Less than three years after his first article appeared, Jews provided the nucleus for both the Amalgamated Clothing Workers Union and the International Ladies Garment Workers Union, the two most powerful clothing workers unions in the United States during the twentieth century.

Following the great American tradition of self-reliance, many Jews started their own businesses, especially in the clothing trades where most of the employees were also Jewish. Hendrick translated that ambitious undertaking into a negative trait:

> in recent years, the Jewish manufacturer has pressed into service thousands of Italians. He has quickly utilized an alien population living on a lower economic plane than himself. In the control of the business he has forced to the wall not only the German, the Irishman, the native-born, but the German Jew.[17]

Hendrick also called the Russian Jew "the most important factor in determining the physical growth of New York. He decides where the people are to live and the form their housing is to take. He does this, not only because he controls the land, but because he also controls the building business."[18] That only a tiny minority of Jews owned land or buildings was irrelevant to the antisemitic case he was constructing.

Most east European Jews in New York, Chicago, Boston, Philadelphia, Pittsburgh, Cleveland, and other major American cities lived at or near the brink of poverty in the first decade of the twentieth century. Hendrick's assessment ignored the actual conditions of Jewish immigrants while inflating the power and authority of some of the more successful entrepreneurs. Hendrick's 1907 essay concluded by asking if the Jew "is assimilable? Has he in himself the stuff of which Americans are made?"[19] There was no doubt about how he would answer.

Six years later, in an even more loathsome assessment, Hendrick tried to frighten his readers into believing that Jews were on their way to conquering the United States. He listed businesses that he thought Jews "absolutely control, and others in which their influence is steadily increasing." Not only did they rule the clothing trades, but "there is not the slightest doubt that in a few years the Jews will own the larger part of Manhattan Island—the richest parcel of real estate in the world," including "the largest part of Harlem, enormous stretches of Brooklyn, thousands of acres of tenements in the Bronx and Manhattan." Moreover, in the New York City civil service, "the Jews are rap-

idly driving out the Irish, the Germans, and the native Americans."
Within the city government Jews had allegedly

> obtained absolute control of . . . nearly all medical and laboratory posi-
> tions—of which New York employs a fair-sized army. They have cap-
> tured a great majority of the engineering jobs. They hold most of the
> minor legal positions. They are the city's searchers, process-servers, and
> law examiners. Most of the municipal office-boys are youngsters from
> the East Side; the stenographers and the type-writers [sic] are nearly all
> Jewish girls; the bookkeepers and holders of minor clerical positions are
> nearly all Jews.

Under a bold heading in the article Hendrick also asserted "PROTES-
TANT AND CATHOLIC CHILDREN NOW TAUGHT BY JEWESSES"; he then
added that "Jews are now the largest single racial element in the public
schools." After observing that a few Jewish men were joining the police
force, Hendrick wondered how long "the Irish will maintain a numerical superiority in this department?"

Hendrick's list of industries and businesses that Jews allegedly controlled included department stores "in all of the large cities," the whiskey business, the mail-order liquor businesses in certain southern and
western states, and the leaf tobacco industry. "The activities of American Jews," Hendrick continued, "extend far beyond the borders of
New York. They control, in particular, one business that reaches into
every part of the country—the business of public amusement." Moreover, without citing evidence, Hendrick stated "another fact which is
a matter of common notoriety—that a majority of the prize-fighters in
New York are really Jews who operate under Irish names." Finally,
before concluding his essay, Hendrick informed readers that succeeding articles by Abraham Cahan, editor of the New York Yiddish newspaper, the *Forwards*, "will make clear why it is that the Jews so easily
surpass or crowd out, at least in business and finance, the other great
immigrating races . . . and why, in the next hundred years, the Semitic
influence is likely to be almost preponderating in the United States."
Hendrick excoriated the "nimble-witted Jew" who he labeled "the
greatest shoestring capitalist in the world," and before another decade
passed he would produce a series on how Jewish socialists were undermining American values. Loathing the Jews as he did, Hendrick used
whatever information he could find, exaggerate, and distort to make
others fear and despise them as well.[20]

No other analysis of the Progressive era rivaled Burton Hendrick's
in terms of vitriol toward Jews, but E. A. Ross' "The East European
Jew" came close. Ross, a celebrated sociologist at the University of Wisconsin, was influenced by the eugenics movement so popular in the
early years of the century, and his commentary on a variety of different
peoples reflected a belief in the immutability of genetic characteristics.

His evaluations of the Jews had the ostensible air of scholarship and objectivity but his sweeping generalizations interspersed fact with fiction. Ross, a practicing academician, made assertions that were often accurate, but in language that suggested negative, underhanded, or diabolic motives. Thus his comment that Jews were well represented "at every level of wealth, power, and influence in the United States" certainly was untrue as regards their impact on state and federal governments as well as major American corporations. But it had sufficient accuracy not to be dismissed out of hand. Similarly, there could be no disputing the fact that between 1888—when the Jewish population totaled about 400,000 in this country—and 1913, one and one-half million more Jews arrived. Nor was there any question that Jewish organizations strongly opposed a literacy test for immigrants. By describing the campaign against such a test as being "waged by and for one race," and indicating that "the literature that proves the blessings of immigration to all classes in America emanates from subtle Hebrew brains,"[21] Ross revealed that he opposed such a stance while denigrating Jews at the same time.

Ross also observed that Jews "rarely lay hand to basic production. In tilling the soil, in food growing, in extracting minerals, in building, construction, and transportation, they have little part." Absolutely true. The east European Jews favored occupations that required brains rather than brawn and while many immigrant Jews labored as painters, printers, tailors, etc., they encouraged their sons to select careers as physicians, dentists, pharmacists, lawyers, and teachers.[72] Moreover, the Ross essay conceded that

> Jewish immigrants cherish a pure, close-knit family life, and the position of the woman in the home is one of dignity. More than any other immigrants they are ready to assume the support of distant needy relatives. They care for their own poor, and the spirit of cooperation among them is very noticeable. Their temper is sensitive and humane; very rarely is a Jew charged with any form of brutality. There is among them a fine elite which responds to the appeal of the ideal and is found in every kind of ameliorative work.[23]

Ross made questionable and inaccurate statements as well. He assumed that "intellectuality" was a Jewish "race trait" and that parents, "however grasping," recognized the worth of schooling. They therefore stressed the value of education to a greater extent than did parents of Irish and Italian ancestry. But instead of recognizing this cultural characteristic as a positive attribute he cited a school principal who explained that while Jews made good students, "their progress in studies is simply another manifestation of the acquisitiveness of the race." Moreover, Ross evoked an "eminent savant" to show that the best Jewish minds were not strong in generalization, but merely "clever, acute, and industrious rather than able in the highest sense."[24]

Ross also accused the east European Jews of being immoral and unethical criminals who were hard to detect. Their "inborn love of money-making" made them unethical businessmen and "insurance companies scan a Jewish fire risk more closely than any other. Credit men say the Jewish merchant is often 'slippery,' and will 'fail' in order to get rid of his debts." In another slur, Ross asserted that "pleasure-loving Jewish business men spare Jewesses, but pursue Gentile girls." It was therefore erroneous to think of Jews as more law-abiding than peoples of other backgrounds because "it is harder to catch and convict criminals of cunning than criminals of violence."[25] And the Ross essay concluded with the statement: "the truth seems to be that the lower class of Hebrews of eastern Europe reach here moral cripples, their souls warped and dwarfed by iron circumstance. Their experience of Russian repression has made them haters of government and corrupters of the police." In Ross' mind Jews were rightly excluded from elite resorts, schools, and private associations not because of bigotry or religion but because of "certain ways and manners."[26]

In fairness, it must be conceded that Ross disdained almost all newly arriving foreigners. He was a member of the American Protestant elite who feared the new immigration from eastern and southern Europe, and as a scholar he tried to frame his judgments of these peoples in what he believed to be a neutral and detached stance. In *The Old World in the New* (1914), he assessed previous generations of immigrants more positively than those of his day. The northern Europeans, those who had populated colonial and ante-bellum America "seem to surpass the southern Europeans in innate ethical endowment . . . discipline, sense of duty, presence of mind, and consideration for the weak are much more characteristic of northern Europeans." Germans particularly impressed Ross as people of integrity. "Quiet, industrious, and thrifty," they loved good music and good drama, were "honest and stable" and "Americanized with great rapidity." The Irish also appealed to him. He found them to be "imaginative" and sensitive to what others thought of them, and also in need of "an appreciative word or glance" in order to continue working toward a specific goal. Not only did they "furnish stirring orators, persuasive stump-speakers, moving pleaders, and delightful after-dinner speech-makers, but they gave us good salesmen and successful traveling-men. Then, too, they know how to manage people." They made excellent executives.[27]

On the other hand, Ross had mixed views of most of the immigrant groups of the early twentieth century. Although Scandinavian immigrants assimilated quickly and provided "an excellent, cool-blooded, self-controlled citizenship for the support of representative government," he noted ruefully that they were not too sociable, "imagination and tenderness [were] traits none too common among" them, and that as a group they had "a tendency toward insanity." All "the Mediterranean peoples" were "morally below the races of northern Europe."

He wrote that "the Greek is full of tricks to skin the greenhorn" and depicted southern Italians as "volatile, unstable, soon hot, soon cool." He minimized their worth because they showed "a distressing frequency of low foreheads, open mouths, weak chins, poor features, skew faces, small or knobby crania, and backless heads." Ross thought of the Slavs, another one of the major immigrating groups in the Progressive era, as "superfecund" with a "small reputation for capacity." He quoted a physician to the effect that the Slavs "are immune to certain kinds of dirt. They can stand what would kill a white man." He believed that they drank too much, and he repeated Slavic proverbs to the effect that "he who does not beat his wife is no man" and "strike a wife and a snake on the head."

It is obvious, therefore, that Ross' antisemitism was part of a broader elite concern with the quality of peoples inundating American shores. It was no wonder that 10 to 20 percent of the immigrants arriving looked "out of place in black clothes and stiff collar, since clearly they belong in skins, in wattled huts at the close of the Great Ice Age." He also worried about the decline of American beauty:

> it is reasonable to expect an early falling off in the frequency of good looks in the American people. It is unthinkable that so many persons with crooked faces, coarse mouths, bad noses, heavy jaws, and low foreheads can mingle their heredity with ours without making personal beauty yet more rare among us than it actually is.[28]

Ross' judgments in *The Old World in the New* reflected the thinking of many other Americans and gave rise only two years later to Madison Grant's *The Passing of the Great Race* (1916), the most celebrated racist tract of the Progressive era.[29] Grant, a vice president of the Immigration Restriction League and an officer of the American Eugenics Society, traced his ancestry to colonial America. Described by John Higham as "the most lordly of patricians," he graduated from Yale, received an L.L.B. from Columbia Law School, and belonged to the Knickerbocker, University, Century, Tuxedo, and Union clubs in New York City.[30] The work of several years, Grant's book provided a summary of American WASP fears and attitudes about contemporary immigrants.[31] Influenced by eugenic theories as well as the racist cult of Anglo-Saxonism that had dominated the thoughts of the American Protestant elite for more than one-quarter of a century, Grant produced the major nativist work of the Progressive era.

The Passing of the Great Race bemoaned the fate of old-stock Americans as they watched this "great" race being diluted by inferior breeds from southern and eastern Europe. Grant divided Europeans into three groups: Nordics, Alpines, and Mediterraneans. Within the first group, people from the northern and western areas of the continent, one found "the white man par excellence" whose brains contained the political and military genius of the world, and who were the guiding lights in the founding, development, and superior stature of the United States.[32] But in the twentieth century the nation welcomed lesser folk like the Alpines from central Europe, a race of "submissive peasants," most of whom were Roman Catholic "represented by various branches of the Slavic nations," and Mediterraneans, "inferior in bodily stamina" to both Alpines and Nordics but superior in intellectual and artistic endeavors. According to Grant, their greatest accomplishments were in the past; as a people they had long since passed their peak of ability.

"The men of Nordic blood to-day," Grant wrote, constituted the entire population of Scandinavia and most of the British Isles; they were "almost pure in type in Scotland and eastern and northern England." Even the Celts were Nordic: "the tall, blond Irishmen are to-day chiefly Danish with the addition of English, Norman, and Scotch elements, which have poured into the lesser island for a thousand years, and have imposed the English speech upon it." The Nordics, were, without question,

> a race of soldiers, sailors, adventurers, and explorers, but above all, of rulers, organizers, and aristocrats in sharp contrast to the essentially peasant character of the Alpines.

Unfortunately, Grant believed, people of inferior breeding were now overrunning our country, intermarrying, and diminishing the quality of American blood. "What is needed in the community most of all is an increase in the desirable classes, which are of superior type physically, intellectually, and morally."[33] His conclusion was obvious. The passing of the great race of Nordics who had built our nation and made it great was a serious cause for alarm.[34]

Grant developed a hierarchy of Europeans based on their alleged racial characteristics. Like Hendrick and Ross before him, however, Grant articulated contempt and abhorrence for lower-class immigrants and displayed a special animus that almost all elites held toward Jews. Although he had little positive to say about the Italian, Slav, or Syrian, Grant thought least of "the Polish Jew . . . with his dwarf stature, peculiar mentality and ruthless concentration on self-interest." He complained that

> the man of the old stock is being crowded out of many country districts by these foreigners, just as he is to-day being literally driven off the streets of New York City by the swarms of Polish Jews. These immigrants adopt the language of the native American; they wear his clothes; they steal his name, and they are beginning to take his women, but they seldom adopt his religion or understand his ideals.

Even the intermingling of groups through intermarriage, Grant concluded, would produce a mongrelized people of a substandard species but the lowest cur would be "the cross between any of the three European races and a Jew" because then you would get a Jew.[35]

The cumulative effect of the writings of Hendrick, Ross, and Grant cannot be exaggerated. Politicians throughout the Progressive era had already sought means of curbing "undesirable" immigration. In 1907 Congress established a commission to investigate the problem and in 1911 a 42-volume report on the subject appeared. Known by the name of its chairman, the Dillingham Commission recommended ways to cut down on immigration (one of them included the literacy test) but President Taft vetoed such a bill in 1913. Hendrick wrote in 1907 and 1913, Ross in 1912 and 1914, and Grant's book appeared in 1916. All three authors provided data to reinforce the views of restrictionists. The outbreak of the war in Europe in 1914 had sharply curtailed the exodus of immigrants from the continent. After the war ended in 1918, and especially in the 1920s, Grant's thesis dominated public discussions about immigration policy. This patrician's analysis of European groups reinforced a desire to keep America "nordic," and minimize the influx and impact of southern and eastern Europeans in this country.

Of all newcomers to the United States in the Progressive era, the public found Jews the least acceptable. Jews were not only non-Christians, but they were seen as uncouth, avaricious, and a threat to the established order because irrational fears existed that they might undermine and take over the reins of economic and political power. More than just words were used to vilify them in the United States. Throughout the era numerous incidents and events reflected society's insulated attitudes. The greatest social disabilities fell on Jews whose backgrounds, accomplishments, and incomes would have placed them among the elite of American society had they been Episcopalians, Presbyterians, or even Baptists. The German or Americanized Jews, who felt the brunt of the existing social and economic discrimination, probably shared many of the assessments of the east European Jewish immigrants that Ross described, yet they were the ones excluded from high society. They felt their accomplishments entitled them to acceptance, and they would in fact have been accepted had it not been for their faith.[36]

The experience of Louis D. Brandeis in Boston illuminates the harsh experiences of America's wealthy Jews. Long the most prominent Jewish citizen of America's Athens, Brandeis suffered not only social exclusion but lack of political opportunity. Vehement opposition from prominent Bostonians forced President-elect Woodrow Wilson to eliminate him from a list of potential cabinet nominees.[37] In 1916, however, political realities necessitated that Wilson choose a leading Progressive for an opening on the U.S. Supreme Court and the President selected Brandeis. The nomination discomforted Brahmins because of Brandeis' progressive leanings; contemporary journalist Norman Hapgood observed, however, "there was more energy put into the fight against him because he was a Jew."[38] Massachusetts Republican Senator Henry Cabot Lodge, among those who tried to obstruct the appointment, received a letter from his nephew, Ellerton James, supporting the nomination in the face of old Boston's prejudice. James acknowledged that Brandeis "is a Hebrew, and, therefore, of Oriental race and his mind is an Oriental mind, and I think it very probable that some of his ideas of what were fair might not be the same as those of a man possessing an Anglo-Saxon mind." Still, "contrary to the usual ideas of Jews I do not think the almighty dollar is very attractive to him."[39] But it was not only conservative Boston that found Brandeis objectionable. The President's choice also displeased many Democratic Senators. Southerners especially grumbled, including one who claimed that he found Jews "positively repulsive."[40] Nonetheless, Wilson prevailed in holding together enough party loyalists to ensure Brandeis' confirmation.

One did not have to be a Supreme Court nominee, however, to suffer from Gentile antisemitism. Among the groups Jews had most difficulty with were the Irish. About five million people from Ireland arrived in the United States in the nineteenth century, and they ranked second only to the Germans among immigrants to this country during that period. The Irish concentrated mostly in the cities of the northeast and midwest and in time produced a majority of local political leaders and policemen in most of the communities where they settled.[41] Since the overwhelming number of immigrant Jews went to New York, Chicago, Boston, Philadelphia, and other cities in the northeastern quadrant of the United States, and since they were at first near the bottom of the socioeconomic scale, they came into frequent contact with Irish police, politicos, and workers. Historically, most of the publicly acknowledged inter-ethnic friction occurred in urban areas where economic groups near the bottom of the rung were in proximity with one another, so Irish and Jews clashed frequently.

The basic values of the Irish contrasted with those of the Jews. The former were Roman Catholics, of a more conservative temper, and educated to believe that all descendants of the killers of the Savior were forever outcasts in society. Their occupational and educational goals were also in conflict as the Jews pushed the Irish out of their jobs in New York City's garment centers at the turn of the century and successfully competed with them for civil service positions within the city government. As a result of these differing experiences, backgrounds, and attitudes, considerable controversy developed between the two groups in a number of the cities in which they lived in adjoining neighborhoods.

In 1914, one Chicago woman of Irish background who married a Jew wrote bitterly about her experiences. Although she loved her husband, she observed, had she known what would happen as a result of their union she would not have married him. She grew up in an antisemitic household and neighborhood. She worked for Jews, and along with co-workers ridiculed their manners and accents. Nonetheless, a Jewish man won her heart. When she first informed her family that she was going out to dinner with a Jew they were incensed. After she married him, however, one of her friends informed her, *"of course,* dear, you cannot expect us to receive a Jew."[42] After the newlyweds rented apartments further unpleasantness resulted. Neighborhood children came, threw dirt and stones at their windows, and chanted

> The rose is red
> The violet's blue—
> Everywhere you go
> There comes a Jew.

Although the young couple moved several times, people in "American" neighborhoods refused to accept them. "We have tried two suburbs," she wrote, "and found the persecution even worse than in 'exclusive' city districts. We have practically ceased associating with any persons who are not Jewish." She praised her husband's friends who "have shown a respect for my religion that my people have been far from showing for theirs." What hurt the woman most was the treatment received by her son from his peers at the predominantly Irish Catholic parochial school to which she sent him. "A fight was a daily occurrence," and whenever he did anything wrong peers blamed it on "the Jew in him." Boys repeatedly taunted him and refused to play with "that dirty Jew." One day the lad came home and wanted to know if "his papa had helped kill Jesus."[43]

Jews were hardly the only immigrants who had difficulty with the Irish. Urban Irish did not take well to newcomers of any background who threatened their turf. As James Richardson wrote, "Italians, Jews, Chinese, and Negroes found that to the Irish beating up newcomers was a kind of sport. Too often the predominantly Irish police force arrested the victim rather than the aggressor or joined in on the Irish side."[44] But the swelling numbers of Jewish newcomers from eastern Europe—and the severity of their economic positions—accelerated Irish-Jewish hostility in the twentieth century. Jews tried to avoid contact with the Irish but their tormenters seemed to enjoy a good fight. Boys and young men especially were likely to be physically abusive and street ruffians who pulled men's beards, overturned pushcarts, or attacked Jewish youth without provocation were everywhere. Particularly feared was the careless use of clubs by policemen enforcing what they regarded as "curbside justice."

A notorious event occurred on July 23, 1911, in Malden, Massachusetts, where Irish teen-agers and young men brutally assaulted Jews. Shouting "Beat the Jews" and "Kill the Jews," the youths used iron bars, wagon spokes, stones, jagged bottles, and sticks to smash store windows, break glass, and beat up whichever Jews they found. The police responded slowly when informed of the melee, arrested few of the culprits, and made light of the incident. Captain Foley of the police department claimed that this was an isolated attack and praised the

Irish youth for being fine citizens. In court the defendants argued that they had been set upon by "old men with whiskers" and that they responded in kind. The Jews, they claimed, were the ones responsible for the violence. The judge did not believe the tale of the defendants yet only four of the sixteen assailants were found guilty.[45]

The Malden incident echoed the first egregious antisemitic episode of the twentieth century where similar Irish animosity toward Jews and official police indifference and brutality combined to create a police riot. It took place during the funeral of Rabbi Jacob Joseph on New York's Lower East Side in July 1902. Jewish immigrants had been streaming into the area since the 1880s but there had been a lull at the end of the 1890s. Then in the first three years of the new century over 175,000 Jewish immigrants arrived in the United States, about a third of whom settled on the Lower East Side. Crowded conditions upset the non-Jews in the neighborhood who felt overwhelmed by the foreigners. Germans began to move out of the community but the Irish were less mobile. Irish-Jewish friction increased as Catholic youths preyed on the Jews and policemen sometimes beat up or arrested Jews for little cause such as speaking loudly in the park.

Then, on July 30, 1902, during the course of a Jewish funeral procession, Irish workers at a printing press factory began throwing debris and metal objects at the bier and its following carriages. Water hoses and buckets of water showered down from factory windows to douse those following the hearse. Angry and hysterical mourners charged the factory demanding a cessation of the disgraceful behavior only to be rebuffed by management who called the police. The entire incident took no more than twenty minutes and by the time the police arrived most of the procession had passed the factory, the bier was on the ferryboat headed to a Brooklyn cemetery, and the fracas had just about ended. But the police inspector in charge, without inquiring about the event, charged the remaining procession of mourners, encouraging his men to do likewise, slashing his club while shouting, "Kill those Sheenies! Club them right and left! Get them out of the way!" The people in the street were shocked by the abominable police activity, but in the thirty minutes it took to completely scatter the crowd hundreds of Jews were clobbered by policemen who had apparently lost control of their senses.

The event outraged the local Jewish community. Protest rallies demanded an investigation by Mayor Seth Low whom Jews had helped to elect the previous year partially because he pledged to reform the abuses of the police department. Low responded appropriately. He appointed a five-man investigation committee which within a month prepared a devastating indictment of the police. The committee completely exonerated the mourners, condemned the police for "marked incivility and roughness," and observed that for a number of years there had been "a complete lack of sympathy between the policemen and the residents of the East Side." Despite the recommendations of Low's committee, the policemen guilty of inappropriate behavior were exonerated by their peers and police harassment of Jews on the Lower East Side continued. While various commissioners came and went police attitudes toward Jews remained constant.[46] In September 1908, six years after the riot at the rabbi's funeral, New York City Police Commissioner Theodore Bingham started a brouhaha when he observed that 50 percent of the city's criminals were Jews. Challenged to substantiate his accusation, he could not supply details and publicly apologized for his previous remarks. Subsequent New York county statistics indicated that only 16.1 percent of total convictions were Jewish and, given the city's ethnic population, Jews were only half as likely to be convicted of criminal activities as were non-Jews.[47]

It was publicly acceptable, however, to highlight accusations of Jewish criminal activities even when the evidence presented did not warrant the claims made. Reformers of the Progressive era assumed that the new immigrants were always overrepresented among the pimps and prostitutes when, in fact, Jews, Italians, and Irish were generally underrepresented. In the Midwest prostitutes were more likely to be Protestant whereas two separate analyses in New York City, in 1909 and 1911, showed that over 75 percent of all women convicted of prostitution were native born.[48] But such figures may not have been available to George Kibbe Turner who in 1909 wrote an essay entitled "The Daughters of the Poor." The article essentially condemned the Jews of Eastern Europe and New York City for dominating the white slave trade throughout the world. Two years earlier he had written a piece on Chicago in which he observed that "the largest regular business in furnishing women . . . is done by a company of men, largely composed of Russian Jews, who supply women of that nationality to the trade."[49] "Daughters of the Poor" resulted in national publicity, great amounts of handwringing among the Jews, and contributed, in part, to passage of the Mann Act by Congress in 1910 that prohibited transporting women across state lines for immoral purposes.

When a New York grand jury questioned Turner under oath about the sources he used for his notorious essay he told the jurors that he had no personal knowledge of the things about which he wrote. "It is by such worthless evidence that the impression has been created in the minds of the people that the traffic in girls is largely in the hands of Jews" a Jewish publication, *The Temple*, noted.[50] But if Turner had no specific knowledge of what he wrote, he was not completely wrong in what he said. Prostitution was a serious urban problem and many immigrant girls and women of various nationalities and backgrounds were enslaved in the world's oldest profession.[51] What clearly set Turner's writing apart was his distorted emphasis on the Jewish factor. There is no evidence available to show that Jews actually dominated or controlled the white slave trade. Nor has there ever been any information that showed a disproportionate Jewish interest or control of prostitution. Historian Harold Hyman notes, "Indeed, a particularly Jewish opposition existed to prostitution. This antipathy is rooted both in the Talmud and in a secular environmentalist view popular among Jewish Progressives, that prostitutes (like many convicts) were less evil offenders than victims who deserved more remedial help than contumely."[52] But even if Jews had not even been involved in the white slave trade, one scholar later wrote,

> the association of Jews with abduction was predetermined by the logic of white slavery. It did not depend upon the existence of Jewish procurers. Indeed, white slavery allegations were similar to the other notorious charges in the anti-semite's arsenal: that Jews were by nature criminal, that they organized widespread conspiracies to corrupt and pollute the Christian world, and that they ritually murdered Christian children in order to obtain their blood for the baking of unleavened Passover bread.[53]

Turner's articles, were, in fact, politically inspired. His essay on the daughters of the poor was meant to show how the politicians in Tammany Hall worked hand in hand with Jewish procurers. Even the subtitle of the essay, "The Plain Story of the Development of New York City as a Leading Center of the White Slave Trade of the World, Under Tammany Hall," focused on the involvement of Tammany Hall in the business rather than on the role of the Jews. Turner's piece was weakened by his use of antisemitic stereotypes and emphasis on the dominance of the Jews in the trade because he lacked the data to substantiate his contentions. "Turner was utterly wrong," a more recent scholar concluded, "to assert that prostitution in New York was primarily a Jewish problem, and that New York was the world centre of white slavery."[54]

It was the continuing popular misconceptions about Jews that contributed to the establishment of permanent American Jewish defense agencies during the Progressive era. The American Jewish Committee (AJC), formed in 1906 by middle- and upper-class American Jewish philanthropic and community leaders, dedicated itself to protecting the civil and religious rights of Jews throughout the world. The self-selected Committee members, mostly of German extraction with roots in the United States that dated back to the middle of the nineteenth century, enjoyed contacts with people in important political positions. They hoped to improve conditions for Jews throughout the world by discreetly speaking to men in power and shunning any publicity for their activities. They had been effective in getting President William Howard Taft to abrogate an 1832 treaty with Russia because some of the provisions allowed that government to discriminate against American citizens of the Jewish faith, and the AJC continually pressured state legislatures to pass legislation promoting equal treatment for all citizens. One of its accomplishments was a New York state law that barred

ıers and managers of public accommodations from engaging in dis-
ıinatory advertising policies.[55]

Another defense agency of lesser stature and accomplishment until
after World War II, the Anti-Defamation League of B'nai B'rith (ADL),
was begun in 1913, seven years after the formation of the AJC. The
ADL, representing the interests of midwestern and southern Jews, most
of whose forebearers arrived in the United States from central Europe
in the middle of the nineteenth century, was initially concerned with
the negative caricaturing of Jews on stage and screen, social and economic discrimination, and the increasingly antagonistic tone American
society had toward Jews. In 1908 the president of B'nai B'rith, the ADL
parent organization, noted how

> our attention has lately been directed to the low ebb of the public's favor-
> able opinion of the Jew. Our petty vices have been magnified and our
> slightest errors exaggerated in glaring headlines. More so than at any
> other time in the last two decades have our black sheep been held up as
> the cynosure of all eyes. Undisguised innuendos and open insinuations
> have been directed at us.[56]

B'nai B'rith's concern with that public impression, and its desire to
promulgate the "correct aspects of Jewish culture and heritage" led to
plans for an antidefamation league.

The unjust conviction of Leo Frank for the murder of a southern
child in 1913 (to be discussed in chapter 9) provided the final spur. In
September 1913 the new president of the ADL announced that evidence
of discrimination often manifested itself in an attempt to influence
courts of

> law where a Jew happened to be a party of the litigation. This symptom,
> standing by itself, while contemptible, would not constitute a menace,
> but forming as it does but one incident in a continuing chain of occasions
> of discrimination, demands organized and systematic effort on behalf of
> all right-thinking Americans to put a stop to this most pernicious and
> un-American tendency.[57]

The Frank case permitted the new ADL to start with a bang, but the
organization had little impact on either Jews or other Americans until
after World War II.[58]

In contrast, the AJC started strong in 1906 and so long as Louis
Marshall served as President of the organization (1912 until his death
in 1929) it remained the major American Jewish defense organization.
During its early years, the AJC dealt with most facets of antisemitism
in American society, and correctly regarded attempts at immigration
restriction as manifestations of antisemitism. After the United States
entered World War I, in April 1917, communiques about wartime discrimination overwhelmed the AJC Executive Committee. Hardly a
meeting occurred in 1917 and 1918 in which some aspect of bias was

not dealt with by this small group. The most frequent objections concerned discrimination against Jews by various governmental agencies and departments.[59]

As in previous wars, and again during World War II, anxiety gripped the American populace. During these stressful times minorities appeared outside the fold. Members of such groups were regarded as uncertain patriots and unsuitable for inclusion in the most important aspects of the defense of our country. Thus the American Army inserted discriminatory advertisements in the New York *World* on November 8 and 9, 1917, blatantly stating its need for "Christian" carpenters. In a letter to Secretary of War Newton D. Baker, Louis Marshall strenuously opposed prejudicial want-ads and Baker quickly issued a directive forbidding such bigotry. Another objection Marshall made about War Department policies occurred after the Provost General noted in the *Manual of Instructions for Medical Advisory Boards:* "Foreign born, especially Jews, are more apt to malinger than the native born." This comment so infuriated American Jewry that Marshall protested directly to President Wilson. Once informed, Wilson ordered the immediate deletion of that "unfortunate expression."[60]

Within the military, as well as at the Annapolis naval academy, Jewish soldiers, sailors, and students suffered from inordinate amounts of hazing, ridicule, and harsh punishments. During the war not a single Jew was promoted to the officer corps from the ranks in Europe and animosity existed toward those Jews who had been trained and commissioned in the United States. For breaching a sanitary regulation at Camp Wadsworth, in Spartansburg, North Carolina, Private Otto Gottschalk received a brutal flogging. After the AJC intervened on Gottschalk's behalf, the officer in charge of the beating was court-martialed. Nineteen Jews entered the Annapolis Class of 1922 in 1918 but only nine graduated. Unmerciful hazings contributed to the resignations of the others.[61] One of the nine survivors was Admiral Hyman Rickover, and his biographers later wrote that "in the social climate of Annapolis of the 1920s, the hazing of a Jew would seem to be inevitable."[62]

During World War I Navy Secretary Daniels also had to review an antisemitc doggerel posted in the main tailor shop of a camp at the Naval National Reserve in Pelham, New York. It read

> Largest mob that the world ever saw
> Trying to beat the Conscription Law
> Jews in front and Jews behind
> Jews of every conceivable kind.
>
> Massed on the steps of the City Hall
> Jews that were big and Jews that were small
> Jews that were fat and Jews that were thin
> Prominent nose and receding chin.

> Socialists, anarchists, slackers and sneaks
> Faces impertinent brazen and weak
> Jews of all station, poor Jews and rich
> Jews that were dirty and Jews with the itch.
>
> Eager to marry and hide behind
> Any old skirt of any old kind
> Pushing and crowding till ready to drop
> Not an Irishman present excepting the cop.[63]

The comic verse not only reflected malevolence toward Jews but also reemphasized the common belief that they lacked patriotic fervor, that they were cowards who shirked their responsibilities, and that as Jews their loyalty could not be relied on. In contrast, the last line emphasized the patriotism of the Irish Americans. The stereotypes so obviously enjoyed in the tailor shop would be a feature of jingles circulated both on and off military bases during World War II.

Another scurrilous verse of the same type appeared in 1917 under the fictitious title *Yiddisha Army Blues* and it, too, repeated the same shibboleths of the previous ditty while adding others that were still commonly believed. This time, however, the Anti-Defamation League succeeded in getting publishers to recall the sheet from music stores and refrain from selling further copies by mail. The jingle's lyrics ran:

> Jake tried to sell his business
> But he couldn't give it away
> So he set the place on fire
> In a business way.
> It's time I saved my money
> Till I get a little start
> Now for the battle-fields
> I'm going to depart.
> My head it starts a-reeling
> I get such a sickish feeling
> When I hear them sing that song
> Over there——
> Oi, Oi vey, I'm shaking in my shoes.[64]

Several episodes during and after the war reminded the Jewish defense organizations of the necessity of constant vigilance against defamation. Most Americans probably would have seen nothing wrong with the statement in the *Junior Plattesburg Manual* noting that "the ideal officer is a Christian gentleman," or the unwillingness of Major George H. Savage of the Quartermaster Department to hire a woman secretary because he preferred not having a Jew on his staff, or even the Department of Interior's appointment of a "Special Collaborator and Racial Advisor on Americans of Jewish Origin" whose duty was to interpret for Jews "the ideals, traditions, and standards of this great nation." But to Jews such activities and actions appeared too frequently and reflected the enormous task they faced in trying to have other Americans accept them simply as people of another religion.[65]

But that was not to be. When the war ended, prejudice toward all religious and ethnic minorities increased and a new wave of nativism stood ready to crest. Jews and others of east European background entering adulthood during the 1920s would face a new set of societal barriers. As a result of the Bolshevik Revolution in Russia in 1917, a fourth dimension of prejudice against Jews—in addition to religious, economic, and racial—came to the fore. The brush of "bolshevism" now tarred them as radicals and socialists. Fears arose in the United States that Jews desired to implant this foreign ideology on American shores and thereby undermine the American way of life.

Thus as the Progressive era and World War I ended, the Jewish battle to prove their loyalty and justify their acceptance by others continued, and had to be fought in an America in no mood to be tolerant. Unlike many other changes in society during the Progressive era racism remained potent and seemed to be getting even stronger than it had been at the turn of the century. Although more Jews had become Americanized and Jewish organizations appeared on the scene to protect their image and fight antisemitism, other Americans did not think more kindly of the Jews or behave more warmly to them. The 1920s was to be a decade in which nativism dominated public concerns and attacks on Jews, Catholics, and immigrants intensified. Paradoxically, Jews, whose socioeconomic status rose, thereby allowing them greater material luxuries, also faced increased private, public, and institutional onslaughts. More than any other Caucasian group, Jews were perceived as inimical to the American way of life.

5

Erecting Barriers
and Narrowing
Opportunities
(1919–1933)

The aftermath of the first World War left Americans disillusioned with internationalism, fearful of Bolshevik subversion, and frightened that foreigners would corrupt the nation's values and traditions. A majority seemed tired of almost two decades of domestic reform and longed for what the 1920 Republican presidential nominee, Warren G. Harding, termed "normalcy." And to most people "normalcy" meant remaking the United States into what it symbolized in the minds of old stock Americans. The Secretary of the Chamber of Commerce of one community in Florida captured an extreme expression of this feeling in 1924 when he advocated expelling all Jews and foreigners from St. Petersburg to make the community "a 100% American Gentile City."[1] More specifically, "Gentile" meant "Protestant." In rural and urban areas alike Catholics and Jews were regarded as outcasts intent on undermining American values rather than as groups longing to accept them.[2]

This intensified desire to keep America for the Americans coincided with the coming of age of the children of turn-of-the-century immigrants who sought to escape the ghetto. In the 1920s Jews, more than the offspring of any of the other southern and eastern European minorities who entered America between the 1880s and the first World War, seemed more prepared to enter the elite Protestant world in terms of schooling, employment, housing, and recreation. Efforts to acculturate, however, were met with strong resistance from members of the middle and upper classes who had little inclination to associate with Jews.

Such disinclinations were not new. Antisemitic traditions had existed in this country for centuries. Nevertheless, the antagonism toward Jews increased alarmingly during the postwar decade. Psuedo-Scientific racist thinking, which became fashionable in the early years of the twentieth century, contributed to this animosity. So, too, did the 1917 Bolshevik Revolution in Russia that had an impact in the United States far out of proportion to any realistic concerns that Americans need have had about socialism and communism replacing the existing political structure. Jews were also attacked for allegedly leading the charge for a domestic socialist revolution that existed only in the minds of the most anxious Protestants and Catholics.[3] Another important source of escalating enmity was the fear that many sons of immigrants who arrived in previous decades would make inroads in established institutions where the Protestant elite reigned.

Throughout the United States there was near universal concern about Jews infiltrating cherished organizations and abodes. Leaders of this opposition provided a rationale that different Americans could use to vent their wrath on the despised Jew who symbolized not only foreign and subversive ideologies but also a variety of unspecified anxieties agitating most Americans. From members of the elite in business, government, or education; popular writers outraged at the alleged mongrelization of American society; or religious leaders seeking to purify the hearts and souls of their followers, came an unending number of charges against the disabilities, characteristics, and alleged malfeasance of Jews.[4]

Displays of antisemitism appeared in connection with the Red Scare during and after World War I. The Bolshevik Revolution found few sympathizers in the United States and there was little reason to assume that the philosophy of those who brought down the Tsar would make inroads in America. Yet a spirit of intolerance carried over from the war into the postwar era and then a series of labor strikes in 1919 frightened many Americans into believing that these disruptions served as a prelude to revolution. When the labor disorders were both punctuated and followed by letter bombs sent to some prominent people around the country a full-fledged "red-scare" ensued. At first the targets were "foreigners" and "anarchists" but those charges eventually resulted in accusations against Jews.

Two articles published in the popular periodical, *Literary Digest*, reinforced the views that the terms "Jews" and "Bolsheviks" were syn-

onymous. On December 14, 1918, the journal reprinted a piece from
World's Work indicating that Jews stood prominently among the leading
Bolsheviks. Then, in March 1919, an essay entitled "American Jews in
the Bolshevik Oligarchy" began, "Jews from New York's east side, so
reports from Russia have run for a long time, are playing a large part
in promoting the present activities of the Bolsheviki throughout Russia." The narrative quoted Methodist minister George A. Simons who
told a U.S. Senate investigating committee that "Yiddish agitators"
from New York City's Lower East Side were creating chaos in the
Soviet Union.[5] In 1919, editorials in the *Minneapolis Journal* and the *Minneapolis Tribune* likened Jews to Bolsheviks,[6] and the former asked, "Are
Russians and Russian and German Jews and others besides psychologically unfit to be citizens of an Anglo-Saxon state or country?"[7] That
same year conservative businessmen in the Greater Iowa Association
spread rumors that Jewish peddlers disseminated Bolshevik literature
along with their wares.[8]

Poles in Chicago also equated Bolshevism and Judaism, denounced
President Woodrow Wilson as a dupe, and labeled the Treaty of Ver-
sailles a product of "Jewish internationalists."[9] By 1924 Chicago's
Polonia, written in Polish, observed:

> Practically the whole world knows that the Jews direct socialism, Jews
> who through socialism are striving to stir up in various countries ferment
> and social unrest which always results in harm to Christian society and
> in profit and gain for the Jews.[10]

Another scurrilous association of Jews, radicalism, and subversion
that emerged from the war came packaged in the form of a book: *The
Protocols of the Elders of Zion*. Concocted by members of the Russian
secret police at the turn of the century, this spurious work, based on
European Christian mythology, purported to show that Jews, agents
of the devil, were secretly plotting a world revolution to undermine
Christian civilization. The document received little attention until after
the Bolshevik Revolution when it soon made its way into western
Europe and then into the United States. For those needing an explanation for the Russian upheaval the book provided a great deal of information, albeit preposterous. Nonetheless, the myth of the Jewish world
conspiracy harkened back to earlier Christian teachings. Also, Americans in previous generations, and Christians for centuries, had been
told of Jewish economic conspiracies and the Jews' alleged control of
the world's gold—hence the argument that they were planning to
undermine Christian governments did not seem so farfetched. Published in London early in 1920, it took only a few weeks before *The
Protocols* appeared in Boston. An American publisher then put out an
edition entitled, *The Cause of World Unrest*.[11]

On May 22, 1920, an article entitled "The International Jew: The
World's Problem" reproduced the essence of *The Protocols* and domi-

nated page one of Henry Ford's newspaper, *The Dearborn Independent*. This piece inaugurated a series of attacks upon Jews that the newspaper ran for 91 consecutive weeks, and then intermittently until 1927. When the essays began, *The Dearborn Independent* had a circulation of about 72,000 copies a week. By 1922 that figure had increased to 300,000. It peaked in 1924 at 700,000, only 50,000 fewer than that of the largest daily paper in the United States at that time, New York City's *Daily News*.[12] The outlandish tales of alleged Jewish vices were read by millions of Americans, mostly in rural areas, who knew little or nothing about Jews except what they had already absorbed from religious teachings, gossip, and *The Dearborn Independent*.

Exactly what motivated Henry Ford to launch an attack upon Jews is unclear. He had grown up on a farm in rural Michigan during the Populist era when there was a general wariness of Jews, especially about their alleged greed and control of wealth. But not until 1915 did Ford publicly express antisemitic thoughts. He took a group of people to Europe on a "Peace Ship" that year with the hope of influencing the European powers to end the war. The mission failed and Ford attributed this to Rosika Schwimmer, a Jewish woman of Hungarian birth who had encouraged him to proceed, and to Jews in general. In his later virulent antisemitic campaigns, Ford would repeatedly return to the ill-fated Peace Ship incident and Rosika Schwimmer's influence on him at that time. As Ford recalled it later, that event introduced him to the connection between radicalism and Jews. By 1918, when he ran as the Democratic nominee for U.S. Senator from Michigan and lost, Ford was attributing almost everything that went wrong with his political fortunes to Jews and Jewish capitalism.[13]

In 1919 Ford purchased *The Dearborn Independent*. "Let's have some sensationalism," in the form of a strong and forceful series on a major topic, suggested a consultant brought in to help improve sales. Given Ford's increasing hostility toward Jews, the publication of *The Protocols* in the United States in 1920 provided him with a vehicle that brought national attention to his fledgling weekly. In 1921 when asked for his rationale for the series on "The International Jew" Ford responded that he was

> only trying to awake the Gentile world to an understanding of what is going on. The Jew is a mere huckster, a trader who doesn't want to produce, but to make something out of what somebody else produces.[14]

The first of the essays on "The International Jew" in *The Dearborn Independent* described the Jew as the world's enigma. Poor in his masses, he yet controls the world's finances. Scattered abroad without country or government, he yet presents a unit of race continuity no other people has achieved. Living under legal disabilities in almost every land, he has become the power behind many a throne. There are ancient prophecies to the effect that the Jew will return to his own land

and from that center rule the world, though not until he has undergone an assault by the united nations of mankind.

> The single description which will include a larger percentage of Jews than members of any other race is this: He is in business. It may be only gathering rags and selling them, but he is in business. From the sale of old clothes to the control of international trade and finance, the Jew is supremely gifted for business. More than any other race he exhibits a decided aversion to industrial employment, which he balances by an equally decided adaptability to trade. The Gentile boy works his way up, taking employment in the productive or technical departments, but the Jewish boy prefers to begin as messenger, salesman or clerk—anything—so long as it is connected with the commercial side of the business.

"In America alone," *The Dearborn Independent* went on,

> most of big business, the trusts and the banks, the natural resources and the chief agricultural products, especially tobacco, cotton and sugar, are in the control of Jewish financiers or their agents. Jewish journalists are a large and powerful group here. "Large numbers of department stores are held by Jewish firms," says the Jewish Encyclopedia, and many if not most of them are run under Gentile names. Jews are the largest and most numerous landlords of residence property in the country. They are supreme in the theatrical world. They absolutely control the circulations of publications throughout the country. Fewer than any race whose presence among us is noticeable, they receive daily an amount of favorable publicity which would be impossible did they not have the facilities for creating and distributing it themselves. Werner Sombart, in his "Jew and Modern Capitalism," says: "If the conditions of America continue to develop along the same lines as in the last generation, if the immigration statistics and the proportion of births among all nationalities remain the same, our imagination may picture the United States of fifty or a hundred years hence as a land inhabited only by Slavs, Negroes and Jews."[15]

These ideas were old but *The Protocols of the Elders of Zion* added an international twist to the myths already perpetuated. What Henry Ford and *The Dearborn Independent* did was to merge these accusations against Jews into one series on "The International Jew" and run the articles for almost two years. Had the topic not won reader approval, it would certainly have been abandoned.

Articles on Jews in *The Dearborn Independent* evoked widespread and enthusiastic comment in the years 1920 to 1924. People from all walks of life, including college professors, illiterates, and large numbers of Christian ministers, sent money to Ford, praised him for his attacks on Jews, and requested additional materials for their own use. In Chicago, many of the leaders of the two major Polish-American groups, the Polish National Association and the Polish Roman Catholic Union, agreed with the positions taken in the series on "The International Jew," and the latter organization's newspaper approvingly publicized Ford's anti-Semitic ravings.[16]

The series also won plaudits overseas. Copies of "The International Jew" were translated into several languages and circulated in Europe and Latin America throughout the 1920s and 1930s. The work is widely credited with influencing the writing of Adolf Hitler's *Mein Kampf*.[17] Hitler kept a picture of Ford on the wall of his office in Munich, praised the automobile magnate in *Mein Kampf*, and later told a *Detroit News* reporter, "I regard Henry Ford as my inspiration."[18] On Ford's 75th birthday, in 1938, Hitler sent personal greetings and bestowed on him the highest honor the German government could grant a foreigner: the Grand Cross of the German Eagle.[19]

Despite widespread acceptance of Ford's writings about the Jews it would be a mistake to conclude that his antisemitism or the repetitious attacks in *The Dearborn Independent* won universal approval. Articles in *The Century, Current Opinion,* the *Independent,* the *Outlook,* and *Harper's Weekly* strongly denounced the *Protocols,* Ford, and the series on "The International Jew."[20] Samuel Walker McCall, Governor of Massachusetts from 1916 to 1919, wrote:

> The so-called "Protocols of the Elders of Zion" have recently been brought forward among us as a basis for anti-Jewish agitation. And this has been done after they have been fully discredited in Europe, where scholars of independence and character had no difficulty in penetrating into their fraudulent character.[21]

The Nation also noted that same wave of antisemitism sweeping the world in 1920 and agreed that "the chief responsibility for the survival of this hoary shame among us in America attaches to Henry Ford."[22]

At its annual convention in December 1920, the Federal Council of Churches (FCC) condemned Ford without mentioning his name. In a circuitous statement members of the FCC indicated that "for sometime past there have been in circulation in this country publications tending to create race prejudice and arouse animosity against our Jewish fellow-citizens and containing charges so preposterous as to be unworthy of credence." The FCC, concerned about "unity and brotherhood," deplored the "cruel and unwarranted attacks upon our Jewish brethren" and earnestly admonished "our people to express disapproval of all actions which are conducive to intolerance or tend to the destruction of our national unity through arousing racial divisions in our body politic."[23] A few weeks later 119 prominent Americans, including Woodrow Wilson and William Howard Taft, signed a statement denouncing what they described as "an organized campaign of anti-Semitism, conducted in close conformity to and co-operation with similar campaigns in Europe," but that declaration, too, did not mention Henry Ford.[24]

Despite the disclaimer, many genteel Americans, including some of those who publicly opposed Ford's attacks against Jews, were also more comfortable socializing with Gentiles and questioned the wisdom of allowing Jews too many opportunities in the United States. Some of the most articulate and influential members of American society also wanted to rid themselves of Jews in their midst—or, more accurately, prevent Jews from encroaching on their established preserves—but they sought this aim less bluntly. The economic success of some of the eastern European Jewish immigrants and the goals that they and their children aspired to in education, social equality, and positions of leadership in society alarmed many members of the American elite.

President A. Lawrence Lowell, for example, thought that Harvard university had a "Jewish problem" because the percentage of Jewish undergraduates had tripled from 6 percent in 1908 to 22 percent in 1922. He suggested that a limit be placed on the number of them who would later be admitted to the university. To be a real American, Lowell believed one had to be a Christian, and he feared that too many Jews at Harvard would result in the institution's losing "its character as a democratic, national university, drawing from all classes of the community and promoting a sympathetic understanding among them."[25]

Not only the Harvard President, but members of the Board of Governors, alumni, students, faculty, and other administrators shared this concern. The subject of what to do with an increasing number of Jewish students had been previously discussed at Harvard as well as at other schools, and it affected significant areas of university life outside the classroom as well. In 1920, for example, a year after Judge Julian Mack became the first Jew to win election to Harvard's Board of Overseers, a colleague on the Board, Jack Morgan, of the J. P. Morgan investment bank, communicated his concerns to President Lowell about another vacancy on that body:

> I think I ought to say that I believe there is a strong feeling among the Overseers that the nominee should by no means be a Jew or a Roman Catholic, although, naturally, the feeling in regard to the latter is less than in regard to the former. I'm afraid you will think we are a narrow-minded lot, but I would base my personal objection to each of these two for that position on the fact that in both cases there is acknowledgement of interests of political control beyond and, in the minds of these people, superior to the Government of this country—the Jew is always a Jew first and an American second, and the Roman Catholic, I fear, too often a Papist first and an American second.[26]

Morgan's views represented those of a broad spectrum of individuals. Few, if any, of those associated with the university wanted, as historian Frederick Jackson Turner wrote, a "New Jerusalem" at Harvard.[27] As one British alumnus put it, "while I am not anti-Semitic in principle, I cannot help feeling, as many others do, that the entire disappearance of the Christian (at least, of the non-Jewish) element from my old University would be a grievous loss to the community. To Harvard men in this world at large it would be an overwhelming tragedy."[28] A student expressed a similar opinion more dramatically: "Imagine having an alumni so strongly Jewish that they could elect their own president and officers! God forbid!"[29]

The Harvard president knew the sentiments of various members of the university community before he made his announcement, and he was not expressing a view that came uneasily to him.[30] Lowell had been an officer of the Boston-based Immigration Restriction League and shared other members' values about the superiority of the Anglo-Saxon culture and their anxieties about it being undermined in the United States by hordes of foreigners with "inferior" backgrounds. Of all of those so-called lesser groups, none seemed more threatening to the reigning elite in America than the immigrant Jews. "The Jews," as Lewis Gannett noted in *The Nation* in 1923, were "the most eager to get the benefit of all that the Anglo-Saxon has to offer him. No other group flocks to our colleges with the same pathetic thirst for learning."[31] As a result, Jews seemed particularly ominous to people who wanted their institutions to remain in traditional hands.

Antisemitism pervaded Harvard's elite groups. In 1920, when Francis Biddle, a scion of a distinguished Philadelphia family, invited Harvard Law School Professor Felix Frankfurter to speak at a dinner of The Fly, one of the most select of the Harvard undergraduate clubs, several alumni declined their invitations because they refused to socialize with a Jew. Until that incident Biddle claimed to have been unaware of "how strongly anti-Semitic feeling permeated Harvard clubs."[32]

Never before had any secular American university openly acknowledged that it wished to limit the number of its Jewish students. Lowell's announcement, however, publicly signaled that the increase in Jewish undergraduates alarmed him, as well as the faculty, students, and alumni of the university. In the same fashion the elevated number of Jews at many other American universities was perceived as an invasion that would ultimately undermine both Christian traditions and the social prestige of schools that housed too many of them. Lowell's position therefore reflected what Dean Frederick Paul Keppel of Columbia University had written in 1914 about the multiplication of Jews at his college. Keppel feared that too many immigrants on campus might make Columbia "socially uninviting to students who come from homes of refinement."[33] Similar concerns bothered administrators at Princeton, Williams, Yale, and the University of Pennsylvania.[34]

The action taken at Harvard inspired other institutions. The Association of New England Deans, meeting in Princeton in May 1918, had already discussed the "problem" of having too many Jewish students[35] but reached no conclusion about how to deal with that dilemma. Once Harvard took the lead, however, many of the nation's most prestigious private colleges and universities, along with less distinguished academies throughout the country, established their own allotments. These schools included Columbia, Princeton, Yale, Duke, Rutgers, Barnard,

Adelphi, Cornell, Johns Hopkins, Northwestern, Penn State, Ohio State, Washington and Lee, and the Universities of Cincinnati, Illinois, Kansas, Minnesota, Texas, Virginia, and Washington. New York University discriminated on its Bronx campus but not at Washington Square. As the 1920s admissions moved from selective to restrictive fewer places became available for Jews.[36]

Many universities found unique ways of handling what they perceived of as "the Jewish problem." The Dean at Colgate wanted only six Jews in the university so that if charges of antisemitism were raised the six could be trotted out to disprove the assertion. At Syracuse University, where a Ku Klux Klan chapter existed, Jews were also excluded from most social organizations and, from 1927 to 1931, were housed separately from Christians. Ohio State segregated Jews in some female dormitories while Gentile students at the Universities of Michigan and Nebraska were advised against associating with Jewish males. One of the questions on the application form for Sarah Lawrence College in Bronxville, New York, asked, "Has your daughter been brought up to strict Sunday observance?" Columbia College's application wanted to know the student's "religious affiliation," whether he or his parents had ever been known by another name, parents' place of birth, mother's full maiden name, and father's occupation. Harvard came up with the idea of geographical diversity, assuming that outside of the major cities in the East and Midwest one would find few Jews.[37] The President of Dartmouth, Ernest Hopkins, summed up the general reasons for these questions and policies when he said: "Any college which is going to base its admissions wholly on scholastic standing will find itself with an infinitesimal proportion of anything else than Jews eventually."[38]

Jews who attended these "selective" universities—and practically all the colleges allocated 3 percent to 16 percent of the slots in the freshman class for Jews—contended with social snobbery and exclusion. Sometimes the experiences were painful, other times merely offensive. Jews allegedly lacked college spirit, set too high an intellectual standard, did not drink, smoke, or participate in athletics, were physically repulsive, too anxious to please, or all the above. Walter Lippmann, who would later become an internationally famous journalist, was always an outsider as an undergraduate at Harvard where few Gentiles were nice to him or cordial to Jews. Antisemitism kept Laura Z. Hobson out of Phi Beta Kappa at Cornell because there was a feeling among the trustees and some of the faculty that "too many greasy little grinds from New York [and] not enough people from other parts of the country" were qualifying on merit for the honor societies. One correspondent of *The Nation* in 1923 thought that Jews were acceptable in class but that otherwise their deportment undermined social prestige. He wrote, that "at the 'prom' or the class-day tea, the presence of Jews and their relatives ruins the tone which must be maintained if social standing is not to collapse."[39]

Journalist Max Lerner remembered his days at Yale: "We were kept out of everything. Not in any formal way, but in how we were treated."[40] One Jewish student believed that "a 'Cohen' has as much chance to be the editor of Harvard *Crimson* as I have to be Pope."[41] Most fraternities shunned Jewish pledges but at Brown, where Jews were absolutely excluded from these brotherhoods, the President of the university refused to sanction a Jewish fraternity on the grounds that it might "kindle the fires of racial antagonism" and "damage" the school's reputation.[42] An antisemitic atmosphere also existed at the U.S. Naval Academy in Annapolis where Jews were harassed, excluded, and psychologically ostracized. In the class of 1922 the editors of the school yearbook humiliated the Jewish cadet who ranked second academically by printing his picture on a perforated page so that it could be removed without defacing the rest of the volume.[43] When asked about his college experiences a half century later, Democratic party chairman and later U.S. Ambassador to the Soviet Union Robert Strauss recalled, "Not many things happened to me at the University of Texas. I discovered I was Jewish, which meant that you were ostracized from certain things."[44]

Universities had even less desire for Jewish faculty than for Jewish students. Before the 1920s few Jews had earned doctorates. With the growth of educational facilities at the turn of the century, which coincided with the huge immigration of Jews from eastern Europe, a number of Jews tried to prepare themselves for academic positions. They found opportunities almost non-existent. This was especially true in the disciplines of English and History where it was thought inconceivable for a Jew to transmit or comprehend the culture and traditions of an American Christian society. When Max Lerner (Yale, B.A., 1923) informed a college instructor whom he was on good terms with that he would like to teach English at a university, the instructor replied: "Max, you can't do this. You can't teach literature. You have no chance of getting a position at any good college. You're a Jew."[45]

A few Jews obtained academic jobs. In 1927 the faculties of Yale, Princeton, Johns Hopkins, and the Universities of Chicago, Georgia, and Texas each included one Jew. There were two each at Berkeley and Columbia, three at Harvard, and four at the City College of New York, "the poor man's Harvard." English departments generally considered themselves the bastions of Anglo-Saxon culture and not until 1939 did Lionel Trilling become the first Jew appointed to a tenure-track position in that field at Columbia University. Diana Trilling later wrote that "it is highly questionable whether the offer would have been made" had her husband borne the surname of his maternal grandfather, Cohen. After Trilling became an assistant professor, a colleague came by to chat and expressed the department's hope that the new appointee would not use his opportunity "as a wedge to open the English department to more Jews."[46]

In reading through the accomplishments of Jewish academics between the wars the uniqueness of their good fortune and the difficulties they had in obtaining their ultimate rank are always stressed. Isidor I. Rabi, for example, graduated at the top of his Cornell class in chemistry in 1920 but could not find appropriate employment. Informed that the academic world was closed, he took a job in a chemical laboratory analyzing mother's milk and furniture polish. He changed jobs frequently during the decade before the patrician chairman of Columbia's physics department befriended him. Rabi obtained a lucrative fellowship, returned to Columbia to study physics, and then received a lectureship to teach mathematical physics. At the time, no one could recall another Jew who had been appointed in the department.[47]

All told, fewer than 100 Jews were on the liberal arts and sciences faculties of American universities in the mid-1920s and not many more a decade later.[48] The tales of the rare Jews who succeeded are usually accompanied by the caveat that they were specifically chosen because of their unusual qualities rather than as a sign that barriers were being dropped. Letters of recommendation for historians Oscar Handlin, Bert Lowenberg, and Daniel Boorstin, for example, bore phrases like "has none of the offensive traits which people associate with his race," "by temperament and spirit . . . measure up to the whitest Gentile I know," and "is a Jew, though not the kind to which one takes exception."[49] A professor of mathematics at the University of Chicago, commenting about a mathematical astronomer, noted, "He is one of the few men of Jewish descent who does not get on your nerves and really behaves like a gentile to a satisfactory degree." Many academics, however, believed that people of the "Jewish race" should be totally barred from the academic world.[50]

Discrimination in university employment was simply part of a larger practice. At the turn of the century most immigrant Jews had little choice but to engage in industrial occupations, peddling, or starting up new businesses. Those who could educate their children in the United States did so, and the offspring, in turn, acquired skills that qualified them for clerical, managerial, and professional positions. But the second generation found impediments beyond their control.

As more Jews sought white-collar jobs, newspaper advertisements indicating a preference for Christians proliferated. Such announcements had appeared intermittently before World War I but their numbers increased thereafter. A sharp rise in 1922 and 1923 coincided with the impact of Henry Ford's series on "The International Jew." The peak year for discriminatory advertising during the decade occurred in 1926, reflecting the cumulative effect of a variety of publicly enunciated antagonisms against Jews. Restrictive advertisements were far more prevalent for women than they were for men. The reason for the discrepancy was probably that 75 percent of the ads were for typists, stenographers, or secretaries, positions then going mainly to women.[51]

Many businesses discriminated whether they publicized it or not. Utilities, banks, insurance companies, publishing houses, engineering and architectural firms, advertising agencies, school districts, major industrial companies, civic bodies for art and music, hospitals, universities, and law firms were among the major culprits.[52] Paul Cravath, senior partner at New York City's prestigious law firm, Cravath, Swaine and Moore, informed students at the Harvard Law School about the kinds of associates he and his partners favored:

> Brilliant intellectual powers are not essential. Too much imagination, too much wit, too great cleverness, too facile fluency, if not leavened by a sound sense of proportion, are quite as likely to impede success as to promote it. The best clients are apt to be afraid of those qualities. They want as their counsel a man who is primarily honest, safe, sound and steady.[53]

Other firms apparently had similar goals. Humble Oil Company in Houston had a rule never to engage Jews. The Indianapolis drug firm, Eli Lilly, employed nearly 4,000 people in the 1930s only two of whom were Jewish. Western Union refused to hire Jewish messenger boys.[54] Actor Kirk Douglas, recalling his boyhood days in a small town in upstate New York, wrote:

> It's tough enough to be a Jew, but it was very tough in Amsterdam. There were constant reminders. No Jews worked in the carpet mills. No Jews worked on the local newspaper. No Jewish boys delivered the newspaper.[55]

During Ford's lifetime, no Jewish physicians served on the staff of the Henry Ford hospital in Detroit and few Jews were on hospital staffs anyplace else in the country. Young women of Jewish descent at the Dental Hygienist School of the University of Minnesota were told that it would be impossible to place them after graduation. Secretaries heard the same story. New York City Congressman Fiorello LaGuardia denounced the State Department for invariably failing Jews who had done well on written examinations in their oral interviews. At the end of the 1920s, when Jews constituted about 26 percent of New York City's population, and were also the best educated group in the community, one study found that 90 percent of white-collar positions went to non-Jews. Another examination of 27,000 cases, made by a "Christian placement specialist," also indicated that 90 percent of the openings discriminated against Jews.[56] The owner of one factory succinctly expressed the prevailing sentiment: "We try to have only white American Christians in our factory regardless of religion."[57]

Both cultural values and limited employment opportunities dictated the kinds of jobs and professions Jews sought. In New York City the major garment workers' industries were dominated by Jews as was a good part of the building industry. Among the professions teaching, social work, pharmacy, medicine, and law also attracted them. In 1930, half of all lawyers and one-third of the physicians in New York City were Jewish.[58] Finally, of course, there were the myriad opportunities in retail businesses and real estate construction that individuals could embark on themselves. In the 1920s Jews predominated among those building apartment houses in New York City. A 1937 survey, moreover, estimated that Jews owned two-thirds of the 34,000 factories in the city, about the same percentage of the 104,000 wholesale and retail establishments, and 11,000 restaurants and luncheonettes.[59] The more successful entrepreneurs not only employed members of their own families but others as well. This pattern of endeavors applied, to some extent, in communities all over America. In southern California the fledgling movie business provided unique opportunities because so many Jews participated in the founding and development of the industry, but throughout most of the rest of the country Jews could be found mainly in family businesses, as independent professionals, and in government service.

A case in point is Samuel Rosenman who graduated from Columbia Law School with highest honors but refrained from seeking interviews with prestigious law firms; he knew that would be a waste of time.[60] Instead he moved into state government where he worked with Governor Al Smith. Franklin D. Roosevelt, Smith's successor in 1929, chose Rosenman as his counsel. He proved so adroit in that position that after Roosevelt won the 1932 presidential election he elevated Rosenman to the state judiciary. In 1937 the President brought Rosenman into the White House where he remained until after Harry S Truman had served out his first year.

Despite Rosenman's status, associations, and accomplishments, he, like Ben Cohen, another Jew who served in the White House during Roosevelt's presidency, was excluded from prestigious private clubs. Unfortunately that also limited access to networks of corporate power. Organizations like the Knickerbocker, Links, and Union Clubs were part of the New York City business establishment and included some of the most important members of Wall Street houses, prominent law firms, and international businesses but few Jews. Jewish industrialist Henry S. Hendricks (1892–1959), who traced his lineage back to the Revolutionary era in New York, was allowed into the Knickerbocker Club but not the Yacht Club to which his father and grandfather had belonged. In Pittsburgh, an exclusively Gentile male elite gathered at the Duquesne, a club later characterized by social analyst Vance Packard as "the *ne plus ultra* of business clubs in the United States"; in Washington, D.C., the men who counted congregated at the Chevy Chase.[61]

The private men's clubs came into vogue in the United States during the middle of the nineteenth century. They included individuals with common, identifiable business interests—in a city like New York these usually included the areas of shipping, banking, and investments. At first an individual's religion was not a factor in membership but it later became grounds for excluding non-Protestants. Five members of the Gratz family, for example, were members of the Philadelphia Club before the Civil War; one had even served as temporary President during the Mexican War. Jesse Seligman, a founder of New York's Union League, resigned in 1893 after his son was blackballed because Jews were no longer admitted. Similarly, Los Angeles associations like the University Club and the California Club took Jews at the end of the nineteenth century but excluded them by the 1920s.[62]

It seems logical that people who did not want to study, work, or fraternize with Jews would have no desire to socialize with them. Hence graduates of those universities that wanted to restrict their presence as classmates, and employers who did not want to hire them, avoided Jews in moments of relaxation. Occasionally, individuals were admitted to restrictive organizations, and in some smaller communities there were too few Jews, or their dues and contributions were necessary to keep a club going, so membership was less exclusive. But the general practice throughout the nation, especially in elite men's and women's clubs, was to bar Jews.

Patterns that developed in the men's clubs appeared in the newer country or social clubs for families that began in the 1880s. While intermittent prejudice existed in the nineteenth century, no consistent policy of exclusion had prevailed. Elite members of the Jewish community were sometimes founders of, or welcomed at, the prominent clubs in New Orleans, Tucson, and Akron, Ohio, but were banned in the Athlestane Club of Mobile, Alabama, and shunned by social groups in Denver. Georgia's Jekyll Island Club, founded in the early 1880s, mostly by wealthy New York and Chicago Episcopalians and Presbyterians of British ancestry, solicited newspaper magnate Joseph Pulitzer, who was born in Budapest to a Jewish father and a Catholic mother, to become a charter member of the club. Listed among the other founders were J. P. Morgan and William K. Vanderbilt. The Jekyll Island historians, however, acknowledge that Pulitzer was "an anomaly among the original members" and identify no others of Jewish descent who belonged to the club.[63]

Club discrimination was common throughout the country. One found it in places as diverse as Cleveland and Columbus, Ohio, Indianapolis, Denver, Los Angeles, Philadelphia, Atlanta, Detroit, New York, Charleston, South Carolina, Newport News, Virginia, and New Orleans.[64] In Muncie, Indiana, Jews were not even acceptable in service clubs like the Elks; in Marietta, Ohio, several men, mostly Methodists, threatened to resign from the Rotary Club if it accepted one of the community's leading Jews; and in Minneapolis Jews were barred from the Boat Club, the Automobile Club, the Athletic Club, the Blue lodges of the Masons, and the Rotary, Kiwanis, and Lions organizations.[65]

There were sad individual stories of discrimination. Many Jews of German origin tried every method to impress their Gentile peers. They mimed the speech and dress of the Protestant elite, donated huge sums of money to favorite Gentile charities like museums, libraries, and universities, and some even became Christians like Otto Kahn. But they were never accepted by those they tried to emulate. Bernard Baruch, a self-made millionaire, a confidant of President Wilson, and head of the War Industries Board during World War I, received a letter from the Palm Beach Bath and Tennis Club, to which he belonged, asking members not to bring Jewish guests. He promptly resigned. His daughter was denied admission to a dancing class that her mother, a non-Jew, had attended when she was a child. A man who chose not to reveal his name wrote a letter to *The Atlantic Monthly* in 1924 detailing the difficulties of his family in a suburban area of New York City. The social life in his community revolved around the club but his children were not invited to the Friday night dances or allowed to go to the skating rink in the winter or the tennis courts in the summer. His letter indicated that he had attended one of the finest universities in the country and that both he and his wife descended from many generations of "cultured Americans." Nonetheless, the religion of his parents prevented social acceptance. In turn, this social discrimination would foster the dehumanization of Jewish-Gentile relations and pave the way for the vicious antisemitic behavior of the 1930s.[66]

Members of exclusive social clubs also preferred to vacation in resorts that barred Jews. The general practice of refusing Jews seems to have begun in the 1880s, although incidents occurred earlier, and steadily increased thereafter. Newspaper advertisements in the 1920s specified "Gentile" or "Christian" patronage, or included phrases like "No Hebrews or tubercular guests received," "Adverse to association with Hebrews," and so forth. Driving through Michigan in 1925 Bostonian Charles Russell saw a sign reading: "A hotel exclusively for gentiles." A Pennsylvania retreat described its location as a place with an "altitude one thousand feet; too high for Jews." Throughout much of New England, the Lake George region in New York state, the Poconos in Pennsylvania, and even at the Camelback Inn in Phoenix, Arizona, restrictions against Jewish guests were enforced. In some places they still exist.[67]

As a result of this near universal prejudice, Jewish resorts developed where kin could feel comfortable. The most popular ones appeared in New York's Sullivan County, about 90 miles northwest of New York City. In later years these Jewish hotels and resorts were sometimes referred to as the "Borscht Belt" because of the clusters of Jews who vacationed there. As third- and fourth-generation Jews matured, in the 1960s and thereafter, and as resort discrimination declined and worldwide travel increased, so too did the need for an exclusive Jewish vacation area. Therefore many of the most famous of the "Borscht Belt" hotels including the most prestigious—Grossingers—were out of business by the late 1980s.

Still another manifestation of widespread discrimination occurred in residential areas. Bigotry and "restrictive covenants" kept Jews out of some of the most desirable neighborhoods in New York, Chicago, Washington, D.C., Los Angeles, Miami, Denver, Baltimore, Boston, Chattanooga, and Cleveland, as well as their suburbs.[68] The Sharon, Connecticut, Chamber of Commerce distributed leaflets in 1922 urging property owners to refrain from selling to Jews.[69] Helen Reid, wife of the owner of the New York *Herald-Tribune*, worried about who might acquire acreage near her home in Purchase, New York. Her concerns apparently revolved around the influences that might be brought to bear on her sons: "I hate the thought of Whitelaw and Brownie growing up with nothing but Jewish neighbors around."[70]

Toward the end of the 1920s apartment house owners in the Jackson Heights section in New York City advertised that their buildings were "restricted" and prohibited Catholics, Jews, and dogs. A legal battle over the exclusions ensued but the court upheld the rights of the property owners to choose their own tenants.[71] Not until the U.S. Supreme Court outlawed restrictive covenants in the 1948 case of *Shelley* v. *Kraemer* would such agreements be unenforceable in courts of law.

And yet even where discrimination reigned Jews still found opportunities to move out of the ghetto and into more agreeable housing. By the 1920s over 160,000 of the approximately one-quarter of a million Jewish people from New York's famed "Lower East Side" had moved north to Central Park West, West End Avenue, and Riverside Drive in Manhattan and into the outer boroughs of the Bronx and Brooklyn. Although kept out of areas like Park Slope and Brooklyn Heights, and sections of Riverdale in the Bronx, they found clusters of new and desirable apartment buildings on Eastern Parkway and Ocean Avenue in Brooklyn and on the Grand Concourse in the Bronx. In parts of what many have called "bedroom communities," as Jews moved in Gentiles moved out. But this did not bother the Jews. Historian Deborah Dash Moore has observed that "because most Jews wanted to be 'at home' in their neighborhood, they were happiest living near other Jews."[72]

In addition to various facets of American life where Jews were discriminated against, a strong national movement arose to limit the entry of European Jews to the United States. It was part of a much broader immigration restriction crusade that received widespread support from Protestants, and many Catholics, of every social class throughout the nation. Most of these people believed that too many immigrants from southern and eastern Europe had already arrived in

the United States, that their increased numbers would subvert American culture and traditions, and that assimilation would be impossible.
The nation already was receiving about a million newcomers a year;
without restrictive legislation, it was assumed that figure might surge
to over 2 million. The vast majority of Americans considered most foreigners "undesirable" additions to the nation.[73] Opposition strong-
holds to restrictions, however, existed in major cities like New York
and Chicago, which housed myriad ethnic enclaves, mostly of Catholics and Jews.

Xenophobia had been part of the national credo since the colonial
era when colonists questioned the wisdom of allowing Scots-Irish and
Germans to congregate among them. Periodically thereafter specific
ethnics like the French in the 1790s, and Irish Catholics and Chinese
workers in the middle of the nineteenth century, aroused the ire of
different groups of Americans. Not until 1882, however, was any
national group totally barred from our shores as were the Chinese, first
temporarily, and then on a permanent basis.

After 1896 a majority of immigrants to the United States came from
southern and eastern Europe. Their religions were Roman Catholic,
Jewish, and Greek Orthodox, their complexions seemed darker than
those of "real" Americans, and increased numbers of Protestants
wanted them excluded from this country. Post World War I disillu-
sionment with foreign entanglements, widespread fear of Bolshevism
and radicalism, and a shift in the policies of businessmen outside of
the Southwest who no longer felt the need for cheap, immigrant labor
thereby allowed a Congressional majority to meet constituents'
demands and restrict the number of foreigners who might enter the
United States. Moreover, unlike Presidents Taft and Wilson, who were
reluctant to alienate minority voters or whose principles led them to
veto legislation curbing the entry of immigrants, in the 1920s Presidents
Harding and Calvin Coolidge had no such qualms. Thus the efforts of
those who wanted to keep immigrants out proved successful.

The writings of some of the most articulate racists of the Progressive era, like Madison Grant and Burton J. Hendrick, contributed to the
popular demand for restriction. So, too, did the views of other writers
like Theodore Lothrop Stoddard, Kenneth Roberts, and Gino Speranza.
Stoddard, a disciple of Grant's, dubbed most immigrant Jews "Asiatics" and a threat to the "nordics"; he later supported Nazi racial laws.
Roberts, a young journalist who won renown as the author of *Northwest
Passage*, and Speranza, a former proponent of free immigration who
wrote a series of articles in 1924 stressing the trinity that America was
Protestant, Nordic, and Anglo-Saxon. In his essays Speranza emphasized the fact that the newer immigrants were "alienizing" America.[74]

The first journalistic series of the decade, the 1920–1921 articles of
Kenneth Roberts that appeared originally in issues of the *Saturday Evening Post* before coming out as a book in 1922, paralleled Ford's series

on "The International Jew" in *The Dearborn Independent*. Although no friend of the Jews, Roberts dismissed the *Protocols* as the "rankest poppycock," and went on to explain that they had been "plagiarized [sic] from Maurice Joly's *Dialogue Aux Enfers*, which was originally written to show that Napoleon planned to dominate the world."[75] Getting past that point, however, Roberts then criticized several of the southern and eastern European groups who hoped to emigrate to America.

He described the Jews of Poland, as "human parasites, living on one another and on their neighbors of other races by means which too often are underhanded . . . , they continue to exist in the same way after coming to America, and . . . are therefore highly undesirable immigrants." Roberts labeled many of them "Asiatics, and in part, at least, Mongoloids," and, unlike Nordics, unqualified to govern themselves. In short, Jews seeking entry into the United States were "unassimilatable [sic], undesirable, and incapable of grasping American ideals." To underscore his point, Roberts emphasized that

> Every American who has at heart the future of America and of the race that made it a great nation owes it to himself and to his children to get and read carefully *The Passing of the Great Race*, by Madison Grant; *The Rising Tide of Color*, by Lothrop Standard; and *Race or Mongrel*, by Alfred P. Schultz.[76]

Burton Hendrick, who questioned the Jews' worth as possible citizens of the United States and fretted about their potential control of the economy in articles written in 1907 and 1913, again showed anxiety about their increasing numbers in a series of articles that *World's Work* published in 1922–1923. As in his previous essays, he discussed Jewish economic endeavors. While noting that Jews were the major factors in the clothing trades, the movie business, and theatre ownership, Hendrick indicated as well that Christians still dominated the national economy. They constituted "an overwhelming majority of the bank presidents and officers," controlled the major industries, and had the most influential positions in newspapers. Happily, according to Hendrick, the list of the wealthiest New Yorkers showed "that the racial stocks which founded the United States one hundred and fifty years ago still control its wealth."

On the other hand, Hendrick feared that the Polish Jews were bringing socialism to America. He labeled New Yorker Morris Hillquit, "the leading exponent of Socialism in this country," and remarked how "the names of Polish Jews so constantly appear in all revolutionary and anti-nationalistic movements in New York." Hendrick lashed out at the most popular Yiddish newspaper in America, New York City's *Forward*, which, he claimed, substitutes "internationalism" for a robust American nationalism, "the solidarity of the working classes" for the American allegiance to the central government, "the dictatorship of the proletariat" for representative institutions.

The fourth article in this series, "Radicalism Among the Polish Jews: Their Destructive Political Activities as Shown in Their Newspapers, Their Votes, and Their Labor Unions," opened with the following paragraph:

> Are the Polish Jews anti-nationalistic in spirit, devoid of patriotism, unsympathetic with the thing known as Americanism, lacking in understanding and appreciation of the principles that control the American system? Are their political tendencies subversive, destructive? Is their attitude, so far as American institutions are concerned, one of restlessness, dislike, contempt, and even of hostility? This is the most serious charge brought against the Polish Jew."[77]

After reading Hendrick, Roberts, Grant, and others like them, one might be hard-pressed to distinguish their basic philosophical thrust from that expressed in Henry Ford's series on "The International Jew." *The Freeman,* in evaluating Hendrick's essays when they appeared together in his 1923 book, *The Jews In America,* called them a postscript to Ford's "anti-Semitic vagaries."[78]

Nonetheless, these writings underscored what most Americans were thinking,[79] and Congress passed restrictive immigration legislation first in 1921 and again in 1924. In 1921, the legislators stipulated a quota of 3 percent for groups based on how many people of that national background had been counted in the census of 1910. By 1924, however, it was clear that the earlier bill still allowed too many newcomers into the country and the base year was set back to 1890, before most of the east Europeans had arrived, the percentage lowered to 2 percent, and all Asians were barred as well. In 1929, formal quotas for different nationalities were firmly established with people from Great Britain, Ireland, and Germany accorded the largest allocations.

The passage of the immigration restriction laws has sometimes been looked on as a reflection of the views of the nativist group, the Ku Klux Klan, which became influential in the first half of the decade. Klansmen and women certainly favored keeping the United States 100 percent American, and by that they meant white Protestants. But their views in the matter of immigration restriction, and many others as well, reflected what most Americans thought—only their methods and uniforms distinguished them from the great majority of their fellow citizens.[80]

This revived Klan of the 1920s promoted the interests of "Nordic" Americans and focused its hatred primarily on Roman Catholics whom they regarded as the prime enemy in undermining Protestant America. They also inflicted severe beatings on errant members of the dominant culture whom they believed violated traditional and sanctified moral codes such as monogamy. To be sure, Klansmen also wanted African Americans to know their place and restricted their organizational membership to Gentiles (although one Indiana Klansman indicated that "any Jew can belong if he believes in the divinity of Jesus Christ"[81]) but, except in parts of the West, Jews were not the prime targets of the group.[82] In states like Louisiana, Arkansas, Texas, and Oklahoma moral and political reform dominated the Klan agenda, Jewish and Catholic stores were not generally boycotted, and the social standing and economic well-being of some of the most prominent Jewish families in the region was scarcely affected.[83] In 1923 the "Imperial Wizard" of the organization, Texas dentist Hiram W. Evans wrote:

> There seems to be a prevalent impression in some sections that the Klan is hostile to the Jew. This impression was not generated and has not been fostered by the organization. Idealistic Americanism is the purpose in which members of the Knights of the Ku Klux Klan are dedicated. Any man—whether native or alien by birth; Gentile or Jew by faith; white or black by race—who so commits himself in allegiance to his country that nothing is reserved, and in devotion to his flag that nothing remains uncommitted, is not the enemy, but is the friend of Klansmen.

Then, in the 1980s, when folklorist Carolyn Lipson-Walker interviewed Southern Jews about their recollections of Klan activities, she was surprised that instead of hearing tales about persecution and hostility, she heard primarily narratives about "Klan cordiality toward Jews."[84]

Examining the countless activities of the Klan during the decade elicits isolated examples to buttress Evans' assertions and Lipson-Walker's findings. In a 1925 local election in Detroit, Klansmen tried to recruit Jews to support their candidate for Mayor on the grounds that Henry Ford favored the reelection of the Catholic, John W. Smith.[85] Twice the Klan in Monroe, Louisiana, refused to campaign openly against Mayor Arnold Bernstein, and in 1924, Klansmen in Fairfield, Illinois, dressed in full regalia before 15,000 well wishers, tendered local storekeeper Emanuel Steiner with a wreath of American beauty roses. Herbert G. Markley, pastor of the Fairfield Presbyterian Church and Secretary of the local chapter of the KKK, accompanied the presentation with these words:

> Mr. Steiner, we are [here] today as your friends. You have lived here 50 years. You have been an honest, upright man. The Knights of the Ku Klux Klan respect and revere you. It is the constitutional right of every man to worship God according to the dictates of his conscience. The Ku Klux Klan never has and never will try to violate that right. You have built up by your honesty, uprightness and integrity a successful business. As a citizen there is no better. You have always been behind every proposition for the community and its welfare. As American citizens the Knights of the Ku Klux Klan congratulate you for the many things that you have done for the flag and for the country. . . . We hope that as you go down the trail you will remember with kindness and generosity the men in the masks.[86]

Despite the accolade, the assertions of Evans, or the findings of Lipson-Walker, no one should think of Klansmen as friends of the Jews. Jews may not have been prime targets of animosity but they were not white Protestants and on that ground alone were suspect. Moreover, while Jews were not generally intimidated by Klansmen in the eastern half of the nation, they were victims of Klan aggression in the West. Even in areas where antisemitism was not particularly severe among Klansmen, as in Oklahoma, Jews were among those attacked. The KKK was also quite active in Pocatello, Idaho, where one of its leaders threatened to expel Jews and Catholics from the city in 1923.[87] And in a speech to a South Portland klavern in Oregon, Klansmen were told that

> Jews are either bolsheviks, undermining our government, or are Shylocks in finance or commerce who control and command Christians as borrowers or employees. It is repugnant to a true Christian to be bossed by a Sheenie. And in some parts of America the kikes are so thick that a white man can hardly find room to walk on the sidewalks.[88]

Many women in the Klan denounced Jewish film producers who allegedly made "lewd" and "immoral" films, and "Bishop" Alma Bridewell White, an evangelical preacher and spokeswoman for the local Klan in parts of Texas and Colorado condemned Jewish men for engaging in lascivious activities with Gentile women, the Jewish fashion industry for foisting immodest attire on Christian women, and Jewish owners of "vile places of amusement" who kept their businesses open on Sundays.[89]

Klan activities were particularly vicious in Denver where Jews constituted about 11,0000 of the city's 250,000 people. Like Indiana, by 1925 kleagles and dragons occupied important state and local governmental offices including the Governorship, the Mayoralty in Denver, and a majority of the seats in the state assembly. Klansmen also controlled numerous county bureaus where antisemitism "ran like a fever." While members of the KKK did not create antisemitism in the state, they capitalized on its existence.[90] Denver's Klansmen included socially and economically prominent individuals in the city. They argued that Jews were unassimilable in American society and that Catholics were subject to a foreign power. The white-clad knights made inroads in Masonic lodges and the ruffians among them drove through Jewish neighborhoods on Friday nights shouting obscenities. A KKK women's fife and drum corps also marched through the streets successfully intimidating members of the Jewish community.[91]

By the 1920s Denver reflected the results of a half century of American nativism. The city had never been hospitable to Jews or any other minority. Anti-Chinese riots broke out in 1880 and later in that decade thousands of Coloradans joined the anti-Catholic American Protective Association. Members of anti-immigrant and nativist organizations abounded in the 1890s, and they joined with the Populists in antisemitic tirades. World War I witnessed book burnings and near lynchings of the "Hun" while individuals of German background in Denver suffered grievously from local animosity. Thus it was not surprising that in the 1920s the Ku Klux Klan found the city fertile territory for growth.[92]

In one regard, Denver typified the Klan nationally. It utilized Protestant churches and their ministers to advance organizational dogma. Just as ministers of the Baptist, Presbyterian, Dutch Reformed, Evangelical, Pillar of Fire, and Methodist churches in New Jersey supported Klan doctrines so, too, did their counterparts in the Highland Christian Church, the First Avenue Presbyterian Church, and the Grant Avenue Methodist Church in Denver stand militantly for the cause. One reason Protestant ministers generally refused to oppose anti-immigration laws was that they shared their congregants' views that the United States needed protection from foreigners and foreign ideologies. Some ministers were even more specific in their targets. In Illinois, for example, one preacher called a Sunday evening mass meeting to discuss "The Problem of the Jew, or How Shall We Get Rid of Him?" Throughout the 1920s and 1930s antisemitic commentaries abounded in Protestant churches and the Protestant press, and Hitler's policies toward Jews in Germany during the 1930s therefore did not seem as shocking or inappropriate to religious people as they might otherwise have appeared had there not been that same underlying prejudice in the United States.[93]

A particular concern of Jews and Gentiles alike was how Protestant and Catholic children were being taught. Some clergymen worried that Jews wanted to prohibit Bible reading in the public schools. Jews and some Gentiles were also disturbed about lessons and textbooks in parochial schools where Jews were repeatedly vilified in texts. Parochial school materials oversimplified the Crucifixion, emphasized Jewish culpability for the suffering and death of Christ, and included suggestions in manuals that teachers substitute the phrase, "wicked Jews" wherever the words "wicked soldiers" appeared. Many Protestant groups still emphasized that Jews had killed the Savior, and this continued well into the 1980s in some places.[94]

Bishop Gilmor's *Bible History*, first used in 1870, describes Jews as "barbarous," "blood-thirsty," wanderers without homes, "strangers amongst strangers—hated yet feared . . . bearing with them the visible signs of God's curse. Like Cain marked with a mysterious sign, they shall continue to wander till the end of the world."[95] In other texts Jewish rabbis were described as "worthy of nothing but 'scathing, stinging words' from Jesus." And in many parochial schools a common catechism went:

Q: Who killed Jesus?
A: The Jews.[96]

A Jewish writer attributed the origin of antisemitism to such teachings. "Basically, in the Christian mind," he wrote, "every Jew bears responsibility for Calvary, precisely as if every Jew repeated over and over again, in every generation, the crucifixion."[97] Another critic, writing in *Christian Century*, noted:

> Careless Sunday school teaching about the Jews has unquestionably played a large part in the production of such anti-semitic feeling as exists in this country. The constant reiteration that "the Jews" did this and that reprehensible thing with reference to Jesus, that they were his critics, his enemies, and finally his murderers, cannot fail to produce an unfriendly emotional tone toward them which will persist long after the specific teachings upon which it is based have been forgotten.[98]

In her study of Johnstown, Pennsylvania, Ewa Morawska found evidence to confirm such views. During the 1920s and 1930s, terms like "Christian morals" and "Church-loving citizens" appeared frequently, while children unselfconsciously called their Jewish counterparts "Christ-killers." Daily prayers in public schools included phrases like "Jesus loves me this I know, 'cause the Bible tells me so." During the Easter season Jews sensed increased resentment toward them, preachers delivered sermons on topics like "The Trial of Our Lord at the Hands of the Jewish Crowd," and churchgoers sang verses of "They Have Slain Him."[99]

American Jews had learned to live with this kind of prejudice, however, and by the end of the 1920s the harsh attitudes of so many different spokesmen seemed to fade away from public view. Ford had curtailed his attacks against Jews by 1927, the universities had put into place their limitations on Jewish students, immigration restriction laws had been passed and ceased being a subject for much public commentary or periodical attention, and sex scandals and corruption among prominent Klansmen in Indiana had caused the decline of the organization's nationwide membership and activities. Therefore antisemitic prejudices were bandied about in fewer causes and concerns receiving public airing. *The American Jewish Yearbook* for 1928 cheered the fact that "the past year witnessed the practical cessation of all organized anti-Jewish propaganda."[100]

Despite the sanguine assessment, however, a pervasive antisemitic atmosphere existed and sometimes provoked violence against Jews. In 1916, at Kings County Hospital in Brooklyn, New York, a Jewish intern had been bound and gagged by Christian interns, put on a train in Grand Central Station, and told that he would be thrown into the East River if he ever returned to the hospital. He did not return, and no Jews served as interns again on the Kings County staff until three appeared in 1925. These men were warned by their Gentile colleagues that they, too, should leave because "this is a Christian institution, and we will tolerate no Jews here." The three continued in their posts but were not allowed to eat at the same table with the Gentiles or play on the hospital's tennis courts; nurses behaved discourteously to them and occasionally refused to carry out professional orders. Then, on June 20, 1927, the men were dragged out of their beds in the middle of the night, bound, gagged, and ducked in a bathtub of ice cold water by some 20 odd other young physicians who worked in the hospital. The victims protested this heinous behavior, the Commissioner of Public Welfare conducted an investigation, and six of the perpetrators were expelled.[101]

Another shocking affront occurred in Massena, New York, reflecting on medieval superstitions that had never quite vanished. On September 22, 1928, two days before Yom Kippur, a four year old disappeared from her home. After a disappointing and lengthy search by townspeople and state police, the Mayor speculated that perhaps she had been killed by the Jews in the community who wanted to use her blood for a religious ritual. The libel that Jews kill Christian children and use their blood for religious purposes may have been an ancient myth but it has had a powerful hold on people's minds.[102]

And Massena was not the first place that such an accusation had been made in the United States. As was pointed out in Chapter 2, on the eve of Yom Kippur, in September 1850, Irish immigrants in New York City spread the rumor that Jews murdered a Gentile girl and used her blood for religious purposes. The next morning a mob of almost 500 men, led by three policemen of Irish descent, ransacked the synagogue. Tales of ritual slaughter also spread in New York in 1913 when a crazed woman of Polish descent killed her husband because she thought he had sold their son to the Jews for use as a religious sacrifice. Other episodes of a similar nature were reported in 1913 in Clayton, Pennsylvania, and in 1919 in Chicago, and in Pittsfield and Fall River, Massachusetts.[103]

The preposterous accusation in 1928 frightened and infuriated the Jews of Massena because they did not know how to defend themselves against it. But before the charge became a massive problem the child was found unharmed in a nearby wooded area. Apologies from the Mayor and the state trooper who had interviewed the local rabbi about the alleged ritual were rejected by the Jewish community. Louis Marshall, head of the American Jewish Committee, had been informed of the event, and demanded an investigation by the Superintendent of the State Police. An inquiry occurred, the state trooper publicly apologized, and his superiors reprimanded and suspended him.[104]

Marshall's quick reaction in the Massena affair contrasted with his more cautious responses to much of the antisemitism exhibited during the 1920s. Discussions of antisemitic incidents proliferated at executive committee meetings of the American Jewish Committee but relatively little action was taken against the antisemites. The ineffectiveness of the AJC was clearly demonstrated in the kinds of endeavors that it did undertake. Statements were issued repudiating the charge that Bolshevism was a Jewish movement, *The Protocols of the Elders of Zion* was declared a forgery, and in December 1920 the members adopted a resolution "to demonstrate to Congress the desirability of Jews as immigrants."[105] How such a statement would affect Congressmen and Senators was never made clear.

After the initial two *Dearborn Independent* articles appeared, Marshall protested to Henry Ford but the AJC President ceased further intervention after Detroit Jews indicated their disapproval of what they regarded as an intemperate deed on his part. (The Jews of Detroit resented any outside interference and feared retaliation of some kind even after the mildest protest.) Marshall also complained about remarks that he regarded as antisemitic on a local New York City radio program,[106] and urged President Coolidge to veto the immigration restriction act of 1924.

Other attempts of Marshall and fellow Jews to combat antisemitic attacks during the 1920s included the promotion and distribution of works like John Spargo's *Jews and American Ideals* (1921), which presented Jews in a favorable light, and newspaper articles describing the Americanizing contribution of the Yiddish press. The Anti-Defamation League of B'nai B'rith conducted a campaign against school use of William Shakespeare's *Merchant of Venice* on the grounds that the depiction of the grasping, vengeful "Shylock" led others to assume that he was typical of all Jews. By 1920 41 cities banned it from their curricula, by 1937 over 300 had done so.[107] While the elimination of the classic from some high school English classes did not change people's attitudes toward Jews, it helped somewhat to alleviate the discomfort of Jewish students.

In 1927 Marshall again intervened with Henry Ford, but this time the auto magnate sought his assistance. Ford's series of antisemitic articles proceeded intermittently between 1921 and 1927. In 1925, however, Aaron Sapiro sued Ford for slander and slurs against him and "his race." The first trial ended abruptly because one juror injudiciously discussed her opinions with a reporter before a verdict had been reached. Just before another courtroom battle was scheduled to begin in 1927 Ford had an automobile "accident." Many suspected that the so-called accident was a ruse because Ford, for reasons that remain unknown, did not want to appear in court. His emissaries subsequently sought a way to get Ford out of his difficulties with the Jewish people and they were referred to Marshall. Marshall agreed to help Ford recant and prepared a letter for the auto magnate to sign, retracting all antisemitic accusations, apologizing for any offense he may have caused, and promising to refrain from publishing any slanderous materials about Jews in the future. In addition, Ford agreed to withdraw "The International Jew" from publication and sever all relations with those responsible for helping him to spread his spurious antisemitic policies in the past. Ford signed the letter, and for the rest of his life kept the agreement.[108]

Unfortunately, the leaders of the Jewish community did little more to combat the upsurge of contemporary bigotry. Jewish defense organizations in the 1920s had not yet achieved the know-how and sophistication that they would develop at the end of World War II and afterwards. Therefore, when the issue of Bible reading in public schools came to the group's attention, the AJC did nothing because Marshall considered it a "local issue."[109] Some Jews questioned whether Marshall and Julius Rosenwald, another prominent Jew and the largest shareholder of the giant department and catalog store, Sears, Roebuck, whose personnel department refused to hire Jews, were too cautious in their efforts to combat existing antisemitism. Nonetheless, German Jewish leaders, led by Marshall and other members of the AJC, continued their tepid reaction.

Summing up the AJC's responses to the antisemitism of the decade, one student wrote that the members assumed that as Jews integrated within society the "Jewish question" would disappear from public consciousness. On the other hand, raising the issue of the "Jewish problem" publicly would create a heightened awareness of the Jew as alien. Therefore the AJC resolved to make few references to antisemitism, hoping thereby to erase the subject from the public mind.[110]

To promote their goals, some American Jews supported immigration restrictions and college quotas for the east Europeans whose presence embarrassed them. They encouraged these limitations while advising newcomers to behave properly and display patriotic fervor to show that they were good citizens. Annie Nathan Meyer, a member of the Daughters of the American Revolution and one of the founders of Barnard, also believed that Jews "could win acceptance, both socially and professionally, through hard work, education, refined manners and avoidance of Jewish separatism and particularism."[111]

The most bizarre critique of coreligionists came from the pen of Walter Lippmann. While not endorsing Henry Ford's attitudes toward Jews, he noted that Ford "is at bottom a rather kindly man." But because Ford was "ignorant and childish" and opposed to war, he was persuaded that wars were made by international bankers and the Jews "happened to be more easily visible to his untrained eye than the non-Jewish international bankers." While tolerant and understanding of the foibles of the auto magnate, however, Lippmann excoriated "the rich and vulgar and pretentious Jews of our big American cities [who] are perhaps the greatest misfortune that has every befallen the Jewish people. They are the real fountain of anti-Semitism." Although Jewish vulgarians hardly differed from their Gentile counterparts, Lippmann labeled them "a thousand times more conspicuous." He also asserted that he wasted no time "worrying about the injustices of anti-Semitism. . . . But the anti-Semitism which has its root in our own weaknesses and failings I do worry about.''[112] The noted journalist did not realize that by blaming the victim he was displacing rage against his own rejection by members of the dominant culture and transferring it to coreligionists whom he considered too weak to fight back and whose feelings failed to elicit the same compassion that he displayed for Henry Ford.

The cumulative bigotry that received so much attention in the 1920s set the stage for more virulent attacks in the 1930s. Social and economic discrimination continued but the world crisis, triggered by the stock market crash in the fall of 1929, helped to usher in lethal onslaughts that neither Jew nor Gentile could have anticipated only a few years earlier. A new magnitude of hatred arose within a few years of Louis Marshall's death in 1929. He was the most prominent spokesman for Jewish rights and causes during the last twenty-five years of his life and the AJC's enunciated positions during his presidency essentially resulted from his "fundamental beliefs that the Jewish position in America was sound, that great public controversies were ordinarily unnecessary, and that Jews could deal more effectively with anti-Semitism by their own example than by attacks on their enemies.''[113]

Future events would severely test Marshall's philosophy but American Jewry would be without forceful leadership. Not until the Holocaust would Jews in the United States and their organizational leaders change strategies in combatting the devastating effects of anti-semitism. In the decade of the 1930s the only friend Jews would find among world leaders was Franklin Delano Roosevelt whose tenure in the White House coincided with the most inhumane antisemitic episode in world history.

6

The Depression Era
(1933–1939)

Initially, antisemitic displays in the United States did not increase with the onset of the Great Depression.[1] But after 1933, when a Nazi-led government came to power in Germany and Franklin D. Roosevelt inaugurated a New Deal at home, the deepening economic crisis contributed to an explosion of unprecedented antisemitic fervor. Fueled also by the rise of Protestant and Catholic demagogues, deeply entrenched Protestant fundamentalism, and the widespread expression of antisemitic attitudes by respectable social and religious leaders, they illustrated how centuries of denigrating Jews culminated in the most savage accusations, and in some urban areas—especially New York and Boston—violent physical attacks. In the 1930s American antisemitism was "more virulent and more vicious than at any time before or since"[2] as rabid antisemites, almost without exception, envisioned an international Jewish conspiracy aimed at controlling the government of the United States. They believed that unless maximum vigilance was exercised, Christian America would be lost.[3]

The ingrained prejudices of respectable people, an attitude confirmed as proper in the 1920s, contributed to the escalated ferocity. Their beliefs and behavior set the tone that allowed for the excesses of the more frustrated and demonstrative Americans. A vignette of how futile it was to try to alter entrenched views occurred in 1931 when one man complained that he had been hearing for years how "the Jews were steadily driving the Gentiles out of business" from people who were doing quite well economically. In his attempt to convince friends and relatives of the inappropriateness of antisemitism he came up against a brick wall. One woman told him that the strength of his argument only fortified existing beliefs:

> John, you mean well and I love to hear you talk, but you have unwittingly reinforced every feeling we have about the Jews and you have splendidly shown up the very thing which makes them so dangerous, and that is that every now and then they are able to fool some good man like you. That is what makes it all so hopeless.[4]

"Nordic" Americans held deeply rooted stereotypes of most minorities in the United States, and Jews were hardly alone as victims of discrimination. But unlike African Americans, no legal segregation kept Jews in an inferior position, they were not banished from communities like Mexicans and their American-born children who were rounded up and sent to Mexico involuntarily, nor were they interned as Japanese Americans on the West Coast would be during World War II. Since the beginning of the twentieth century Jews adjusted to existing discriminatory conditions by establishing their own businesses, working in civil service or for coreligionists, attending less prestigious schools, and living mostly in neighborhoods with people of their own background. Although Jews were socially and culturally marginalized by the prejudices of the dominant society, for the most part they maintained a higher standard of living than either WASPs or members of other ethnic groups. Moreover, even during the 1930s American-born children of east European Jewish immigrants continued to move upward much faster than children of other immigrants who arrived in the United States in the late nineteenth and early twentieth centuries.[5] Nonetheless, in the 1930s a majority of American Jews probably lived close to what we now call "the poverty line," victims of both the depression and also a pervasive intolerance. After the stock market crashed in October 1929, the depressed economy hurt the poorest Jews but all of them suffered from apprehensions steadily reinforced by antisemitic rumblings in the non-Jewish world.[6]

In the United States, as in Europe, antisemitism had been growing for at least two generations but political antisemitism was acceptable in the European nations to an extent that it never was, or would be, across the Atlantic. It spread rapidly after the breakup of the German and Austro-Hungarian empires, and throughout the 1920s riots, pogroms, and indiscriminate assaults on Jews occurred in the new states of eastern and central Europe. Jews came under attack by classmates at universities in the Ukraine, Latvia, Poland, Hungary, Romania, Austria, Greece, and Germany while in England antisemitism seemed rife immediately after the end of World War I. Its societal manifestations diminished by 1926 but it remained strong in the universities.[7]

After 1929, coincident with worldwide depression, conditions worsened for Jews throughout Europe. Nazism gained strength in Germany and German Jews visiting the United States reported that the growth of a Nazi party "had been stimulated, to a great extent, by the anti-Jewish publications which had been sponsored by Henry Ford in the United States."[8] In eastern Europe governments began to bar Jews from employment, small cooperatives eliminated Jewish shopkeepers and artisans, and universities restricted Jewish admissions. The precarious situation of the Jews in Romania may be seen from the fact that some Romanians wore swastikas even before July 11, 1930, when every Jewish home in the village of Balaceano was smashed after the signal for the attack came from the church steeple. A year later 54 Jewish families had their homes burned down in Salonika, Greece.[9] As outbursts continued, a British publication stated in July, 1932, that

> a crusade of anti-Semitism has been raging from the Rhine to the Vistula, and from the Baltic to the Aegean Sea, during the past six months, with a vindictiveness that almost surpasses all previous manifestations of anti-Jewish hatred since the end of the war.[10]

Six months later Adolf Hitler was sworn in as Chancellor of Germany.

As the depression worsened in the United States, Hitler's attacks on Jews as the root causes of the world's economic and social problems no longer seemed so outrageous to genteel bigots. As frustrations intensified, people eagerly blamed Jewish businessmen, who allegedly controlled the money supply, for the economic crisis. Throughout the country, those who had easily accepted myths about the nefarious qualities of Jews and who had imbibed the Populist ethos about the malevolence of "Wall Street" bankers and international financiers had no difficulty in believing the carelessly hurled economic charges.[11]

Jew hatred permeated the United States. Hostility pulsated in small towns and large cities, in fashionable social circles and exclusive boardrooms, and even on the floor of Congress.[12] A Mount Vernon, New York, man wrote in 1933:

> before we see this Hitler flareup end, it would not surprise me to have it reach America and have the blessing of the very men who have been damning Hitler now. Because when the Jew finally reaches that point when he will be satisfied at nothing but complete control of money, business, society, and government, well, goodbye Jew. And it is only a question of time until the ever-recurring pogrom becomes necessary.[13]

Maud Nathan, a New York Jewish social reformer and socialite, complained that "the prejudice against Jews, of which one was not conscious fifty years ago, has become so serious that today it is a burning question and is frankly spoken of as 'The Jewish Problem.' "[14] A Pennsylvania Congressman declared that the Jews have all the gold and money while the Gentiles have only "the little slips of paper," and financier Jack Morgan, son of J. P. Morgan, told a friend that he did not like Hitler "except for his attitude toward the Jews, which I consider wholesome." Author Laura Z. Hobson recalled being at a dinner party where the discussion turned toward Hitler and Germany. One of the guests remarked:

> "The chosen people ask for it, wherever they are."
> "Oh, come on," another man said. "Some of my best friends are Jews."
> "Some of mine are, too," Hobson heard herself saying, slowly.
> "Including my mother and father."[15]

It is not clear from Hobson's description of the event whether other guests sympathized with her or the bigot.

The types of people with whom Hobson dined also abhorred many of Franklin D. Roosevelt's closest advisors and the programs they devised to pull the nation out of the depression. New agencies were created that gave the federal government unprecedented authority to regulate and control business and agriculture while providing direct relief to individuals throughout the nation. To facilitate the implementation of a massive reform agenda the President ordered subordinates to find the most talented people available, recommended to Cabinet officials that Felix Frankfurter, Dean of the Harvard Law School, be used as a source for lawyers, and reached out to individuals and groups not heretofore well represented in the government to help propel his New Deal.[16] As a result of the President's actions more minorities and women achieved responsible positions in the federal government than ever before.

Of all these groups, Jews, both relatively and absolutely, benefitted the most.[17] As Governor of New York, Roosevelt worked comfortably with Jews and as an administrator sought to bring talented people into his political orbit.[18] After he moved to the White House, Felix Frankfurter and Supreme Court Justice Louis D. Brandeis stood out among his closest advisors on governmental policies. Within his administrative circle, the President felt extremely comfortable with Sam Rosenman, his former chief assistant as Governor of New York, whom he finally brought to Washington in 1937. And at different periods Ben Cohen and David Niles were also highly regarded subordinates. Some agencies, like the Securities and Exchange Commission and the Departments of Agriculture, Labor, and Interior, had many Jews in high positions.[19] For the first time in American history, therefore, the federal government provided significant numbers of opportunities for Jews.

Within months of Roosevelt's taking office in March 1933 rumors spread that Jews were running the government. During the summer inquiries poured into the *Kiplinger Washington Newsletter* about how

many Jews were employed by the new administration and what influence they possessed. Anti-Roosevelt Americans frequently blamed Jewish advisors for New Deal policy failings and antisemitic verbiage
became both common and obsessive. In October 1934 a New York
woman wrote to the President:

> On all sides is heard the cry that you have sold out the country to the
> Jews, and that the Jews are responsible for the continued depression, as
> they are determined to starve the Christians into submission and slavery.
> You have over two hundred Jews, they say, in executive offices in Wash-
> ington, and Jew bankers run the government and [Bernard] Baruch is the
> real President. This is the talk that is heard everywhere.[20]

Many Washington politicians as well as some of the nation's wealthiest
people had already expressed similar views.[21] In June that year, *Boston
Herald* editor Frank Buxton confided to former Secretary of State Bain-
bridge Colby, "I was amazed at the intensity with which highly intelligent men argued that Jews were controlling the President."[22] Eight
years later journalist W. M. Kiplinger specifically rejected that perception when he wrote that while "men who are Jews occupy very influential positions, there is no such thing as a 'Jewish influence' in Washington."[23]

   Nevertheless, for the first time in American history the religious
heritage of the President and some of his advisors became an issue of
public discussion. Millions of Americans believed that Jewish influence
over Roosevelt was responsible for the administration's "Jew Deal."[24]
Thus, so long as Roosevelt occupied the White House, tales circulated
that he not only favored Jews in making appointments but that he, too,
was of Jewish ancestry. An extensive amount of antisemitic literature
flooded the 1936 presidential campaign including an essay by the Reverend Gerald Winrod of Kansas, one of the better known fundamentalist Protestants who had started the Defenders of the Christian Faith,
stating that the President was descended from Rosenbergs, Rosen-
baums, Roosenvelts, Rosenblums, and Rosenthals.[25] (The litany was
reminiscent of the *Katholische Volkszeitung*'s description of about 40 people with the name of Rosenfeld attending William McKinley's second
inauguration in 1901.) A doggerel also circulated at that time informing
readers that President Roosevelt allegedly told his wife:

> You kiss the niggers,
> I'll kiss the Jews,
> We'll stay in the White House
> As long as we choose.[26]

   Christian religious groups and publications also contributed to the
growing critiques of Jews. *The Christian Century*, "the nation's most
prominent liberal Protestant weekly magazine,"[27] had long been analyzing "The Jewish Problem." In the 1930s it published more than

twenty articles and editorials apparently sympathetic to the plight of the Jews yet at the same time urging them to assimilate and become "real" Americans.[28] The editors acknowledged in May 1933 that

> the Christian mind has never allowed itself to feel the same human concern for Jewish sufferings that it has felt for the cruelties visited upon Armenians, the Boers, the people of India, American slaves, or the Congo blacks under the Leopold imperialism. Christian indifference to Jewish suffering has for centuries been rationalized by the terrible belief that such sufferings were the judgment of God upon the Jewish people for their rejection of Jesus. If it is God's judgment, why should Christians interfere, and why should they sympathize?

Yet *The Christian Century* itself assumed the same stance. It urged Jews to convert because "Israel needs Jesus to complete its own life."[29]

In April 1936 the magazine intensified its efforts and began publishing a series of pieces virtually demanding conversion of the Jews to Protestantism, blaming them for their sorry predicament in the United States, and warning, in veiled terms, of dire consequences should they refuse to comply. While continually attacking antisemitism per se, *The Christian Century* reinforced the views of respectable bigots who thought of themselves as tolerant. The magazine argued that in a dynamic society it was difficult for two religions to be mutually tolerant. A May 13 editorial told every Jew he

> will never command the respect of the non-Jewish culture in which he lives so long as he huddles by himself, nursing his own "uniqueness," cherishing his tradition as something which is precious to *him* but in the nature of things cannot be conveyed to others, nor participated in by others.

On June 9 another editorial proclaimed:

> The simple and naked fact is that Judaism rests upon an impossible basis. It is trying to pluck the fruits of democracy without yielding itself to the processes of democracy. In a dynamic society a national culture cannot help seeking the unity of all its component elements.

Then, on July 1, came the denouement. While denying any antisemitic feelings, the editors of *The Christian Century* concluded that Jews "must be brought to repentance—with all the tenderness, in view of their age-long affliction, but with austere realism, in view of their sinful share in their own tragedy."[30]

Such pronouncements from one of the nation's premier Christian journals alarmed a great many Jews. Sober-minded and intelligent individuals ran *The Christian Century;* nonetheless, these responsible Christian liberals were committed to bringing nonbelievers into the fold. Their views could not be dismissed as the rantings of the lunatic or fundamentalist fringe. *The Christian Century* did not speak for Protestant America but it reflected what millions of ordinary middle-class

Protestants thought—and for that reason its comments were frightening to Jews.

Fundamentalist Protestants also thought that Jews should convert to Christianity. More so than in the 1920s, fundamentalists were increasingly alarmed about the effects of modernism, evolution, and the spread of communism. At a 1931 Atlantic City conference, evangelical Christians considered the question "The Christian Approach to the Jew—especially in North America." Speakers agreed that mistreatment of Jews by Christians in the past had prevented Jews from knowing "Jesus in His true character, mission and power," but they argued that Christianity was the fulfillment of Judaism and that Jews "must be brought back to faith in God. . . . All sinners need salvation."[31] Many fundamentalists could not understand the reluctance of Jews to embrace Christ. As Dr. Louis Evans of Pittsburgh told a 1935 audience in Princeton, "Christ is the universal need. He is no more sectarian than sunshine or rain."[32]

Fundamentalists also believed deeply that America must continue as a "Christian nation founded on God's word" and that communism threatened that position.[33] Their perspectives on communism and Jews allowed them to see Hitler and his policies in a different light. A later day assessor of articles in *The Alabama Baptist* in the 1930s concluded that the editors saw Jews as a deicide people, interested only in material gain, greedy and hypocritical, and sympathetic to, if not actually members of, the Communist Party. This observer wrote that from the point of view of Alabama Baptists, "a man, even Hitler, cannot be all bad if he does not drink, smoke, or allow women to use cosmetics." Many fundamentalists also interpreted Hitler's actions toward Jews as God's "rod of correction" for "Jewish sin and unbelief."[34] After *Kristallnacht* in 1938, when Germans went on a rampage against Jews in their country, *The Alabama Baptist* opined, "While we are not party to Jewish persecutions, we believe this era of persecution can be used as an opportunity to preach repentance to Israel."[35]

William Dudley Pelley, the son of a Protestant minister, stood out as the most prominent fundamentalist antisemite, a man convinced that he was divinely inspired to lead a mass movement against the anti-Christian conspirators in America. In 1932 he had a vision that something important would occur on January 30, 1933, and after Hitler became Chancellor of Germany that day Pelley interpreted the accession as a God-given sign. He then began to think of himself as the American Hitler. In February 1933 Pelley founded the Silver Legion, generally known as the Silver Shirts, which offered the same venomous message in the United States as did the leader of Germany. Like Hitler, Pelley used materials from Henry Ford's "International Jew" and *The Protocols of the Elders of Zion* in many of his various publications, including *Liberation*.[36]

Several of Pelley's views perplexed rational people. He claimed to

have spent seven minutes in Heaven during which he allegedly conversed with God. Pelley also described an international conspiracy with
300,000 to 400,000 European Jews coming to the United States to spearhead an assault on the American government.[37] And he also stated that
he had "proof—pressed down and overflowing—that the New Deal
from its inception has been naught but the political penetration of a
predominantly Christian country and Christian government, by predatory, megalomaniacal Israelites and their agents."[38] Pelley's assertions
provided the raison d'etre for his group, the Silver Shirts, which one
1933 observer dubbed "the most important native anti-Semitic organization in the United States."[39] During the 1930s the Silver Shirts
enjoyed especially strong support in the South, the Pacific Northwest,
and California.[40]

Other antisemitic groups in the 1930s, several of which were also
headed by fundamentalists, included the Friends of New Germany that
became the German-American Bund, the Defenders of the Christian
Faith, the Knights of the White Camelia ("We're for Christ and the
Constitution"), the Industrial Defense Association, the American
Nationalist Confederation, the James True Association, and the
National Union for Social Justice. In fact, from 1933 through 1941, over
100 antisemitic organizations were created, as contrasted with perhaps
a total of five in all previous American history.[41] The men who headed
the most important of these new organizations included Pelley, the
somewhat deranged son of a Christian minister, the Reverend Gerald
Winrod, who spearheaded the Defenders of the Christian Faith and
who praised Hitler for saving Germany from "Jewish communism,"[42]
and Father Charles Coughlin of the National Union for Social Justice,
the only Catholic but the most notable hate monger of the decade. These
zealots cloaked their views with religious imageries, defended the anticommunist (thus antisemitic) policies of Adolf Hitler, and charged that
the New Deal programs of Franklin D. Roosevelt emanated from the
Jewish conspiracy fantasized in *The Protocols of the Elders of Zion*.[43] Pelley, who issued a stream of antisemitic pamphlets under a variety of
titles from his headquarters in Asheville, North Carolina, told a House
of Representatives Committee in 1940:

> I don't hold any hatred toward any Jew in the United States. I feel exactly
> as the Nazi Party in Germany felt in regard to Germany, regarding the
> Jewish element in our population, yes sir.[44]

During World War II rabble rousers like Pelley, Winrod, and
Coughlin would be deemed fascist extremists by the American government and silenced under the provisions of the 1917 Sedition Act
that prohibited speech aiding the enemy. Yet in the 1930s they received
support not only from religious people who accepted the truthfulness
of Jewish responsibility for a communist conspiracy to undermine the
government of the United States[45] but also from fearful men and

women suffering from and frustrated by the depression. The rantings of the demagogues resembled medieval accusations against the Jews as well as views popularized earlier in the twentieth century by men like Burton Hendrick, Madison Grant, and E. A. Ross. In later years, Protestant religious journals, and the fundamentalist press in particular, reported Hitler's persecution of the Jews but they generally failed to convey the horror of Nazi pogroms, were indifferent to the plight of the Jews, believed that the Jews were partly responsible for Nazi attacks, or saw the tragedy as the fulfillment of God's judgment. Many of their readers subscribed to all of the above views.[46]

Catholic, as well as Protestant, hostility toward Jews also intensified in the 1930s. Fears of communist subversion, the rise of Hitler, and the depression in the United States did not create Catholic antisemitism but exacerbated existing prejudices and led to vituperative expressions of antipathy toward Jews. Catholics and Jews differed on a wide variety of domestic issues as Catholics generally favored, and Jews opposed, state aid to parochial schools, religious instruction in public schools, censorship, and bans against divorce and birth control. In New York City politics, Irish Catholics and Jews differed over the election of liberal-progressive Fiorello La Guardia as Mayor in 1933. Nationally both groups backed Roosevelt but many members of the Catholic hierarchy pulled away from the President in the middle of the decade because they thought his New Deal "communistic." Father Charles Coughlin was among the leaders of those who sought to thwart Roosevelt's reelection in 1936 when Jews embraced the President.[47] Catholics in general, and the church in particular, favored order, authority, and conservatism while the more liberal Jews valued intellectual exploration, socialism, and even communism.[48] In response to a 1939 nationwide poll question, "If you had to choose between Fascism and Communism, which would you choose?" 66 percent of Catholics chose fascism, 67 percent of Jews chose communism.[49] Moreover, a tradition of hostility toward Jews existed in Catholic America that for decades had led to assaults on Jewish children and adults by Irish youth in New York, Boston, Pittsburgh, Jersey City, Philadelphia, and other cities.[50]

The American government's international policies widened the rift between Catholics and Jews. Catholics opposed Roosevelt's recognition of the Soviet Union in 1934, criticized the administration and non-Catholic Americans for their apparent indifference to outrages against Catholics in Mexico, and expressed relatively little opposition to Hitler's policies. Jews, on the other hand, generally supported Roosevelt's positions in foreign policy and saw Hitler as the devil incarnate. A small minority even believed that Stalin, who before the Nazi-Soviet pact of 1939 was the only European head of state to denounce Hitler's Germany and antisemitism, would lead the world to a socialist nirvana.[51]

But it was the Spanish Civil War that had the most profound effect on Catholic-Jewish relations. The war began in July 1936 when a military coup, led by General Francisco Franco, attempted to overthrow the popularly elected, left-wing Spanish government. Civil war ensued, with the Soviet Union aiding the established government and the Catholic Church, along with Nazi Germany and fascist Italy, backing the rebels. To American Catholics who supported church positions it seemed obvious that the anti-Franco movement was "dominated by Jews."[52] The Brooklyn *Tablet*, the official archdiocesan newspaper, observed in March 1937, "It is rather galling to find vociferous and misrepresentative Hebrews championing Stalin and Caballero while they denounce Hitler."[53] When a few Catholic periodicals like *Commonweal*, the *New World*, and the *Catholic Worker* tried to give some perspective to the Franco-Loyalist conflict they "were doomed to failure from the start. There is no parallel in all our history," a Catholic scholar later wrote, "for the rabid abuse [these publications] brought down on themselves."[54]

The war brought smoldering Catholic-Jewish animosities into the open, as both Catholic periodicals and priests forcefully expressed antisemitism. Two leading diocesan newspapers, *The Boston Pilot* and the Brooklyn *Tablet*, were especially strong in their opposition to Jews, and Catholics were advised to read only their own publications and shun those published or dominated by Jews.[55] In 1937, the *Catholic Transcript* of Hartford informed readers that "the Jews . . . are hated because they are too prosperous, too successfully grasping. . . . They are the richest men in the world."[56] In Cincinnati, the editor of the diocesan *Catholic Telegraph-Register* repeated much of what appeared in other Catholic periodicals but in a more venomous fashion. Over a period of eight years, from 1936 through 1943, he accused Jews of being pagans, asserted that the Soviet government was predominantly Jewish and that the leader of Loyalist Spain was also a Jew, accused many Jews of being communists and atheists, and claimed that antisemitism was a result of the immoral business ethics of Jews. After World War II began in Europe, and many Americans feared that the United States might be dragged into the conflagration, the *Telegraph-Register* editorialized in August 1941:

> Today there is a highly organized minority of the war party in our country. It is vocal, it is tyrannical in smearing patriotic Americans who disagree with its program. Jewry in certain aspects is a very highly organized minority. It possesses great wealth and extraordinary influence. . . .
> Jewry seems committed to a war program for our country. Jewry seems concerned with the things of a passing day. Yet, if this country goes to war, we predict that opposition to Jews will gain uncontrollable momentum. They will be blamed, in large measure, for influencing the officials and agencies favoring war.[57]

Catholic hatred of Jews was later commented on by a woman who received a parochial school education in the New Deal era. Abigail McCarthy, former wife of the unsuccessful 1968 aspirant to the Democratic presidential nomination, Eugene McCarthy, recalled that she was taught that the Jews had rejected Jesus and they in turn were the rejected people. The students did not genuflect during the prayer for "the perfidious Jews" on Good Friday, either. She asserted that whole immigrant groups arrived in the United States with the same smoldering and venomous antisemitism that flourished in the nations that they left behind. McCarthy also characterized her fellow Catholics as "racists," who have been that way "throughout our history in this country—and there are anti-Semites among us. If truth were known there are probably more anti-Semites than there are racists. Some of us are Klansmen at heart."[58]

By 1938 conditions were ripe for a charismatic Catholic to vocalize these views to the general public. A Detroit priest, Father Charles Coughlin, seized the opportunity and in articulating positions favored by many in the Catholic hierarchy, he became one of the most controversial and revered figures of his time. As one of his biographers put it: "He was Christ; he was Hitler; he was savior . . . he was demagogue."[59] Coughlin had already established a reputation as a national critic of malevolent and predatory economic forces and his mellifluous and appealing radio voice won him millions of followers in 1933 and 1934. During the early years of the New Deal he also made references to money lenders and international financiers who kept Americans in the throes of depression and in 1935 he complained about "the Tugwells, the Frankfurters, and the rest of the Jews who surround" the president. By 1936 this pastor of the Little Flower Church in Royal Oak, a Detroit suburb that had earlier been a bastion of the Ku Klux Klan, was friendly with prominent antisemites like Henry Ford and Joseph P. Kennedy. He then joined, and became a leader of, the unsuccessful campaign to prevent Franklin D. Roosevelt's reelection.[60]

In 1937 Detroit's newly appointed Edward Cardinal Mooney tried to curtail Coughlin's public activities but the Vatican intervened on his behalf. A few months later the charismatic priest turned his criticism sharply against the Jews, and developed the largest national following of any demagogue in American history. Coughlin's emergence as the United States' most prominent and vocal Jew-hater coincided with the German *anschluss* of Austria in March 1938. Capitalizing on known Catholic animosities and the growing reluctance of western European and North American countries to welcome Jewish refugees, Coughlin sensed that attacking Jews would enhance his status among Catholics and those who sought easy explanations for their woes. He received copies of the fraudulent *Protocols of the Elders of Zion* from the men who had supplied them to Henry Ford some years earlier, and in July Coughlin's newspaper, *Social Justice*, began reprinting selections of the forgery.[61] In early November 1938 *The New Republic* charged the priest with being "cynically aware that he is peddling falsehood," and asserted that there was "almost no editorial difference between the Nazi weeklies and Coughlin's *Social Justice*."[62]

But criticism from the liberal press hardly bothered Father Coughlin and both American and European events may have spurred his activities. Observers sensed much greater antisemitic fervor in 1938 on both continents. During the fall, Americans read about the worst antisemitic political campaign in Minnesota history as Jews, while being labeled Communists were also accused of gangsterism, radicalism, and controlling the state government. Farmer-Labor Party Governor Elmer Benson had appointed the first Jewish regent in state history which upset many voters, while rumors also spread that the state's first lady and most of the Governor's aides were also Jewish. Campaign rhetoric obviously impressed the voters as all but one member of the Farmer-Labor Party, including the Governor, lost on election day.[63]

A much more shattering event occurred the following week in Germany. On the night of November 9–10, 1938, *Kristallnacht*, Germans brutally attacked Jews throughout the nation, burned their synagogues and destroyed their businesses, invaded their old-age homes and children's schools. German police also arrested at least 20,000 of the victims. Never before had there been a pogrom of that magnitude and the civilized world reacted with revulsion. Over 90 percent of Americans who knew of the event disapproved of German actions. President Roosevelt declared, "I myself could scarcely believe that such things could occur in a twentieth-century civilization."[64]

Only ten days later, on the night of November 20, Father Coughlin gave the most incredible radio address of his career. Coughlin minimized German barbarities on *Kristallnacht* as he launched into an expansive antisemitic diatribe. He blamed Jews for imposing communism on Russia, and noted that Germans correctly believed that Jews were responsible for the economic and social sufferings of the Fatherland ever since the signing of the Treaty of Versailles. "Nazism," Father Coughlin told listeners, "was conceived as a political defense mechanism against Communism and was ushered into existence as a result of Communism."[65]

To support his contentions Coughlin used counterfeit documents disseminated by the Nazis. Twenty-four of twenty-five "quasi-Cabinet members" of the Soviet Union, and 56 of the 59 members of the Communist Party in Russia in 1935 were alleged to be Jewish. The radio priest also claimed to have before him a quotation from the September 10, 1920, issue of *The American Hebrew* in which the editors allegedly claimed that the Bolshevik revolution of 1917 "was largely the outcome of Jewish thinking, of Jewish discontent, of Jewish effort to reconstruct." Coughlin expressed some sympathy for the 600,000 German Jews persecuted by the Nazis but immediately followed that observation with his opinion that Nazism could not be eradicated until Jewish leaders in synagogues, finance, radio, and the press, attacked communism and coreligionists who showed any sympathy for it.[66] Coughlin also complained that the American press and the American government were more upset about the treatment of Jews in Germany than they were about the attacks on Catholics and their property in Mexico and Spain. The radio priest accused the American press of "muzzling the truth" about the horrors in these countries where Catholics were the victims, especially Spain which was the world's "battleground of Communism versus Christianity." Coughlin began his peroration with the observation that "Thanks be to God, both the radio and the press at length have become attuned to the wails of sorrow arising from Jewish persecution." He then advised his audience that "Gentiles must repudiate the excesses of Nazism. But Jews and gentiles must repudiate the existence of Communism from which Nazism springs."[67]

Officials at New York City radio station that carried the talk, WMCA, were incensed by the speech. They had screened the presentation earlier and had pointed out several errors of fact that needed correction. Spokesmen for the priest assured officials that the changes would be made but they were not. Coughlin submitted this, and subsequent, radio presentations to the archdiocese for prior censorship and he received permission to deliver his talks; he may have thought that the objectionable passages had been eliminated. A half century later a Catholic theologian observed:

> the astounding fact of the matter is that ecclesiastical censors named by [his supervisor, Cardinal] Mooney had previewed and passed Coughlin's anti-Semitic broadcasts as not offending against Catholic faith or morals. It was a pre-Vatican II Catholicism in which anti-Semitism did not arouse moral outrage and institutional considerations outweighed virtually all else.[68]

But at station WMCA antisemitism and erroneous dissemination of information did arouse moral outrage. Immediately after Coughlin finished his talk an announcer came on the air and stated: "Unfortunately, Father Coughlin has uttered certain mistakes of fact,"[69] and then proceeded to enumerate them. In contrast to what Coughlin had said, the Bolshevik Central Committee was not dominated by Jews, Jewish bankers had not financed the Bolshevik Revolution, no British white paper had ever accused the American Jewish banking firm of Kuhn, Loeb of having aided the Communists, *The American Hebrew* had never stated that the Jews were responsible for the Bolshevik Revolution, and so forth. It later turned out that most of Coughlin's "facts" came from Nazi publications like *World Service*.[70]

A few prominent Catholics and periodicals took exception to the distortions in Coughlin's presentation. They included the *Michigan Catholic, Commonweal,* and George Cardinal Mundelein of Chicago who announced that Coughlin spoke for himself and not the church. Monsignor John Ryan accused Coughlin of being guilty of great distortions of fact, and of promoting antisemitism in the United States,[71] while an editorial in *Commonweal* denounced Coughlin's "tendentious radio talk of November 20," which "gained him . . . accolades from the inspired German Nazi press." The editorial condemned Coughlin's

> cavalier disregard for pertinent historical testimony, his insensitivity to the consequences of his acts on German and Italian Jews, his all too pious acceptance of propaganda from a party whose Fuhrer proudly boasts his machine is based on huge lies.[72]

Catholic World scolded Coughlin, though not by name, in an editorial undermining the thrust of his speech while reinforcing many Catholic beliefs that laid the groundwork for the charges. These words, in turn, made the radio priest's accusations more palatable:

> Anti-semitism is unjust, brutal and opposed to the teachings of Christ. If simple-minded people have sometimes thought that they needed to avenge our Savior for the treatment He received from His own people, they are very badly mistaken.
>   Of course, it is true that the Jews were rejected by God as the nation through which salvation was to come to the world, when they called down upon themselves and their children the blood of Christ. It is probably true that many of the hardships that have befallen God's chosen people are the punishment of the Heavenly Father, who wants to bring them back to Him whenever they "grow fat and kick." But neither Hitler nor Mussolini or any other individual has a special mandate from God to carry out this punishment.[73]

Despite the mild chastisement of the *Catholic World,* Coughlin's views apparently touched the right chord among millions of Americans discontented with their own lives, fearful of communism, and frustrated by the enduring depression and the failure of Roosevelt to turn the economy around. No American Catholic had ever before achieved such commanding attention and approval in the United States.[74] A Roseford, Illinois, priest voiced an opinion undoubtedly shared by millions of others: "If Father Coughlin be not the most useful citizen in America, he is among the first in that distinguished category."[75] In December 1938, 45 radio stations carried his weekly address that 3.5 million Americans listened to regularly; another 15 million had heard him at least once. Two-thirds of his loyal followers and more than half of those who tuned in occasionally subscribed to his views while polls showed that the lower the economic class, the larger the percentage of people who approved the radio priest's views. Among the lowest classes of Irish Catholics, and in the Irish middle classes, Coughlin had enormous support. His office received approximately 80,000 letters a week, 70 percent of which came from Protestants, and it took 105 staff members to read them.[76]

Perhaps because of Coughlin's exhortations, perhaps because of another downturn in the economy in 1938, perhaps because of the increased and cumulative effect of Nazi policies that led to an augmented number of German Jews trying to get into the United States, or perhaps because of a combination of factors, observers throughout the United States sensed an increased ardor in both antisemitic feelings and expressions. According to George N. Shuster, a prominent Catholic layman who deplored the bigotry of so many of his coreligionists and who would later become president of New York City's Hunter College, Coughlin's

> utterances stiffened the backs of all those who in one way or another were friendly to Hitler and Mussolini. It gave anti-Semitism the same religious inlay which played so great a role in the Austria whence Hitler came; it raised the issue of fascism vs. democracy; and it carried the virus of race hatred to regions where it had never previously been known.[77]

Thus, as Shuster pointed out, not only were Catholics aroused by Coughlin but his continuing attacks in print and on the air reflected sentiments heard throughout the United States. A Chicago business-man with "a reputation for liberalism" indicated that "the Jews are the cause of all our troubles in this country and I wish that every one of them could be deported."[78] In Minneapolis, a streetcar conductor obvi-ously believed the same thing. Forced to move a stalled auto off the track so that his trolley could proceed, and without knowing who the driver was, he muttered, "That dirty Jew, we ought to have a Hitler here."[79] One Catholic priest in Akron, Ohio, claimed, "The Jews here are at the bottom of most of our troubles and will someday suffer for it." A 68-year-old rubber plant worker in the same city told an inter-viewer, "I don't know what's the matter with me, but I hate the sight of a Jew. They control the money of the United States." He then added, "I'm like Hitler when it comes to the Jews. They would all leave the country if I had the power. I get mad when I start talking about them."[80] And in St. Louis, where the local chapter of the Friends of New Ger-many was generally known as the "Hitler Club," antisemitic slogans proliferated. Some parks and pools displayed signs readings: "Restricted to members and Gentiles only."[81]

Prominent officials throughout the country received letters from citizens reflecting this intense animosity toward Jews. On November 18, 1938, only eight days after *Kristallnacht* and two days before Father Coughlin's sensational radio address attacking Jews, Idaho Senator William Borah spoke out against relaxing our immigration laws to increase the number of refugees admitted to the United States.[82] As a result of the talk hundreds, perhaps even thousands, of Americans inundated his office with messages of congratulations and apprecia-tion. This mail not only supported Borah's stance regarding refugees but also included ferocious attacks on Jews.

The letters may have repeated tired shibboleths yet they mirrored the bigotry so many Americans had already revealed to pollsters in 1938. (At least half of all respondents that year thought Jews were par-tially or entirely responsible for Hitler's treatment of them while four separate inquiries resulted in anywhere from 71 to 85 percent stating that they opposed increasing immigration quotas for refugees.[83]) With great self-assurance correspondents mentioned a variety of inaccuracies including the "fact" that the President's family had been known as "Rosenfelt" in Holland, that Jews controlled the economy, that they prospered while others starved, that they were "the scum of Europe," and that millions of Americans were "put off their farms" and "lost their homes during the Jewish New Deal depression." A man from Fond du Lac, Wisconsin, praised Borah's "stand regarding the Jew. We don't want any part of them in this country, at least no more of them. There are far too many of them now." A New Yorker wrote, "Please Senator Borah dont allow anymore jews to come into the U.S.A. We have enough trouble here without bringing in the greatest trouble making race in the history of the world." Another New Yorker claimed that

> the Jewish problem has been with us for over 2000 years since the time of Pontius Pilate. The Jews have been ejected from and are a problem to any nation they inhabit. The Jews themselves are a very large measure responsible for this. Why don't they ask themselves why they are anathema all over the world? Germany is not the first nation nor shall it be the last confronted with this dilemma.

A resident of Washington, D.C., complained that he

> had twenty-five years experience with the Jew and have yet to find a good one. You probably know more about them than I do. If there is to be a comparison between the German and the Jew, there is not a good quality possessed by the German that is in a like manner possessed by the Jew. The German is clean, law-abiding, ethical in business, patriotic, and American. The Jew is dirty, lawless, unethical, unpatriotic and un-American.[84]

"There must be some reason for all this dislike of the Jews," another of Borah's correspondents speculated.[85]

The harsh antisemitic sentiments expressed in November and December 1938 escalated during the winter. As part of his crusade to eliminate communism Coughlin called for the establishment of a united Christian Front in the May 23, 1938, issue of his newspaper, *Social Justice*.[86] His call met with an enthusiastic response from his Irish Catholic admirers in some of the largest cities in the United States. Embittered and unnerved by the depression and its downward turn again in 1938, displaced from seats of political influence and power in New York City, and upset by the apparent potency of militant elements in the labor movement and elsewhere who championed radical changes in society as well as the Loyalist cause in the Spanish Civil War, Irish Catholics were ripe for the demagogic appeals made by men like Coughlin. They provided the backbone for Christian Front chapters in the sections and cities where more than 85 percent of all American Jews lived: Brooklyn, New York, Boston, Philadelphia, Baltimore, St. Paul, Minneapolis, Chicago, St. Louis, Detroit, Pittsburgh, Cleveland.[87] In all of these places Christian Fronters "were likely to be anti-Semitic, bellicose and vulgar."[88] Christian Fronters who received the most attention resided in Boston and Brooklyn where about 90 percent of the participants were Irish Catholics. In both areas the local diocesan newspapers, *The Boston Pilot* and the Brooklyn *Tablet*, staunchly supported Father Coughlin. Christian Front meetings were not exclusive, and members of the German American Bund, Protestant War Veterans, Christian Labor Front, Christian Order of Coughlinites, Crusaders for Social Justice, Crusaders for Americanism, American Nationalists, and the Christian American League often appeared at gatherings, many of which ended with the Nazi salute.[89]

At these rallies Christian Fronters were called on to "liquidate the Jews in America." Copies of *Social Justice* and the Brooklyn *Tablet* were sold, Coughlin was celebrated, and Jews were assailed as communists, international bankers, and war mongers. Speakers referred to the President of the United States as "Rosenfelt" or "Rosenvelt" or some similar sounding Jewish name, praised Franco as "that great Christian general who drove the reds out of Spain," and championed Hitler as "the savior of Europe." One critic assailed most Christian Front speeches as "plain, unvarnished incitements to murder," while another described the organization as a "savage anti-Semitic movement."[90]

From 1939 through 1942 roving gangs of Christian Fronters picketed, and placarded obscene stickers on, Jewish-owned retail establishments, desecrated synagogues, and indiscriminately attacked Jewish children and adults on the streets of cities like New York and Boston where sympathetic policemen of Irish background allowed the outrages to continue. In New York City more than 400 members of the police force also belonged to the Christian Front. On one occasion a Jew passed by an antisemitic rally where the speaker of the moment proclaimed that the only good Jews were in a cemetery. The passerby retorted that he was a good Jew and very much alive. Immediately a policeman on duty arrested the Jew for disturbing the peace. An incredulous judge could not believe what the patrolman had done but the officer explained that he arrested the "culprit" on specific orders from his captain, Michael McCarron.[91]

The Christian Fronters were variously assailed as storm troopers for Coughlin, young thugs, and neo-Nazis but they were hailed in working-class Irish Catholic neighborhoods for helping thwart the spread of communism. Leaders of the group included priests like Thomas E. Malloy, called the "Bishop of the Christian Front," and other prominent clerics and Catholic laymen who admired Father Coughlin. These men commanded respect from fellow Irish Catholics, especially since neither the archbishops of Brooklyn, New York, or Boston publicly criticized the radio priest or his followers. Church leaders even failed to endorse the newly formed Christian Committee Against Anti-Semitism in April 1939 on the ground that it would divide loyal Catholics.[92] The Brooklyn *Tablet* acknowledged that some members of the Christian Front were "anti-Semitic. Well what of it? Just what law was violated?"[93] One student later wrote that

> the Irish character of the Front seems ... to have been due to the well known loyalty of the Irish to the Catholic Church. The devout and even militant Catholicism of the Front organization's leaders, and the support given the Frontists by *The Tablet* all testify to the fact that the Front was primarily a Catholic movement.[94]

The activities of the Christian Fronters and Coughlin's continuing attacks on Jews, both on the radio and in print, also triggered a noticeable upsurge in antisemitism in the New York area.

Christian Fronters and their cohorts also provided most of the audience for a spectacular display of antisemitic fervor that the German American Bund put on at New York City's Madison Square Garden on February 20, 1939. Ostensibly a celebration of George Washington's birthday, the arena was filled with 19,000 animated people, hundreds of Nazi flags, 400 men who looked like storm troopers in their Nazi uniforms, and 1,745 New York City policemen. Throughout the arena massive banners proclaimed: "Stop Jewish Domination of Christian America!" and "Wake Up America! Smash Jewish Communism." Mention of Father Coughlin's name resulted in prolonged cheers and applause while shouts of "Heil Hitler" punctuated the evening.[95]

The combination of Coughlinites, Christian Fronters, and Nazi sympathizers celebrating in unison alarmed concerned Americans. A *Nation* reporter wrote on April 1, 1939,

> this account of anti-Semitism in New York City today could not have been written a year ago. These things now in abundant open manifestation were not happening then. But in 1939 anti-Semitism in New York has ceased to be whispered and has become an open instrument of demaguery, a vast outlet for idle energies.[96]

Journalist James Wechsler attributed the rise of this antisemitic furor to "Coughlin's personality, Coughlin's speeches, and Coughlin's propaganda." The Christian Front, he added, provided "the dynamic core of the movement. It calls the mass meetings, floods the city with leaflets, and rallies the crowds under its own signature." *The Christian Century* added, "Father Coughlin ... is thoroughly Hitlerish in outlook, in method and in the effect he produces."[97]

A spate of other articles in 1939 reinforced the impression of rising hostility toward Jews. In June, historian Henry Pratt Fairchild observed that in this country "some degree of hostility or dislike of Jews in general is very widespread, even among the most broad-minded and kindly disposed of his associates." The Catholic *Interracial Review* noted in July that "the rapid spread of anti-Semitism has caused well-justified concern." And in November, a man with a Jewish wife, living in the suburbs of an eastern city, admitted:

> my work has taken me into about every large city in the United States and into many small cities, too. My face and name bring me into contact with Christians at their clubs, bars, and homes. And each time I travel I grow more conscious of the rising tide of anti-Semitism that is moving across the country.

A New England Episcopalian minister added that he had lived "over forty years in a country where 'something would have to be done about the Jews.' I've been surrounded by and part of a passive anti-Semitic multitude of Christians." But there had been "no concerted effort. No *active* desire to do something." Yet in 1939, he emphasized, there were leaders. There were active and organized efforts to do something about the Jews and it frightened him.[98]

As anguished as the New England minister was, his apprehensions did not compare with the fears and anxieties of American Jews who desperately wanted to reverse what appeared to be the unrelenting growth of antisemitism in the United States. A 1940 article in the American Jewish *Congress Bulletin* noted that "at no time in American history has anti-Semitism been as strong as it is today."[99] To be sure, American animosity did not compare in severity with what was happening to Jews in Europe, but American Jews did not know what calamity might next occur in the United States. Psychologist Kurt Lewin advised American Jews how and when to inform their children of the situations that they might encounter. "The basic fact is that [your] child is going to be a member of a less privileged minority group, and he will have to face this fact." Do not try to avoid a discussion of the subject of antisemitism because "the problem is bound to arise at some time, and the sooner it is faced, the better." The child might not be called a "dirty Jew," Lewin wrote, until about the fourth grade, he or she could be expected to be invited to parties of their Gentile peers until adolescence when the invitations would cease, and both boys and girls would not have to face the existence of discrimination in colleges and jobs until the end of their high school years.[100]

Jewish community leaders admonished coreligionists to remain circumspect in their public behavior, to draw no attention to themselves as Jews, and to disassociate themselves from any group considered foreign to American society.[101] "Conservative Jews," a 1938 article in *The Nation* indicated,

> faced with the insanity of anti-Semitism, are tempted to abandon rationality themselves and accept as their own criteria of behavior the prejudices that operate against them. A radical Jewish labor leader or public official is looked upon not merely as a wrong-headed fellow, but as a

menace to the race; a Jew who militantly espouses even the cause of free speech is considered a person of dubious judgment.[102]

The hostile attitudes exhibited toward Jews by so many Americans had devastating psychological effects on individuals who could not bear their minority status. "The fact is," Lewis Browne wrote in 1939, "we Jews as Jews can't do anything. Working by ourselves, we are utterly impotent. For, being a minority, we cannot act; we can only react. All of which means simply this: if the Jewish problem is ever to be solved, it will have to be done by the Gentiles." One Jewish writer even claimed that "the vast majority of Jews do not remain Jews by choice. Basically, the Jew hates his Jewishness, and bewails his fate."[103]

Thousands of Jews abandoned their heritage. A man who chose to remain anonymous indicated that he found it impossible to be both a Jew and an American. He therefore converted and claimed that he enjoyed being a Christian.

> I am now raising children who need never learn to endure snubs, who will never be tempted to retaliate against cruel discrimination. From this pleasant sunshine, I look back with horror at the somber world in which my race-proud kin persist on their ancient and unhappy courses. Life is good. I never regret my step.[104]

That same author wrote how embarrassed he had been by other Jews who spoke English badly, who used gestures to emphasize their points, and who interspersed Yiddish words or expressions in their speech.[105]

The vast majority of Jews, however, suffered silently or bewailed their fate. Untold numbers modified or altered their names and several reporters at *The New York Times* believed that publishers Adolph Ochs and Arthur Hays Sulzberger were so sensitive to antisemitism that they encouraged newcomers to use initials instead of their given name of Abraham in bylines. Thus readers noted stories by A. H. Raskin, and in later years by A. H. Weiler and A. M. Rosenthal, without becoming aware of their Jewish-sounding given names. (New York *Tribune* editor William O. McGeehan had tried to influence one of his sportswriters to do the same thing in the 1920s but Jesse Abrahamson refused to become A. Bramson.) In the 1930s many Jews changed their names to increase their economic opportunities and comfort levels. Milton Levine became Milton Lewis in the hopes that it would advance his journalistic career and the young Mel Israel gladly accepted the advice of CBS radio officials who told him that keeping his name would hinder professional advancement. Israel then changed his surname to Allen and went on to a lengthy career broadcasting New York Yankees baseball games.[106]

In the 1930s a Jewish name was not only a hardship for those people trying to move into mainstream America but was also a vestige of Old World and immigrant origins from which they wanted to distance themselves. Many Jewish college students changed their names just before they graduated; some people waited a bit longer. A study of name alterations in Los Angeles in the 1930s indicated that the largest number of people who applied for changes were married and prosperous Jewish males who lived in mixed Jewish and Christian neighborhoods. The study provided no analysis or explanation as to why this particular group stood out, but a safe assumption is that being identified as Jewish narrowed their social and economic opportunities. People in Hollywood, for example, often changed their names for business reasons.[107] One Hollywood writer/director, Abraham Polansky, admitted:

> when I arrived at Paramount as a contract writer, another Jewish writer told me to change my name. He told me it sounded Jewish and that movies were seen all over America. I didn't change my name and nothing happened. But many actors did. Americans wanted to see Americans.[108]

During World War II the National Jewish Welfare Board surveyed Jews in the armed services and found that over 50 percent of the returned questionnaires came from people with generalized American surnames like "Smith" and "Brown," who presumably wanted to hide their backgrounds or avoid discrimination. By the late 1940s 46 percent of those people who sought name changes in Los Angeles were Jewish although Jews constituted only 6 percent of the city's population.[109]

Other Jewish responses to the antisemitism of the 1930s were reflected by the reaction of some high-status Jews after the death of Justice Benjamin Cardozo of the U.S. Supreme Court in July 1938. Secretary of the Treasury Henry Morgenthau, publisher Arthur Hays Sulzberger of *The New York Times,* and other well-connected Jews urged Roosevelt not to appoint Felix Frankfurter as Cardozo's replacement because his name, *Time* magazine asserted, had "come to symbolize Jewish radicalism in the New Deal." They feared that with Brandeis still serving, "putting a second Jew on the Court would play into the hands of anti-Semites at home and abroad."[110] On this occasion, however, the President chose to take a symbolic stand and wrote to Judge Julian W. Mack:

> I feel it is peculiarly important—just because of the waves of persecution and discrimination which are mounting in other parts of the world—that we in this country make it clear that citizens of the United States are elected or selected for positions of responsibility solely because of their qualifications, experience, and character, and without regard to their religious faith.[111]

Frankfurter had been a valued advisor to the President for several years and although Roosevelt waited until January 1939 before selecting him, the nomination sailed through the Senate with little difficulty. During Frankfurter's confirmation hearings, however, Senator Patrick McCarran of Nevada asked him, "Are you a Communist?" and then followed up with "Do you believe the doctrines of Karl Marx?" After Frankfurter indicated that he was neither a communist nor a sympathizer with Marxist thoughts, the Senate Judiciary Committee unanimously recommended his approval. Significantly, many Senators preferred not having their stance recorded and floor confirmation came on a voice vote.[112]

Members of the public showed less hesitancy in expressing their views of the appointment. Letters came into Senator William Borah's office accusing Frankfurter of being "a dangerous radical" and "a Jew of frankly Communistic activities." One writer wanted "a person, who first of all is a native born American of pioneer stock—and a devout Christian," another indicated that "a Jew has no right in our courts. We want white men there," and a third claimed that his "objection to Jews in our government is . . . based . . . upon moral grounds, which were established by the Bible, and the teachings of Jesus Christ."[113] William Pelley of the Silver Shirts claimed that the appointment of Frankfurter to the Court doubled his business and George Shuster wrote that "the remarks of various editors" of Catholic journals "on the appointment of Felix Frankfurter to the Supreme Court would make up a rather harrowing little anthology."[114]

Thus, as the decade ended, Jews were more uneasy about the future than they had been ten years earlier. *Fortune* magazine reported in February 1936 that incidents and expressions in the 1920s had aroused concerns but in the 1930s made important Jewish leaders, "men who had previously looked to the future with complete confidence," fearful. "The apprehensiveness of American Jews has become one of the important influences in the social life of our time." *Fortune*'s concern with the rumors led it to examine Jewish power in the business world. Its investigation found that a majority of Jews were poor or living on the margin of poverty in the 1930s but they had to endure accusations that they controlled the banks and monopolized economic opportunity in America. The journal's lengthy analysis of American Jewish business interests led to the conclusion that there was absolutely "no basis whatsoever for the suggestion that Jews monopolize U.S. business and industry." Typically, those who held such beliefs refused to relinquish them.[115]

American Jews knew of existing antisemitism but before 1933 it had been mainly religious, intellectual, verbal, social, and economic. There had also been sporadic attacks on children and adults in a number of cities in this country. But in the 1930s the intensity of antisemitism, the appeal of hate organizations, and the popularity of demagogues combined with an escalation of serious physical abuse, especially in the cities of the northeast and midwest where more than 85 percent of all American Jews dwelled, to have an absolutely chilling effect. And the worst thing was that the hatred seemed to be accelerating. For the first time in American history Jews feared that their attackers might acquire the kind of political influence and respectability that antisemites had in Europe and achieve similarly devastating results. "There is no method by which the present gravity of anti-Semitism may be measured," one Jew wrote just before the United States entered World War II, "the present virulence of anti-Semitism is undefinable, its future unpredictable."[116]

Public opinion polls came into vogue toward the end of the 1930s and their assessments, however crude, reinforced the sense of Jewish insecurity in America. In 1938 at least 50 percent of Americans had a low opinion of Jews, 45 percent thought that they were less honest than Gentiles in business, 24 percent thought that they held too many government jobs, and 35 percent believed that the Jews in Europe were largely responsible for the oppression that had been heaped on them. To the question "Should we allow a larger number of Jewish exiles from Germany come to the United States to live?" 77 percent of the respondents said no. In sum, about 60 percent of those polled had negative impressions of Jews, most finding them greedy, dishonest, and aggressive.[117]

Looking back decades later to assess the status of the Jews in the 1930s provides one with a distorted picture. On a per capita basis, although Jews ranked among the more prosperous Americans economically, the majority of them were poor. It was evident that the second generation of east European immigrants had advanced in status from that of their childhood years but they faced an antisemitism among Gentiles that was deep seated, more unpleasant than a serious threat. Yet American Jews, seeing what Hitler had done in Germany and now personally experiencing the impact of discrimination and racial rhetoric in the United States, seemed more fearful about the future than ever before. For many, there was no light at the end of the tunnel. In both Protestant and Catholic periodicals their faith had been attacked and their raison d'etre questioned. The President of the United States may have brought many Jews into the federal government but he seemed unwilling or unable to cope with the rising tide of antisemitism that Jews saw all about them. A quiet sense of desperation engulfed American Jews who had witnessed several decades of increasing attacks on them from almost every major segment of society. Ironically, although the depression led to increased manifestations of antisemitism, the return of prosperity during World War II did not mitigate its effects. Polls and other measuring rods used to deduce the quantity and intensity of antisemitism during the war suggested that the situation of the Jews in the United States continued to be precarious. It seemed that the very accomplishments of the Jewish Americans were hindrances to their being fully accepted, or at least comfortably tolerated. A sense of foreboding continued to spread among Jews in the United States as their country itself entered an era of maximum mortal danger.

7

Antisemitism
at High Tide:
World War II
(1939–1945)

In retrospect it becomes apparent that the rising tide of antisemitism in America paralleled increased national involvement with European affairs. As the possibility of war increased, Franklin D. Roosevelt ended innovative New Deal initiatives in 1938 and shifted his interests toward foreign policy. The Germans had marched into Austria and Czechoslovakia in 1938 and on September 1, 1939, less than ten days after the Nazi-Soviet pact of August 1939 shook the world, Germany invaded Poland. World War II began as Great Britain and France entered the fray on the side of Poland on September 3, 1939.

In the United States there was considerable concern and sympathy for the British and French, but an equally firm desire to remain out of the war. In addition to the large number of radicals, pacifists, and supporters of the Soviet Union in the country, Americans of German, Italian, and Irish descent opposed going to war against Germany or on the side of the British. The Germans had a natural concern for their country of origin while the Irish, who supported Hitler in his battle against communism, also hated the British who had kept Ireland in subjugation for centuries and refused to grant it independence until 1921.[1] Italian Americans were proud of Mussolini's reforms in Italy and, since he had aligned his country with Germany, they were reluctant to fight against Hitler.

To promote and increase isolationist sentiments the America First Committee formed in July 1940 and won the support of some of the people associated with antisemitic rantings, including Henry Ford and Father Coughlin. America First's goal was to keep the United States out of the war. Although it had only about 850,000 members, two-thirds of whom lived within 300 miles of Chicago, America First claimed 15,000,000 followers.[2] The movement at first boasted of respectable leaders like Robert Wood of Sears, Roebuck and Co., R. Douglas Stuart, a graduate of Yale Law School and the son of a vice president of Quaker Oats, and some of the "finest citizens" of St. Louis but Coughlinites and Nazi sympathizers eventually figured prominently. Monsignor Edward A. Freking, a Coughlin supporter and editor of *The Catholic Telegraph-Register*, served as a member of the Board of Directors of Cincinnati's America First Committee while in Brooklyn the America First Committee was "little more than the Christian Front by another name." The chairman of the Terre Haute, Indiana, chapter thought that " 'Jews were now in possession of our Government' and were causing the 'panic and wars that ruin everybody.' " A woman in the Kansas America First chapter insisted that both Franklin and Eleanor Roosevelt were Jewish and so were 90 percent of the New Dealers. She also believed that Prime Minister Winston Churchill was half-Jewish, and stated that the English government was a "Jewish government." United States Senators Burton K. Wheeler, Democrat of Montana, and Gerald Nye, Republican of North Dakota, spoke before meetings of America Firsters. On August 1, 1941, Nye denounced Jewish producers in Hollywood for allegedly making movies designed to win support for American entry into the European war. A month later he helped launch a Senate inquiry into the interventionist proclivities of the movie makers.[3]

Shortly after the Senate investigation of Hollywood began, flying hero Charles A. Lindbergh, who only a few years earlier had accepted a medal from Adolf Hitler, made the most significant speech in the short life of the America First committee. Speaking in Des Moines, Iowa, on September 16, 1941, he blamed Jews, the British, and the Roosevelt administration for pushing the United States toward a war with Germany. He specifically excoriated Jews. "Their greatest danger to this country lies in their large ownership and influence in our motion pictures, our press, our radio and our Government."[4]

Reactions to Lindbergh's talk were immediate and harsh. Letters to the President roundly denounced the man. One correspondent asked, "How near must we be to actual warfare before Mr. Lindbergh can legitimately be silenced," another called the former flying hero "the chief 5th columnist in the United States and Hitlers [sic] right hand man," while an advertising man in St. Louis thought up the slogan, "TO HEIL MIT LINDBERGH."[5] Respectable publications, which recognized American antisemitism but were also aware of how distorted and inappropriate Lindbergh's comments were, condemned the man and the speech. "No public utterance by a figure of prominence in American life in a generation has brought forth such unanimous protest from the press, the church and political leaders as did Charles Lindbergh's resort to racial and political prejudice in his Des Moines speech," the Davenport, Iowa, *Times* reported.[6] Editorials in otherwise isolationist newspapers like those owned by William Randolph Hearst, and Robert McCormick's *Chicago Tribune*, disassociated themselves from Lindbergh and his presentation. Emporia, Kansas, journalist William Allen White, called the speech "moral treason," and reproved its author: "Shame on you, Charles Lindbergh, for injecting the Nazi race issue into American politics";[7] denunciations from Protestant and Catholic leaders took up almost three pages of New York's *Journal-American*. A contemporary survey found that 93 percent of the nation's newspapers condemned Lindbergh's talk while historian Wayne Cole later wrote, "It would be difficult to exaggerate the magnitude of the explosion which was set off by this speech."[8] For the America First Committee, the talk proved a blow from which it would never recover. This was recognized by one of its opponents, who, in a letter to the President, wrote that Lindbergh

> entirely without intending to do so, finally rendered a service to his country. He placed the America First group in a position which is very difficult indeed. They cannot, without serious loss, either own or disown him. They might follow Hitler's precedent in the case of [Rudolf] Hess—*they may claim that he has been insane for months.*[9]

Lindbergh's words found some admirers, however. A Connecticut woman who supported him thought she saw

> a great deal of hypocrisy . . . by smug citizens in our midst who sounded off to condemn Lindbergh on the basis of a hasty reading of two or three sentences lifted from his speech. Many such citizens practice anti-Semitism every day of their lives—wealthy snobs who would shun a Jew socially as they would shun a leper.[10]

But no matter what people thought of Lindbergh and his explosive accusations the Japanese attack on Pearl Harbor three months later mooted the question. The next day, December 8, 1941, the United States went to war with Japan and two days after that Germany declared war on the United States, thereby ending the issue of whether the nation should participate fully in World War II. As nearly all Americans stood firmly behind their government and the armed forces, the Senate investigation of movie makers disintegrated. Indeed, the Roosevelt administration utilized Hollywood's film industry personnel to help promote wartime unity and create propaganda features.[11]

But the spirit of common interest that marked the fervor of millions of Americans also augmented suspicions against outsiders; entry into war increased intolerance. World War II, a writer noted in 1944, "aggravated every group antagonism in America."[12] The worst victims of this hostility were Japanese Americans on the West Coast who were taken from their homes and interned on interior wastelands. But other minorities suffered as well. The Detroit Race Riot in June 1943 witnessed vicious attacks on African Americans; Mexican American teenagers were unjustly convicted of murder in the notorious "Sleepy Lagoon" case in Los Angeles in August 1942; and the Los Angeles Zoot Suit riots the following June once again highlighted the venom harbored by members of the dominant culture toward Mexicans.[13]

Jews, too, were victims. Antisemitism was more widespread in the United States during the war years, and even as the war began, in December 1941, one analyst wrote, "Today, as never before in their history, American Jews are apprehensive over their security. A great anxiety weighs on them." Movie actor Kirk Douglas later recalled that in the early 1940s he moved easily among people who did not know he was Jewish. Listening to them speak, he heard "the things that in their nightmares Jews speculate non-Jews say, and . . . I found out, they do."[14] Overt hostility toward Jews registered itself in opinion surveys. One recurring poll question asked whether Jews had too much power in the United States; 36 percent of those polled in May 1938 answered affirmatively but by 1945 58 percent held such beliefs. A February 1942 survey asked: "What nationality, religious or racial groups in this country are a menace [threat] to Americans"; 24 percent of the respondents indicated the Japanese, 18 percent chose the Germans, and 15 percent identified the Jews as the malignant group. By June, 1944, almost a full year before the war ended in Europe, 24 percent now identified the Jews as the greatest menace, with the Japanese at 9 percent and Germans reduced to 6 percent.[15]

In November 1942 a poll asked American high school students which of the following groups, if any, would be their last choice as a roommate. The percentages of their responses were:

Swedes	5%
Protestants	4
Negroes	78
Catholics	9
Jews	45
Irish	3
Chinese	9
Makes no difference	5
Don't Know	3

That same month *Fortune* asked factory workers: "Which of the following groups would you least like to see move into your neighborhood?" They answered:

Swedes	3%
Protestants	2
Negroes	72
Catholics	4
Jews	42
Irish	2
Chinese	28
Makes no difference	13
Don't Know	5[16]

The answer to the question, "Have you heard any criticism or talk against the Jews in the last six months?" graphically illustrated the escalation:

	Yes	*No*
1940	46%	52%
1942	52	44
1944	60	37
1946	64	34[17]

Undoubtedly the diatribes of Father Coughlin in the early 1940s contributed to the negative assessment of Jews since between 1940 and 1942 his weekly newspaper, *Social Justice,* with a circulation of over 1 million copies, published 102 antisemitic articles, 88 of which endorsed the America First movement and its leaders. Coughlin also denounced Winston Churchill, and in March 1942 *Social Justice* carried a story accusing Jews of starting World War II.[18] The President decided to break the priest's appeal and ordered Attorney General Francis Biddle to do something. Biddle asked Postmaster-General Frank Walker to invoke the Espionage Act of 1917, which allowed the government to "suspend or revoke" mailing privileges for those undermining the war effort. Moreover, presidential advisor Leo T. Crowley flew to Detroit to speak with Coughlin's superior, Edward Cardinal Mooney, who had previously silenced the radio priest in 1937. At that time unnamed sources in the Vatican "encouraged" the Cardinal to countermand his order. But with the United States at war, and the President determined to prevent utterances subversive to the war effort, the Cardinal again took action. Mooney thus informed Coughlin that unless he ceased all non-religious activities he would be suspended from the priesthood. Coughlin agreed to withdraw from his non-religious activities and, in effect, ended his public career in April 1942.[19]

Nevertheless, his legacy remained. Father Michael A. McFadden, pastor of Most Pure Heart of Mary Church in Shelby, Ohio, wrote to a Kansas City newspaper in April 1942 after the cessation of publication of *Social Justice:* "Father Coughlin has done more good than all the Knights of Columbus for the Church in America."[20] Christian Fronters also honored Coughlin by continuing his work. A New York newspaper reported in October 1943 that in Boston roving gangs of "Christian Front Hoodlums" terrorized Jewish youths so fiercely that Massachusetts Governor Leverett Saltonstall, heretofore uninvolved with the problem, ordered an investigation in November. The findings confirmed the accusations and proved especially devastating to the police when the committee discovered that Jewish youths had been regularly assaulted for a number of years while police officers either abetted the attacking hoodlums or ignored them. After reading the report the Massachusetts Attorney-General labeled the perpetrators of the assaults "primitive and stupid," and stated that "the case establishes an all-time record for brazen disregard and abandonment of equal and impartial law enforcement." He advised the Governor to reorganize the police department and Saltonstall thereupon prevented the reappointment of the Boston police chief.[21]

Explanations for the assaults on the Jewish youths varied but *The Christian Science Monitor* was not alone in attributing them to "the poison put out by the Coughlinites and the Christian Front."[22] A White House staff member thought the Irish Catholics in Boston were so viciously antisemitic that "so far as the Jew is concerned, it would be a simple matter to organize them into the Ku Klux Klan."[23] A scholar who analyzed the conditions that gave rise to such hostility observed that "a seething sense of perceived social inferiority pervaded South Boston throughout the twentieth century" and that "explicit anti-Semitism attempted to compensate for many ills of the Irish ghetto—alcoholism, low rates of socioeconomic mobility, and a sense of defeatism and failure. This mood facilitated the frightening outbursts of anti-Semitic activities in 1942 and 1943."[24] Christian Fronters also thrived outside Boston. Irish and Jewish youth regularly fought during World War II in Brooklyn, the Bronx, Manhattan, Minneapolis, Providence, Rhode Island, on the boardwalk in Bradley Beach, New Jersey, and in a variety of other places. Grossly negligent police often winked at the actions of the perpetrators and rarely protected the victims.[25]

Either by their reticence or approval, teachers in Catholic schools also facilitated the attacks on Jews even though they did not specifically call for them. A particularly sad event occurred in November 1943 in Brooklyn where George Marooney, Jr., aged 12, beat up a four year old Jewish boy. When questioned, the sixth grader replied that his teacher at St. Francis Parochial School said it was permissible to beat up Jews.[26] Most officials of the church, however, abetted the assaults by their silence; refusing to denounce or caution followers against such practices encouraged their commission. Decades later, one woman, recalling her early 1940s schooling in Long Beach, California, noted, "We were taught in Catholic school to love everyone. . . . Yet Catholics were saying the Jews killed Jesus. We're supposed to be against Hitler, yet we were talking anti-Jew all the time."[27]

Although Irish Catholics attracted the most negative publicity for their antisemitism, several Protestant leaders of the 1930s were equally venomous in their attitudes and, like Father Coughlin, were also forced to curtail their activities during wartime. Gerald Winrod, who accused Jews of starting the war, and William Pelley of the Silver Shirts, were ordered to cease publication of their respective journals, *Defender* and *The Galilean*, in 1942. Both Winrod and Pelley were investigated for sedition and undermining the war effort and Pelley, in August 1942, was sentenced to 15 years in jail. Other antisemitic publications that had their mailing privileges suspended because of seditious activities included *X-Ray, Money, Publicity*, and *Beacon Light*.[28]

After the fall of Pelley, the Reverend Gerald L. K. Smith emerged from a secondary role in the 1930s to become the most prominent American spokesman of antisemitism. With Winrod's help, he inaugurated his journal, *The Cross and the Flag*, in 1942 and then embarked on a campaign to sort out the "anti-Semitic problem." Smith, who considered Henry Ford the best American representative of Christian faith, hard work, and patriotism, had also been a member of the Silver Shirts in 1933, and through the early 1970s ranked among the nation's most vituperative bigots. During a rally in St. Louis in 1940, Smith accused Jews of intimidating and coercing members of other political parties who challenged their power, and called for the establishment of a Christian Nationalist Party. During the war Smith characterized B'nai B'rith's Anti-Defamation League as a "Gestapo organization," railed against newspaper columnist and radio broadcaster, the "Jew Walter Winchell," and opposed the "Jew-infested" United Nations. Smith's long career of bigotry peaked in 1944 when he ran unsuccessfully for the presidency.[29]

A number of women's groups also emerged during World War II to promote antisemitism. The parent organization, "We, the Mothers," headquartered in Chicago, but constituent members were also called "Mothers of America" in Minneapolis, "Mothers of Sons" in Cincinnati, the "United Mothers of America" in Cleveland, and the "Crusading Mothers of Pennsylvania." *Life* magazine characterized the Mothers' groups as "an arsenal of rumor and petty gossip directed at the American war effort." Anti-Bolshevik and antisemitic, they distributed materials claiming that the Jews were responsible for the war, hosted speakers like Gerald L. K. Smith, and trembled as a spokeswoman announced that there were "200,000 Communist Jews at the Mexican border waiting to get into this country. If they are admitted they will rape every woman and child that is left unprotected." The Cincinnati Mothers of Sons Forum met twice a month in a downtown hotel and "learned" that in the war of religions taking place in Europe, Jews were persecuting Christians. When the city's *Times-Star* published an editorial protesting the massacre of European Jews, the local Mothers group rebuked the editor.[30]

Antisemitism also permeated many other American psyches during the war.[31] "The barest scratching of an economic or political reactionary," reporter Stanley High wrote in 1942, "almost unfailingly produces an anti-Semite."[32] Newspapers in Anderson, South Carolina, placed the blame for the war on the Jews and residents in the town ostracized the few Jews who lived there. In 1945 a New England publication spoke out against such bigotry: "We feel *The Christian Register* has an obligation to struggle against the 'genteel,' but equally vicious, anti-Semitism found even among Unitarians." A sociologist tried to synthesize the national attitude. Americans' condition for liking the Jew, he argued, "is that he cease being a Jew and voluntarily become like the generality of society. . . . the Jew is our irreconcilable enemy within the gates, the antithesis of our God, the disturber of our way of life and of our social aspirations."[33] Hardly a surprise, Kenneth Leslie, editor of *The Protestant*, reported that the journal's textbook commission found works used in nursery schools through graduate seminaries to be rampant with antisemitism. "Some of these books," he observed, "are characterized by the hypnotic effect with which on page after page through hundreds of pages the suggestible reader is whipped into a frenzy of hatred."[34] In 1943, in a joint effort to promote tolerance and undermine prejudice, 1,951 Protestant ministers pledged to carry out the program of the Textbook Commission of *The Protestant* to eliminate the "anti-Semitic poison in their textbooks," but their efforts came to naught. Reverend L. M. Birkhead, director of Friends of Democracy, acknowledged in 1944 that antisemitism "had become so prevalent and is so widely preached from the pulpit, the press and Congress that the Church is too weak to deal with the problem."[35] Not until the 1970s would significant alterations appear in religious school texts in the United States, lessening the effect of centuries of antisemitica propagated by Christian teachings.[36]

Antisemitism existed in Congress both before and after the Roosevelt era but in the late 1930s and during World War II several Republican and Democratic Senators and Representatives shamelessly articulated their prejudices. Senators Nye, Wheeler, and Robert Reynolds of North Carolina openly aligned themselves with known antisemites; Senator Arthur Capper of Kansas commended the editor of Coughlin's *Social Justice* for doing a "great job"; Senator Paul McNary of Oregon resisted dealing with Jewish administrators and asked a functionary in the Department of the Interior whether there were any "white men" with whom he could discuss his state's affairs. A decision of the Office of Price Administration (OPA) so displeased Senator Hugh Butler of Nebraska that he complained to OPA Director Chester Bowles: "I'll wager that some Goldberg, or some other Berg . . . acted upon it." The most outspoken bigot in the Senate in the 1940s, Theodore Bilbo of Mississippi, rarely minced words when conveying his prejudice. "So far as Jews are concerned, we have Jews in my State," he explained in

1945, "and some of my best friends are Jews." But there was a certain class of "kike" Jews in New York, Bilbo declared, who wanted to "cram" the Fair Employment Practices Commission "down the throats of the American people."[37] On the House side of the Capitol Congressman Jacob Thorkelson of Montana worried in 1939 about a "conspiratorial group of international financiers, mostly Jewish in faith and communistic in principles," whom he believed were trying to drag the United States into battle while Fred C. Gartner of Philadelphia supported the German American Bund, and Hamilton Fish of Dutchess County, New York, allowed the Silver Shirts to distribute copies of *The Protocols of the Elders of Zion* under his congressional frank. In 1941 John Rankin of Mississippi denounced "Wall Street and a little group of our international Jewish brethren" for trying to get the United States into war. Once the war began he punctuated his speeches with the terms "kikes" and "niggers," insisted that Jews were Communists, and attacked Hollywood because it was so "Jewish." Congressman John A. Flannagan of West Virginia asserted on the floor of the House that he did not want "any Ginsberg" to lead his son in battle; and Martin Dies of Texas reminded colleagues that antisemitism was not a crime.[38]

Congressional criticism of Jews reflected contemporary fears and concerns. During the war rumors spread across the United States that Jews lacked patriotic fervor, that they evaded the draft, and that they shunned wartime service.[39] A professor of foreign languages at CCNY, who would later be accused of allowing antisemitism to affect his professional evaluations of students and colleagues, admitted having joked, "The Battle Hymn of the Jews is 'Onward, Christian Soldiers, we'll make the uniforms.'" A few years later, while intoxicated, this same man stated, "The best damned thing that ever happened to America was when that Jew in the White House, Rosenfeld, died."[40]

By 1943 hostility toward Jews in the United States had grown enormously and had assumed "unprecedented proportions," especially, but not exclusively, in the populous states east of the Mississippi and north of the Ohio Rivers. A 1943 Office of War Information report indicated widespread animosity toward Jews in half of the 42 states it surveyed and described intense antisemitism and "unreasonable hate" particularly among the middle class in Pennyslvania.[41] Cleveland was also rife with venom, the atmosphere in Detroit's factories reeked with antisemitism,[42] and a professional man living in Minneapolis told an interviewer:

> Anti-semitism is stronger here than anywhere I have ever lived. It's so strong that people of all groups I have met make the most blatant statements against Jews with the calm assumption that they are merely stating facts with which anyone could agree.[43]

From Los Angeles, a "third-generation American of Irish descent" wrote of his alarm at the openly expressed anti-Semitic convictions of a large portion of the citizens of this community. . . . Recently I have been told that "all the Jews stay out of the army, or if they get in, they are given commissions— the President is a Jew—you can't get a defense contract unless you are a Jew—Jews own 80% of the nation's wealth—Jews got us into the war— the W[ar] P[roduction] B[oard] is controlled by Jews"—and so on.[44]

Throughout the country, moreover, anonymous groups and individuals distributed "millions" of antisemitic leaflets in war plants, airplane factories, post offices, police stations, and other public buildings.[45]

Civilian attitudes carried over to the military, which grew from 190,000 in 1939 to about 13 million by the end of World War II.[46] Since the servicemen and women generally shared the views of their compatriots at home their antisemitism mirrored the prevalent sentiments of the larger society. The Navy used Harvard University as one site for its fliers' training program and many young men who arrived were openly racist. One sent a note to Delmore Schwartz, then teaching poetry at the university, which included the message: "Fuck the Jews!"[47] If some officers and GIs were highly antisemitic in practice, in fairness to the armed services it must be acknowledged that lectures and courses on ethnic and religious tolerance were given. Official prejudice toward those who were of different backgrounds was deplored, and high-level officers, more sensitive to public relations than some of their subordinates, generally did not condone bigotry when it came to their attention. Yet at the same time both the War and Navy Departments opposed the issuance of President Roosevelt's Executive Order #8802, prohibiting defense contractors from discriminating on the basis of race or creed when fulfilling government projects. At the highest levels of both the War and Navy Departments there were fears that integration would lower or "destroy" the morale of the white fighting forces. Placing Jews among Christians in the armed forces increased the likelihood of bigotry expressing itself but nothing could be done about that since religious segregation was unconstitutional. Therefore Secretary of the Navy Frank Knox and Secretary of War Henry Stimson issued orders forbidding the circulation of antisemitic publications at all naval and military posts.[48]

Nonetheless, attitudes could not be inhibited and vicious expressions from the mouths of military personnel abounded. Both Jews and Gentiles heard wounding comments that remained vivid in memory forty years later. Some Christian soldiers stated boldly: "I would kill all of you Jews," and "We're out to kill Jews tonight." A Jewish seaman, whose religious heritage had hitherto been unknown to his buddies, found them distant and cold after he rebuffed a fellow junior engineer for saying that Hitler had done a great thing when he killed the Jews. A Jewish woman who served in the Women's Army Corps (WACs) recalled that there was a sergeant in her unit who muttered "You can't beat the niggers or the Jews" every time he passed her desk. She also met women from places like Oklahoma and Georgia who had never before seen a Jew, but whose stereotyped impressions of them included the belief that all Jewish men were pawnbrokers. They were surprised to find out that the Jewish WAC's father had a different occupation. Playwright Neil Simon later wrote of a buddy's reaction when he found out that Simon was a Jew. " 'You're Jewish?' one of the men he served with queried. 'I said yes.' He said: 'I thought you'd look different, your nose would be different.' " A Hawaiian inducted into the army later recalled that some of his new buddies asked him not to talk to three of the men in the unit.

> I asked why. They said, "They're Jews." I said, "What's a Jew?" They said, "Don't you know? They killed Jesus Christ." I said, "You mean them guys? They don't look old enough." They said, "You're trying to get smart?" I said, "No. It's my understanding that he was killed about nineteen hundred years ago."[49]

Some of the reading matter provided by American military agencies reinforced existing bigotry, and, being distributed from official sources, lent weight to their alleged authenticity. In 1943 the government presented 350,000 servicemen and women with copies of a Bible that included phrases like "Israel's Fall the Gentile's Salvation" and "The Jews are a synagogue of Satan."[50] Until April 1945, when sufficient numbers of protests resulted in its elimination, one of the correspondence courses conducted by the United States Armed Forces included statements reading: "The genuine American is essentially Nordic, preferably Protestant . . . the Jew is an offensive fellow unwelcome in this country,"[51] "The Gentile fears, and with reason, the competition of the Jew in business and despises him as a matter of course," and "Socially, they are only too anxious to attach themselves to the Christians. Surreptitiously they have wormed their way into the clubs in spite of the ostracism and insults designed to exclude them."[52]

Jewish chaplains in the services found that a number of their Christian peers also harbored strong prejudices. Some of the ministers denounced Jews as "Godless Communists," believed that there were too many of them in the government in Washington, and thought that "all of you Jews are good business executives."[53] One Christian chaplain refused to put an ADL publication on the library shelves because it "made Jews out to be perfect patriots." This same librarian refused to circulate another pamphlet because it stated that the Romans had killed Jesus. The chaplain defended his position by saying that "anyone who knows anything at all knows that it was the Jews who were guilty."[54] A Catholic chaplain told a Jewish colleague that "Father Coughlin is the greatest Catholic priest in the world. I would kiss the ground he walked on!"[55] When one Jewish chaplain sought assistance from his Protestant and Catholic associates to get the base commander to alter an order that all the men had to participate in the Easter Service, he could not get any support. The Catholic chaplain said to him, "Why shouldn't all men regardless of creed or race pay homage to our Savior? Don't they all come out when a famous general goes by our area? And, as to the Jewish men, they have a choice of either the Catholic or Protestant services."[56] In Germany, General George S. Patton refused to have Jewish chaplains at his headquarters. When Rabbi Roland Gittelsohn wrote about his life in the armed forces during World War II, he stated, "The reader will not be blamed if he reads of these experiences incredulously. I myself find it hard to believe them now, nearly two years after they occurred."[57]

Antisemitic incidents were so widespread during the war that they reappeared in some of the most revealing literature of the late 1940s. Several postwar novels included Jewish characters, always portrayed as thinking of their Jewishness in terms of antisemitism and aberration; there were almost always bigoted comrades who mouthed the sentiments prevalent in the towns they left behind. The most prominent of these novels included Norman Mailer's *The Naked and the Dead,* Irwin Shaw's, *The Young Lions,* Martha Gelhorn's *The Wine of Astonishment,* and Ira Wolfert's *An Act of Love.*[58]

Savagely clever verses also served as "literary" reminders of venom some military people harbored toward Jews. The most frequently repeated antisemitic doggerel of the war years reflected on the patriotism of the Irish and its alleged absence among Jews. Called "The First American," it appeared on military and naval bases and in a variety of places throughout the United States. In each of its several versions the heroes had Irish names while the villain always appeared to be a Jew. One rendition went:

> The first American soldier to kill a Jap was Michael Murphy
> The first American bomber to sink a battleship was Captain
>   Colin Kelly
> The first American to prove the effectiveness of a torpedo
>   was Captain John Bulkley
> The first American flier to bag a Jap plane was John O'Hare
> The first American Coast Guard to detect a German spy was
>   Ensign John Cullen
> The first American to be decorated by the President of the
>   United States for bravery was Lieutenant Patrick Powers
> The first American to get new tires was Abe Cohen.[59]

The names of the individuals in the numerous variants of the ditty changed, as did the specific attributes, but the heroes were always called O'Hara or Murphy or Flannegan while the first one to get a new set of tires or a defense contract would always be a Finkelstein, Goldstein, Lipshitz, or some other obviously Jewish person. The editors of many of the Army and Navy publications were insensitive to the insidious nature of the doggerel. Sometimes officers appreciated it, sometimes they did not, and it was an individual matter as to whether the editor was called down for its publication.[60]

The military also picked up on the persistent though erroneous rumor that Jews evaded the draft. A parody of the Marine Corps Hymn circulated in different branches of the armed services and reflected this sentiment:

> From the shores of Coney Island
> Looking eastward to the sea
> Stands a kosher air-raid warden
> Wearing a V for victory.
>
> And the gentle breezes fill the air
> With the hot dogs from Cohen's stand
> Only Christian boys are drafted
> From Coney Island's stands.
>
> Oh, we Jews are not afraid to say
> We'll stay home and give first aid
> Let the Christian saps go fight the Japs
> In the uniforms we made.
>
> If the Army or the Navy
> Ever gaze on heaven's scene
> They will find the Jews are selling shoes
> To the United States Marines.
>
> So it's onward into battle
> Let us send the Christian slobs
> When the war is done and victory won
> All us Jews will have their jobs.
>
> If your son is drafted don't complain
> When he sails across the pond
> For us Jews have made it possible
> All of us have bought a bond.
>
> And when peace has come to us again
> And we've licked that Hitler louse
> You will find a Jew a-ruling you
> In Washington's great White House.[61]

The poems and ditties of World War II displayed not only the inventiveness of the writers but also the depth of antagonism extant in the dominant culture. "A Christmas Poem" echoed the worst slurs about Jews' allegedly compulsive concern with material wealth. Its concluding lines reflected the writer's insensitivity to German atrocities and suggested as well widespread American agreement with Nazi policies.

"A Christmas Poem"

Oh little town of Bethlehem
For Christmas gifts see Abraham
At Christmas get a tree of pine
And buy the balls from Silverstein.
Down the chimney Santa drops
With toys bought from Isaac Blatz.
Ring out the old, ring in the New
And help fatten up some dirty Jew.
Hark! The herald angels sing,
Buy from Katz a diamond ring.
Our savior was born on Christmas day
Levy gives you six months to pay.
Peace on earth, good will to many
Hock your things with Uncle Benny.
Silent Night! Holy Night!
Damned if I don't think Hitler's right.[62]

Similarly, the "Parable of the Shekels" reflected almost all the grotesque ideas Americans harbored toward Jews. Allegedly written by a marine and circulated on military bases, "The Parables" read, in part,

I. And it came to pass that Adolph, Son of Abitch, persecuted the tribes of Judea and there was war.

II. And when the war was four years, many tribes came to the help of the Jews, but the Jews took up arms not.

III. They took up arms not lest in so doing they would take from their pockets their hand and it would come to pass that they would lose a scheckel.

IV. And the Gentiles came up in great multitudes from all the lands to fight for the Jews and the Jews lifted up their voices and sang "Onward Christian Soldiers." We will make the uniforms.

V. And the Jews lifted up their eyes and beheld a great opportunity and they said unto one another "the time has come when it is good to barter the junk for pieces of silver" and straightaway it was so.

VI. And they grieved not when a city was destroyed for when a city is destroyed there is junk and where there is junk there are Jews and where there are Jews there are money.[63]

It would be redundant to detail every verbal antisemitic thrust made by service personnel during World War II. Suffice to say the expressions reflected deep-seated feelings held by Americans in civilian life that could neither be countered nor controlled by military personnel even when they desired to do so. A student who examined the antisemitic doggerels noticed that they had the widest circulation in those camps where there seemed to be a nucleus of men who in civilian life had been inoculated with anti-Semitic propaganda and were merely continuing their civilian activities. The Boston area was a particularly sore spot. Frequently, the places where the doggerel had the widest circulation were induction centers and other points where men newly arrived in the army were stationed.[64]

Some of the nation's best educated men and women endorsed the tone expressed in the crude doggerels but couched their views in more acceptable language. In an unsigned article in the first issue of the *Journal of Clinical Psychology* (January 1945), the author acknowledged the dearth of trained personnel in his profession but warned against accepting too many graduate students from "one racial group." He explained that

> because of long racial experience with suffering and personality problems, certain groups of students show an unusual interest and propensity for psychological science. . . . While disclaiming racial intolerance, it nevertheless seems unwise to allow any one group to dominate or take over any clinical specialty as has occurred in several instances. The importance of clinical psychology is so great for the total population that the profession should not be exploited in the interests of any one group in such manner that the public acceptance of the whole program is jeopardized.

The writer's blatant bigotry (and none of his Jewish readers needed to be informed which "racial group" he was discussing) was attacked by several of the new journal's correspondents. In the next issue the author indicated that "the wording of this paragraph was admittedly unfortunate," but he did not retract the sentiment.[65]

That same year other well educated Americans expressed concern over the "dangers" of accepting unlimited numbers of dental students from "one racial strain." Dr. Harlan M. Horner, Secretary of the Council on Dental Education of the American Dental Association, surveyed the dental students at Columbia and New York University and concluded too many of those enrolled seemed to have come from "one racial strain." Although the President-elect of the American Dental Association repudiated Dr. Horner's conclusions, his observations were not totally ignored. In 1946 the dean of one of New York City's dental schools told a New York City Council investigating committee that he had been repeatedly "ordered" by the university's president to reduce the number of Jewish students. And just before the war ended in August 1945 Ernest M. Hopkins, retiring President of Dartmouth College, affirmed that "Dartmouth is a Christian college founded for the Christianization of its students."[66]

Given the intensity of feelings against them, it is not surprising that American Jews, victimized for more than a generation by escalating hostility, were psychologically battered. In January 1945 two different pieces appeared that reflected the hopelessness so many Jews felt. One, by Christian author Philip Wylie, described how every single Jew,

> from the day of his birth until he dies, lives behind a giant, psychological 8-ball—kept there by Gentiles: the hideous handicap of prejudice.... From morning to night, even in free America, no Jew has an exactly fair chance, and no Jew can tell when next the edge of undeserved, unexpected insult will cut him. ... Were you a Jew, even for a day, you would remember the effect of that hard, transparent wall all the rest of your life.[67]

The other lament, written in January 1945 by David Cohn, a Jew who felt pinioned by Gentile attitudes, appeared in the *Saturday Review of Literature.* In words that practically repeated Lewis Browne's dirge of 1939 and which also captured the prevailing reality for America, Cohn wrote:

> Anti-Semitism has not yet become a national menace to the United States. It is only at the stage where it is a personal tragedy that humiliates, frightens, and embitters individuals. It is obviously beyond the capabilities of the minority of Jews—the group against whom it is directed—to stop the progress of anti-Semitism here. They are too weak and too few. The greater part of the task of up-rooting anti-Semitism must be done— if at all—by the Gentile majority with the Jewish minority cooperating.[68]

The sense of impotence that so many Jews felt was reflected in the way the United States dealt with the problem of European refugees before and during the war. In several American circles there were strong sentiments for the government to help save some of Europe's Jews, and New York Senator Robert F. Wagner and Congresswoman Edith Rogers of Massachusetts proposed bringing in 20,000 European refugee children in 1939. Because public opposition was strong, President Roosevelt refused to endorse the idea and the bill failed to receive Congressional approval. Plans had also been proposed to the President for resettlement of refugee Jews in Alaska, in Latin America, and in sections of the United States. Roosevelt, however, was unwilling to alienate significant segments of the public or the Congress for what he considered a divisive issue. Only overwhelming encouragement from a broad segment of the population and/or Congress would move him to reconsider his stance.[69]

The charge that Roosevelt failed to do enough to save the Jews of Europe from extermination has been investigated frequently by historians since the late 1960s. Roosevelt always dealt with the Jews as an issue in the context of a broader agenda. His primary goal was to end the war as quickly as possible, his British allies were pressuring him to do nothing to help Jews escape from Europe, and he wanted to retain Congressional good will to insure support for a United Nations organization after the war. He also had to work with a recalcitrant Congress and was determined to avoid the errors made by Woodrow Wilson who tried, unsuccessfully, to force the League of Nations down the throats of a Republican-controlled Congress in 1919.[70] The country had elected an extremely conservative group of men and women to the House of Representatives in 1942 and shortly after it took office in January 1943 the President saw some of his New Deal domestic legislation scrapped. When he sought permission from Congress to permit him to allow people into the country who would not qualify for admission under existing legislation, the members emphatically rejected his request.[71]

By 1943 indisputable information had reached this country about Hitler's plan to exterminate all of Europe's Jews. That year the national Democratic and national Republican Clubs requested legislation authorizing temporary admission of all possible victims of Nazi persecution. Resolutions to this effect in both the House and the Senate died without coming to a vote. In November 1943, however, bipartisan declarations were adopted in both houses of Congress calling on the President to formulate and effectuate a plan of action to save the Jews in Europe. Responsible members of the Senate Foreign Relations Committee also urged Roosevelt to do something to thwart the Nazi slaughter of Jews. Then, at the beginning of 1944, long-time friend and Secretary of the Treasury, Henry Morgenthau, Jr., confronted Roosevelt with evidence of American indifference to the massacre of European Jews.[72]

Ever since his unsuccessful attempt to pack the Supreme Court in 1937, Roosevelt responded positively to divisive issues only after he had received sufficient pressure to act. Now he "succumbed" to Morgenthau's advice and set up the War Refugee Board to help save as many European Jews as possible. But, aware of national prejudices, the President tried to avoid controversy by having subordinates delete any mention of Jews in his statement establishing the new agency.[73] Then, in May 1944, a former law partner asked whether he was aware of the fact that "prominent American Christians" were planning to present him with a petition requesting that he set up refugee havens in the United States for homeless European Jews. Roosevelt responded that it sounded like a great idea but suggested that the petitioners insert the word "temporary" before "refugee havens."[74] They did and he responded positively. But the President would agree to establish only one temporary and separate camp for refugee Jews in an inaccessible part of New York state, from which they would be returned to Europe at the end of the war. Despite Roosevelt's caution in adopting any rescue policy, he was nevertheless attacked by reactionaries like Senator Robert Reynolds of North Carolina and Ohio Governor John W. Bricker for sanctioning the admission of people into this country who were not authorized to be here by existing legislation.[75]

Reynolds and Bricker spoke not only for themselves but also for millions of other Americans who feared the entry of more Jews into the

United States. Strangely enough, even many American Jews were ambivalent about admitting refugees. Millions of the Jews in the United States had either assimilated or were in the process of doing so, and they feared that an influx of coreligionists might result in deepened antisemitic sentiment and increased violence against them. Some Jews also refrained from trying to change the immigration laws because they correctly sensed that any revision would be downward. In 1944 and 1945 national conventions of the Daughters of the American Revolution, the American Legion, and the Veterans of Foreign Wars called for suspension of all immigration to the United States for periods of five to ten years after the war ended.[76] These groups obviously equated the term "refugees" with "Jews" and assumed that too many would be coming into the country.

Thus many Jews in the United States had a sense of desperation. While they were not generally the victims of physical attacks, except in urban, Irish, working-class neighborhoods, they knew that their opportunities were circumscribed and that most Gentiles were either apathetic about their troubles or hostile to them. They perceived as well a general unconcern about the fate of their European coreligionists. Several decades later Alvin Johnson, former President of the New School, recalled that "our academic world was as indifferent to what happened to German [Jewish] Professors as cows in pasture are to the one the butcher takes away."[77] Historian Barbara Tuchman was even harsher in her recollections. With few exceptions, she wrote, "the Gentile world ... would fundamentally have welcomed the final solution."[78] Many contemporary observers could have endorsed Tuchman's assessment. In July 1945, two months after the war had ended in Europe, an American reporter wrote, "One of the most extensive anti-Semitic campaigns this country has ever experienced has developed in the United States since the beginning of the war."[79]

In the 1970s and 1980s some Jewish critics condemned coreligionists who lived in the United States during World War II for failing to take more vigorous actions in regard to rescue.[80] These contemporary critics, knowing only how successful Jews and Jewish organizations have been in influencing legislators during the past fifty years, assume that Jews in this country have always been capable lobbyists. That is untrue. Jewish influence in Washington began with the New Deal, and Jews who did have the ears of important politicos, like Felix Frankfurter and Sam Rosenman, were often loathe to discuss issues that they defined as being of concern only to Jews. Ironically, during the years 1919 to 1939, when Jews had the least influence, they were widely perceived by antagonists to be enormously powerful. During the past half century, however, when they have had a greater impact on the course of national and international affairs, the general public has been less aware or involved with their activities.

Clearly, those who look back on the Holocaust and reevaluate what

might have been done by coreligionists in another era are using contemporary perceptions to judge individual and group behavior during
World War II. Yet when one recalls the experiences of American Jews
in the United States in the period between 1918 and 1945, a better question might be "What could they have done?" How could a disliked
and distrusted minority move a hostile majority to embrace a cause?
How could an "alien race" force a powerful president to ignore overwhelming public sentiment and attempt to save the lives of people most
Americans loathed? After all, as historian Henry Feingold has accurately pointed out, "Roosevelt's anti-refugee policy was based on a
broad popular consensus."[81] "The plain truth," historian David
Wyman has also recognized,

> is that many Americans were prejudiced against Jews and were unlikely
> to support measures to help them. Antisemitism had been a significant
> determinant of America's ungenerous response to the refugee plight
> before Pearl Harbor. During the war years, it became an important factor
> in the nation's reaction to the Holocaust.[82]

Contemporary observers acknowledged the same phenomenon. In
August 1944 newspaperwoman Dorothy Thompson wrote to her novelist friend Laura Z. Hobson that "anti-semitism will never be halted
in America because the majority of people like Jews. They don't. . . .
They don't like the race as a whole."[83] The British embassy in Washington informed Whitehall in 1945 that antisemitism in the United
States was "an ever present problem for every American Jew."[84] And
in June 1945 pollsters found that 58 percent of Americans, the highest
percentage ever recorded, believed that "Jews have too much power
in the United States."[85]

By the end of the war, leaders of various American Jewish groups
realized that they could no longer remain passive in the face of rising
hostility within the dominant culture. Silence, patriotism, and attempts
to behave in a manner pleasing to Gentiles failed to mitigate other
Americans' antisemitic feelings. The malevolence of the growing anti-
semitism in the 1930s had galvanized Jewish organizations to an extent
that no previous expressions of animosity had ever done. American
Jews recognized that they would have to retaliate against the growing
manifestations of bigotry and somehow curb, and hopefully reverse,
the vicious trend.[86]

Before World War II, two of the major defense associations of the
future, the American Jewish Congress and the Anti-Defamation League
of B'nai B'rith, were inexperienced, timid organizations without long-
range programs. They reacted rather than proposed. They wrote letters
protesting antisemitic actions, but basically emphasized patience and
caution.[87] Until the 1940s, moreover, they had neither the personnel

nor the funds to wage major campaigns against antisemitism. The American Jewish Congress, which began during World War I to promote the interests of a Jewish state in Palestine, and the Anti-Defamation League of B'nai B'rith, which was founded in 1913 by mid-western German Reform Jews desirous of promoting goodwill between Jews and gentiles, mushroomed in the early 1940s. The former constituted the voice of east European Jewish Zionists in the United States and directed most of its efforts toward helping coreligionists in eastern Europe in the 1930s. Members of the ADL wanted to be treated with greater respect and dignity; letter writing comprised most of their work before the advent of Hitler. In 1930 the ADL had a $30,000 budget and employed three people. By 1938 there were two hundred full-time workers; in 1941 the ADL budget totaled $741,490.[88]

The most prominent American Jewish defense group, the American Jewish Committee, founded in 1906 to promote the civil rights of Jews throughout the world, switched its focus after 1936 from promoting Jewish rights to fighting antisemitism. Its executive committee also discussed how members might counter the widespread whispering campaign that there was a Jewish-communist plot to overthrow the government. In 1939, therefore, the AJC embarked upon an intensive campaign to curb bigotry in America. Efforts were made to stress the un-democratic and un-American nature of antisemitism and to attribute growing hostility toward Jews to Nazi propaganda. By 1943, the other two major defense groups, the ADL and the American Jewish Congress, were also deeply involved in efforts to eradicate American bigotry as were the newly established local community relations agencies that began during the 1930s in the cities of Cincinnati, Cleveland, and Columbus, Ohio, Indianapolis, St. Louis, Philadelphia, Richmond, Los Angeles, and Houston.[89]

The strategy of the new defense agencies emphasized cooperation with Gentile groups in non-sectarian committees designed to aid refugees or to promote tolerance. Jews offered to provide the professional staffs and most of the financing if prominent Gentiles would grace the organizational letterheads. In addition, pamphlets and flyers were prepared correcting common misunderstandings about Jews. They distributed them under their own auspices as well as under the imprint of friendly Protestant groups like the American Friends Service Committee and the American Council of Voluntary Agencies for Foreign Services. Jewish agencies also tried to project a positive image of Jews without answering specific charges or trying to assail antisemitism head-on since by doing so they would be repeating and republicizing the original slurs.[90] Their goal was to depict Jews as "real" Americans with all the virtues and diversities one found throughout the land. In 1945 one author described how Jews wanted themselves presented. He wrote that when discussing the Jews it was important to stress their

bravery, their athletic ability, their "all-rightness," and their assimila-
tive qualities. They

> are Republicans and Democrats, like everybody else. A few of them are
> Communists—as are a few Irishmen, Italians, and a few everything else.
> They are divided many ways over their own Zionist question. Through
> thousands of years, armies of Jews have gone to battle against each
> other—as loyal citizens of warring nations. Human beings who profess
> one religion have, indeed, seldom been so divided as the Jews and sel-
> dom shared the blood of so many different peoples and nationalities.
> That is the way it really is.[91]

Jews who did not subscribe to the more subtle attempts to down-
play antisemitism, and who were also desperate to change the opinions
of the Gentiles, tried another tack. They stressed the contributions of
Jews to civilization, and many even attributed the source of hostility to
the new Jewish immigrants coming into the country. Large numbers
of American Jews wished, therefore, that the migration would cease.[92]
But, as one astute contemporary observer noted,

> the intensity of anti-Semitism bears little relationship to the degree of
> cultural or social assimilation Jewish communities had achieved. Nor
> have apologetics, the refutation of anti-Semitic propaganda, or a recital
> of Jewish contributions to civilization succeeded in stemming the tide of
> anti-Jewish prejudice.[93]

On the contrary, anything said and written about the Jews, either positive or negative, exacerbated expressed hostility toward them.

The intensification of antisemitic trends in the United States during
World War II combined with the knowledge that Hitler was extermi-
nating Europe's Jews drove leaders of the American Jewish community
toward an unprecedented agreement of almost perfect unity. In 1943
every major American Jewish organization except the American Jewish
Committee united under the umbrella organization of the American
Jewish Conference to work for the Zionist goal—the establishment of
a Jewish state in Palestine. But even as the American Jewish Conference
endorsed the Zionist program the American Council for Judaism was
created to offer an alternative position. Composed mostly of assimilated, Reform, middle- and upper-class American Jews, mostly of German descent who lived in the South and the West, the American Council for Judaism worked to undermine Zionism.

As most American Jews joined under the Zionist umbrella, they
also tried their hand at attacking antisemitism in a multifaceted way.
The annihilation of European Jewry convinced American Jewry that
they could no longer remain passive about disturbing echoes in the
United States. They had to combat them. As a result, Jewish organizations established the National Community Relations Advisory Council (NCRAC) in March 1944 as a coordinating and clearing unit for

domestic defense groups. Components of NCRAC included national agencies like the American Jewish Committee, the American Jewish Congress, the Anti-Defamation League of B'nai B'rith, the Jewish Labor Committee, the Jewish War Veterans, the Union of American Hebrew Congregations, and about twenty local community relations agencies from around the country. The NCRAC established committees on legislative information, community consultation, antisemitism, discrimination in educational institutions, intercultural and scientific research projects, and organized its various members to work together for the common good. Along with the local and national groups, the NCRAC probed, inquired, and conducted research into the causes of antisemitism and the means of combatting it. To lay the base for a more tolerant society and reduce prejudice against Jews, individual agencies like the American Jewish Committee, the American Jewish Congress, and the Anti-Defamation League also carried on broad educational, social action, and community programs.

These Jewish agencies assumed the position that antisemitism, along with other forms of bigotry, threatened American democracy and therefore posed a problem for the entire society, not merely one segment of it. Thus Jewish leaders decided to employ the social sciences to analyze prejudice and develop a cure for it, mobilize public opinion against intolerance, and utilize the courts and legislative bodies to eliminate those discriminatory practices that could be controlled by law. Laws alone would not eradicate prejudice but they would indicate that the government opposed bigotry and discrimination.[94]

Thus as the war neared its end, American Jewish groups were ready, as they had never been before, to deal aggressively and forthrightly with the problem of antisemitism in the United States. The sober recognition of the Holocaust galvanized them. They were not only scared but determined. Their fortitude coincided with renewed economic prosperity, a willingness on the part of some returning veterans and other Americans to curb the worst manifestations of ethnic prejudice, and the steps taken by a new President, Harry S Truman, to promote greater equality and respect for all the people.

8

The Tide Ebbs
(1945–1969)

A remarkable metamorphosis occurred in the United States in the two decades following the end of World War II. After more than half a century of increasing animosity toward the Jews, antisemitism in the United States suddenly began to decline. Given what apparently had been an escalation of bigotry during the war, the transformation in public rhetoric and behavior afterwards was so swift that careful observers were at a loss to explain the changes. Antisemitism, of course, did not disappear between 1945 and 1969 but as a less socially acceptable aspect of American life it waned significantly. *Life* magazine commented on how much ground antisemites had lost in a December 1, 1947, editorial; less than a month later the Ohio JCRC issued a report stating that antisemitism was "really weak and not recently articulated." And the ADL remarked about how little antisemitism appeared in the various election campaigns of 1948 compared to the prejduices manifested in 1940 and 1944. Laypersons and scholars both acknowledged the change by 1955. *Look* magazine believed that Hitler had made antisemitism disreputable, cracks about Jews that used to be taken for granted were fewer in number, and for Jews, antisemitism in America was downgraded from a problem to an irritant. In 1956 sociologist Herbert Gans wrote that "organized anti-Semitism is already confined mostly to the lunatic fringe"; several months later historian Oscar

Handlin observed that the position of the Jews in the United States was far better than even the most optimistic individuals could have hoped for two decades earlier.[1]

Beyond these impressions, almost every survey of antisemitism taken after 1946 showed its decline. Polls commissioned by the American Jewish Committee in 1950, 1951, 1953, and 1954 revealed increasingly favorable attitudes toward Jews.[2] To the question "Have you heard any criticism or talk against the Jews in the last six months?" the responses in different years highlighted the decline, with the most significant change occurring from 1946 to 1951.

	Yes	No		Yes	No
1940	46%	52%	1953	21%	79%
1942	52	44	1954	14	86
1944	60	37	1955	13	87
1946	64	34	1956	11	89
1950	24	75	1957	16	84
1951	16	84	1959	12	88[3]

It seemed apparent that a new atmosphere replaced wartime apprehensions about minorities. For the first time since the 1920s Americans felt optimistic about the future and confident about both their own and society's economic well-being. Husbands, fathers, brothers, and sons rejoined their families free of the fear of death and enthusiastic about a job "well done." A peacetime focused on building and growth resulted in increased numbers of marriages, a prolific birth rate, and a demand for new housing. The normal aspects of everyday living consumed the attention of young adults and they had less time to seek ethnic scapegoats. In a sense, the postwar period, unlike that after World War I, offered wonderful social and economic opportunities and more Americans concentrated on those rather than on the alleged culpabilities of the minorities in their midst.

A third change occurred for some servicemen and women during World War II but only became apparent afterwards. Many Americans resolved to do something about bigotry in America. Thousands of veterans had lived with intolerance in the armed forces and hoped to reform the prejudiced nation that they had left behind. The war brought together a variety of peoples from all over the United States, and from foreign cultures as well. In August 1945, just before the Japanese surrendered, the Army publication *Yank* asked soldiers, "What changes would you like to see made in post-war America?" A majority of GIs interviewed agreed that "above everything else, the need for wiping out racial and religious discrimination" was their major hope. Sociologist E. Digby Baltzell also acknowledged that he and many of his fellow naval officers returned home from World War II "far less willing to tolerate the traditional, often dehumanizing, ethnic snobberies of our pre-war years."[4]

To what extent the knowledge of Hitler's slaughter of six million Jews contributed to the desire to curb bigotry is impossible to state but after 1945 millions of Christian Americans became more cautious in expressing negative reactions to Jews.[5] Moreover, the decline in prejudicial remarks aimed at Jews was part of a general lessening in all American bigotry. Catholics, Asians, and African Americans were also the beneficiaries after 1945 of a less hostile atmosphere toward people of different backgrounds.[6]

If tribute for the waning of prejudice is due both to Jews and Gentiles, the single most pointed example came from the White House. Roosevelt's successor, Harry S Truman, selected blue-ribbon panels to investigate fair employment practices, higher education, and civil rights in the United States, and these commissions invariably discovered widespread racism in this country. The President's Commission on Higher Education specifically recommended that college and university applications remove all questions pertaining to religion, color, race, and national origin, while the Committee on Civil Rights reported that in many northern institutions "enrollment of Jewish students seems never to exceed certain fixed points." Truman ended racial segregation in the armed forces and proposed a civil rights bill in 1948. A year later he told the National Conference of Christians and Jews:

> I have called for legislation to protect the rights of all . . . citizens and to assure their equal participation in American life, and to reduce discrimination based on prejudice.[7]

It was clear that prejudice was no longer acceptable when several Hollywood films of the postwar era analyzed the consequences of bigotry. Singer Frank Sinatra starred in *The House I Live In*, a 1945 short that made a special plea for racial and religious tolerance. Later in the decade *Home of the Brave* and *Pinky* dealt with the poignancy of the victims of racism. Not since D. W. Griffith's *Intolerance* had Hollywood producers directly acknowledged and criticized the intense intolerance that existed in the United States.[8]

It was within the context of movies exposing racial and ethnic prejudice that Hollywood in 1947 produced two major films on American antisemitism: RKO's *Crossfire*, based on Richard Brooks' novel *The Brick Foxhole*, and *Gentleman's Agreement*, taken from the novel of the same name by Laura Z. Hobson. The producers of these films, Dore Schary of *Crossfire*, and Darryl F. Zanuck, of *Gentleman's Agreement*, received public kudos for examining the existence and consequences of antisemitism. Each of the films was nominated by the Motion Picture Academy for the year's best picture; *Gentleman's Agreement* won the Oscar.

The honor was merited; of the two works, *Gentleman's Agreement* made the greater impact. The book and movie unmasked those who tried to hide their bigotry under the guise of gentility and conformity. The picture reached millions at a time when adult Americans regarded the movies as more important than television, and the novel remained a best seller throughout most of 1947. Elliot Cohen of *Commentary* did not like the book but called the movie "a moving, thought-provoking film, which dramatically brings home the question of antisemitism to precisely those people whose insight is most needed—decent, average Americans." Critic John Mason Brown was excited that both films dared "to speak publicly for the first time on a subject which movie goers have long spoken privately."[9]

Although the two films highlighted the insidiousness of bigotry, some of those closely associated with their preparation missed the message. One assistant to *Crossfire* director Edward Dmytryk thought the movie pointless. "There's no anti-Semitism in America," the underling remarked, "if there were, why is all the money in America controlled by Jewish bankers?" A stagehand on the set of *Gentleman's Agreement* told writer Moss Hart that the film had a wonderful moral. Gregory Peck had played a white Christian masquerading as a Jew and therefore the stagehand decided that he would never again be rude to a Jew "because he might turn out to be a Gentile."[10]

Other observers fortunately showed greater insight. Editorials, articles, and cartoons exposing the vile nature of prejudice appeared in magazines like *America, Collier's, Life, Look,* and *Seventeen.* Bruce Bliven wrote an eight-part series in *The New Republic* detailing almost every facet of antisemitism in America. Carey McWilliams argued in *A Mask for Privilege* that antisemitism continued because the elite in society needed it to buttress their positions. Two years later the first two volumes of *Studies in Prejudice,* sponsored by the American Jewish Committee, were published and three more volumes followed in 1950. Contemporaries regarded the series highly but later generations of scholars questioned many of its assertions. The most well known, *The Authoritarian Personality* by Theodor W. Adorno and associates, examined the personality traits of more than 2,000 subjects and concluded, erroneously, that antisemitism is most likely to occur among rigid and repressed individuals who were most susceptible to fascist teachings.[11]

The avalanche of movies, studies, and critical assessments of antisemitism both aroused and reflected national concern. Labor, educational, and women's groups inaugurated programs for the eradication of antisemitism and intolerance, while many individuals of all faiths devoted themselves to activities designed to promote racial and religious harmony. Well-directed civil rights groups like the NAACP, Congress of Racial Equality (CORE), the Japanese American Citizens League, and others also contributed to the evolving changes that were taking place. But no ethnic group had defense agencies as well organized and as well financed as those of the Jews. The American Jewish Committee, the American Jewish Congress, the Anti-Defamation League, and NCRAC stood head and shoulders above all others in conveying their message. Jewish defense organizations used community relations, academic analyses, and legal action to lobby for significant alterations in employment, education, and immigration policies. Attempts were also made to curb discrimination in housing, resorts, and social clubs.

In a remarkably civil division of labor one organization usually assumed leadership for each issue. Stephen Wise and the American Jewish Congress led the way in efforts to abolish legal discrimination in higher education. The American Jewish Committee, assisted primarily by Lessing Rosenwald of the American Council for Judaism, stood in the forefront of the movement to alter existing immigration statutes so that more European Jewish refugees might be welcomed. Other Jewish agencies participated in these endeavors, but to obtain the desired goal, they cooperated without anyone trying to receive too much of the credit; nor did groups split hairs on strategy with the "losers" going their own way.

Employment was one of the first problems addressed since for more than two decades job opportunities for Jews had been narrowing. During the 1920s, when discriminatory practices first became a significant problem for Jews, many immigrants and their children accepted the situation as normal—they had known the same thing in Europe. During the Great Depression, when organized antisemitism increased, Jewish defense agencies focused instead on the rise of Hitler and the refugee exodus from Germany. By the 1940s, however, the time seemed ripe for domestic action. Hundreds of thousands of American-born Jews were reaching maturity and, along with those returning home from the war, needed education, jobs, and housing. Resistance to antisemitism in the job market therefore became imperative.

The Jews' initial victories came in the field of employment discrimination. In New York state, the Ives-Quinn bill went into effect in July 1945 and prohibited racial and religious bias in the selection of employees by nonsectarian organizations. By 1949 fair employment practices legislation existed in several other states including New Jersey, Massachusetts, Connecticut, Rhode Island, Oregon, and Washington. Colorado passed its bill in 1951 and strengthened it in 1955.[12]

Passage of laws did not lead to immediate change. In 1945 Bess Myerson became the first Jew crowned "Miss America" but because of her religion received fewer bookings and invitations from American industrial firms than the Gentile women who previously held the title. An early 1946 analysis indicated that employment discrimination had increased. In cities like Denver and St. Louis no discernable change appeared while in Cincinnati the terms "Gentile" and "Gentiles Only" still adorned classified advertisements. But in states where nondiscriminatory legislation passed, the groundwork was laid for fairer hiring practices. A perusal of Manhattan commercial employment agencies in December 1949 showed that 64.2 percent of them accepted discriminatory job orders compared to 88.4 percent that had done so in December, 1946.[13]

Job discrimination subsided in the decades following the end of World War II but measuring the degree proves elusive. Employment agencies in Arizona, Georgia, Illinois, Missouri, Florida, Louisiana, Nebraska, California, and New York, among others, still accepted orders specifying gender, racial, or religious preferences. Without seeing agency records, however, it is impossible to know how many or what percentage of their total requests were discriminatory. During a two-week period in January 1951 the California state employment agencies in the Los Angeles area received notice of 5,535 job openings, about two-thirds of which indicated that members of one or more ethnic groups would not be acceptable. Only 17 percent of the total orders included Jews among those deemed undesirable.[14]

Anecdotal tales and files of Jewish defense agencies suggest that there were fewer racial or religious specifications on the part of prospective employers during the 1950s than there had been in the pre-World War II period. In Chicago, only one in four employers discriminated compared to approximately 90 percent that did so before World War II. The relatively few complaints to Jewish defense agencies from employment applicants also leads one to conclude that discrimination was becoming a negligible problem. In Minnesota, for example, Jewish community agencies received so few letters about inability to find work that by the 1960s they stopped making it a priority item on their social action agendas.[15]

On the other hand, in some professions and industries barriers remained high. In the 1950s and 1960s dental licensing boards in Arizona and New Mexico were accused of discriminating against Jews. In Los Alamos, New Mexico, in 1950, of the six dentists who tested for a license, only the three Jews failed. Well into the 1960s national surveys disclosed that Jews constituted only about 1 percent of high-level executives in banking, heavy industry, communications, transportation, and public utilities. In the late 1950s, of 2,000 management people in U.S. Steel, only nine or ten were Jewish; in Philadelphia's six major banks, 6 of 1,028 employees may have been Jews. In advertising, aside from Jewish-owned firms, Jews succeeded primarily in creative rather than client work.[16]

In the field of law doors opened slowly. Experience with some of the New Deal bureaus and departments provided some Jews with unique qualifications that paved the way for academic appointments. Participation in agencies during the 1930s like the Labor Relations Board or the Federal Communications Commission meant that these individuals had special assets to offer law schools interested in developing new specialties based on recently developed governmental policies.[17]

Jews also began to enter prominent law offices to a greater extent than in previous decades. A number of elite firms recruited only those law school graduates whose families were listed in the *Social Register*.[18] On the other hand, a survey of the 1951 law graduates of the University of Chicago, Yale, Harvard, and Columbia disclosed that 20 percent of the Jews, 40 percent of the Catholics, and 54 percent of the Protestants received offers from one of their top five choices of employers.[19] That 20 percent of the Jews secured positions with desired firms was a vast improvement over what was the norm before World War II. But that percentage fell considerably short of the 54 percent that their Protestant classmates achieved and showed how far American society would have to go to eradicate deeply entrenched views.

For the years 1951 to 1962, not only did a higher percentage of Christians than Jews get the law jobs they sought, but, on average, Jews earned less than their Gentile peers. Moreover, many firms willing to hire Jews limited their numbers because they did not want to be known as "Jewish firms." Nevertheless, by the 1960s Jewish lawyers who had already won acceptance moved easily in elite professional, corporate, and governmental circles, and the doors seemed to be opening wider.[20]

If employment barriers came down at the highest levels, social acceptance proved more difficult to obtain. Throughout the country important business contacts were made in the exclusive men's business and social clubs; the more selective these were the less likely they were to accept Jewish members. Before World War II almost all of them barred Jews and by the early 1960s about two-thirds, still a hefty percentage, continued doing so. Many of those clubs that accepted Jews welcomed only the most distinguished. Among the elite of these restricted organizations were the Links, Knickerbocker, and Union clubs in New York City, the Philadelphia Club, the Detroit Club, the Pacific Union in San Francisco, the Duquesne Club in Pittsburgh, the Chicago Club, the Piedmont Driving Club in Atlanta, Eagle Lake in Houston, the Milwaukee Club, the St. Louis Country Club, and the Commonwealth Club of Richmond, Virginia.[21]

Members and guests at these clubs networked with one another. Exclusion from membership, therefore, resulted in limited economic opportunities because many firms depended on these social contacts to obtain clients. Social antisemitism among America's business leaders contributed to employment bias, and kept Jews out of executive positions in insurance, banking, automobile manufacturing, utilities, oil, steel, and heavy industry. On the other hand, firms that would not employ Jews in policy-making capacities were sometimes less reluctant to do so in areas of research, actuarial, or creative slots where intellectual prowess rather than social pedigree served as the driving criterion.[22]

Social snobbery in the past had always led to residential restrictions but polls conducted after World War II showed a declining percentage of Americans who would be upset by having a Jewish neighbor. In 1950 69 percent of Americans claimed that they would have no objection to living near Jews; by 1954 80 percent shared that opinion. Nevertheless, as one moved up the social scale the discomfort with Jews nearby increased. Metropolitan Baltimore, Chicago, Cleveland, Detroit, Los Angeles, Philadelphia, and New York contained some of the most restricted residential areas. (San Francisco, on the other hand, was notably free of this type of bigotry.) Bronxville, a high-income suburb of New York City, kept Jews out until after the New York State Commission for Human Rights intervened in 1962.[23] A 1961 survey of Bloomfield Hills, Michigan, home to some of the highest level employees in the automobile industry, concluded: "None of the hundred top auto executives is a Jew and no Jewish families live in Bloomfield Hills proper."[24]

Throughout the 1950s and into the 1960s, homeowners in prosperous Grosse Pointe, Michigan, calculatedly enforced discriminatory policies; *Time* magazine described the system as "Grosse Pointe's Gross Points."[25] The Grosse Pointe Property Owners Association hired private detectives to secretly investigate prospective buyers. After the detectives gathered the requisite data, the Grosse Pointe Property Owners Association distributed points based on the fairness or swarthiness of the individual's complexion, occupation of the breadwinner, ancestry, education, reputation, style of dress, religion, and ethnic background. Ethnicity determined the number of points necessary for acceptance.

Group	Points Needed for Acceptance
WASPS	50
Poles	55
Greeks	65
Italians	75
Jews	85

Jews, however, received fewer credits than Gentiles for categories such as education, reputation, income, etc., and therefore needed a larger number of favorable attributes to reach the mandated 85 points. Thus a research physician, a direct descendant through his mother of Carter Braxton who had signed the Declaration of Independence, was refused approval because having a Jewish father necessitated that he make up for this "deficiency" with additional points in the other categories. When a former resident tried to sell his house and was refused permission to do so he sued the association thereby bringing national and international attention to the practice. After lengthy hearings the state of Michigan ordered the Grosse Pointe Property Association to end its point system.[26]

As housing restrictions eased, and as people began to travel longer distances in the United States and abroad, upscale resorts began to reexamine their guest policies. The ubiquitous Holiday Inns, founded in 1952 and built alongside virtually every new turnpike and interstate highway, became an agent for social change. As these new recreational and hotel facilities proliferated, vacationers had wider options and many older establishments could no longer afford the luxury of turning Jews away.[27] Moreover, forward looking groups and travel agencies began to frown on bigotry, while discriminatory hotel and resort practices attracted unwelcome notoriety. In 1950 forty travel agents in New York City, whose agencies accounted for more than half of all the travel bookings in the city, pledged to avoid recommending restricted resorts and hotels. The 1954 decision by the National Association of Attorneys General to switch its annual meeting from Arizona to West Virginia because Phoenix's Camelback Inn refused to accept Jewish guests became, in retrospect, a watershed event. By 1957, only a quarter of all American resorts still maintained restrictions against Jews, and five years later such practices were all but extinct.[28]

Perhaps more significant were changes in education. As noted earlier, most prestigious American colleges and universities had imposed quotas on Jews during the 1920s thereby severely restricting their educational and employment opportunities. On the undergraduate level the barrier was humiliating but in the professional schools it was debilitating. Good liberal arts educations were available at many colleges, but denial of admission to medical and law schools meant the curtailment of career opportunities. From 1920 to 1940 the percentage of Jews in Columbia University's College of Physicians and Surgeons fell from 46.94 percent to 6.45 percent. The percentage of CCNY graduates admitted to any medical school dropped from 58 percent to 15 percent, and not one graduate of either Hunter or Brooklyn Colleges in New York City entered an American medical school until 1946. A revealing *American Mercury* analysis of the subject in October 1945 pointed out that although three of every four non-Jewish applicants were accepted to a medical school in the United States, only one of four Jews was successful. A similar type of discrimination existed in law schools. In June 1935 25.8 percent of all American law school students were Jewish; in 1946 the proportion was 11.1 percent.[29]

Because more than half of America's Jews, and some of the nation's most prestigious law and medical schools—like Cornell, Columbia, and New York University—were in New York state, Jewish organizations, especially the American Jewish Congress, campaigned for a law barring nonsectarian colleges and universities from discriminating against student applicants on the basis of race, religion, or national origin. Similar efforts were made in New York City.[30]

Attempts to enact legislation in the city proved successful. The New York City Council investigated the community's private universities, found widespread bias, and discovered that conditions had worsened since the previous decade. Within a week of the report's publication, Mayor William O'Dwyer announced that he favored withholding public funds from institutions of higher education that discriminated. On March 12, 1946, the City Council adopted a resolution endorsing the principle that tax exemptions should not be granted to nonsectarian colleges and universities that employed racial or religious criteria in selecting students for admission.[31]

On the state level a movement began not only to bar discrimination but for the establishment of a state university. Jewish groups believed that anti-bias legislation alone would not meet the needs of those seeking entrance to colleges and universities. Veterans were already crowding the state's institutions of higher education and adolescents about to graduate from high school faced keen competition for existing places. In 1946, Governor Thomas E. Dewey appointed a committee to explore the possibilities of establishing a state university. Two years later the group recommended that it be done, and that any enabling legislation prohibit racial and religious discrimination in student selection.[32] An anti-discrimination law was defeated in 1947 because of strong opposition from the Catholic Church. The Church allegedly opposed the bill on the grounds that it infringed on the parents' right to educate their children and was "formed after a Communistic pattern." In 1948, however, a similar measure passed after the governor, who anticipated receiving the Republican presidential nomination that summer, backed it strongly. Within a year Massachusetts and New Jersey also approved legislation barring discrimination in nonsectarian schools.[33]

By the end of 1949 the Chancellor of the New York State Board of Regents declared that virtually every one of the state's nondenominational colleges and universities had eliminated questions from their applications for student admissions that dealt with race, color, or religion. The following year a national conference on higher education adopted a resolution branding the quota system, already on the wane, as "undesirable and undemocratic."[34] There were other signs, moreover, that schools were being influenced by the changing tone in society. In 1948 Jewish undergraduates at Yale protested being automatically assigned other Jews as roommates. As soon as they raised the issue, the practice ceased.[35]

Just as antisemitism declined among the nation's elite colleges, so too did it diminish in professional schools. An examination of application blanks for 39 major medical schools in the United States in 1946 disclosed that all wanted to know the applicants' religious preferences or affiliations; 10 asked for the religion of the applicants' parents; 15 requested information on parents' race; 11 inquired whether the family name had ever been changed. The necessity for the New York anti-discrimination law may be seen from the statistics of individuals applying to Cornell Medical School in 1950. There were 69 Protestant applicants of whom 24.6 percent were accepted; 59 Catholics, of whom 15.3 percent were accepted; and 144 Jews, of whom 4.9 percent were accepted. But because of the nondiscrimination law, the changing atmosphere in the country, and the low birth rates of the 1930s, the percentage of Jewish medical students in New York state leaped from about 15 percent in 1948 to approximately 50 percent by the middle of the next decade.[36]

An exceptional case that aroused concern occurred in Georgia. From 1948 to 1958 15.4 percent of the non-Jewish, compared with 64.6 percent of the Jewish, students at Emory University Dental School in Atlanta failed or had to repeat a course. This was a unique situation since Jewish students throughout the country were generally as successful academically, if not more so, than others in their classes. After investigations and complaints by the Anti-Defamation League the disproportionate failure rate of Jewish students ceased.[37]

Another aspect of academic life that underwent reexamination was college fraternity policies. In November 1949 the Dartmouth College Interfraternity Council urged similar groups in twelve northeastern states to support "a real step" to eliminate restrictive clauses in fraternity charters that barred members because of race or religion. Two weeks later the National Interfraternity Conference, meeting in Washington, voted 36–3 to recommend that view to all its constituents. Within the next two years changes in fraternity rules regarding minority members were apparent in places like Amherst, Brown, Harvard, the Massachusetts Institute of Technology, New York University, Ohio State, Rutgers, Swarthmore, and the Universities of Chicago, Colorado, Connecticut, Massachusetts, Michigan, Minnesota, and Washington. The editors of *The Christian Century* observed, however,

> a fact worth noting is that most of the agitation for this lowering of racial and religious bars has come from the student members of the fraternities, and that the changes have been brought about in the face of determined opposition from older alumni. Most of the older alumni, we expect, are "pillars" in their local churches.[38]

The caustic remark about the "pillars" of the church rang true. Although many individual Christians devoted themselves to promoting the cause of tolerance, there was less evidence of institutional concern. In 1949 a former editor of the *Catholic Worker,* a liberal publication, wrote that he had "never heard a single sermon on the evil of antisemitism, before, during or since the war. And an experienced priest tells me that he can recall no instance of hearing the sin mentioned in the confessional." Thus, the writer continued, "when one considers the enormity of the Nazi blood-carnival, it never ceases to be amazing that the Christian conscience has been so slightly disturbed." A similar indifference existed among many Protestant denominations as well.[39]

The lack of interest in the victims of the Holocaust carried over after the war into a general indifference for the welfare of those who survived it. At first, there seemed little hope that American immigration legislation might be altered to help bring some of the survivors to the United States. Before World War II attempts to alter the quota system to aid European refugees failed since neither the President, the Congress, nor the nation saw any pressure to bring Nazism's victims to the United States. After the war, not even the leaders of organized Jewry made any public pronouncements about the necessity of bringing some of the war's survivors to the United States. Mail to Congress explicitly stated widespread attitudes. "The word 'refugee,'" a New York woman wrote, "is synonymous with *Jew,* and the *latter* is synonymous with Red!" A Texan informed his Senator that lowering the barriers would result "in a flood of Jews coming to the United States. We have too many Jews." "If it was left to me," another New Yorker observed in 1950, "I'd admit all of the D[isplaced] P[ersons] except the Jewish D. P.'s. I'd let in the Catholics, the Protestants, and those in between—but no more Jewish boys." When on April 1, 1947, Representative William Stratton of Illinois proposed a bill to allow 400,000 displaced persons into the United States above and beyond quota limits, a political advisor told him that he was risking his political career. "Most people feel that [the displaced persons bill] will make it possible for several thousands of Jews to be moved into this country. . . . Now, nobody in Illinois, outside of the Jews, wants any more Jews in this country."[40] With perspectives like these, and with the Zionists clamoring for a Jewish homeland, it is understandable why Jews and other Americans urged Great Britain to open the gates of Palestine to the Jewish survivors of the Holocaust. The British government, however, would not act quickly, and after endless delays and inquiries, President Truman announced that he would recommend to the Congress that it make special provisions for accepting some of Europe's refugees.[41]

The President's announcement was greeted coolly in Congress but it galvanized leaders of the American Jewish Committee and American Council for Judaism. Their campaign began in November 1946 and by June 1948 Congress passed a bill to bring 205,000 displaced persons to the United States. Unfortunately, the wording of the law discriminated against Jews and favored former Nazis. However, the methods used by the Displaced Persons Commission that implemented the law tended to undermine Congressional intent since more Jews came into the United States than Congress had anticipated.[42]

The passage of the Displaced Persons Act marked the last legislative victory for American antisemites. And, significantly, it opened new doors for American tolerance. The principle of aiding refugees was written into American law as a way of getting around fixed immigration quotas. Two years later, in 1950, a majority in Congress beat back efforts to retain the discriminatory provisions of the 1948 DP act, and produced new legislation that eliminated all real and implied antisemitic barriers written into the previous bill.[43] In subsequent years legislation of this type would be passed again to help a variety of people including anti-Communist Cubans, Vietnamese, and Russian Jews.[44]

Thus, by the end of 1950 the major areas of overt antisemitism that concerned Jews before and during World War II were already in the process of change. An obvious sign of decreased hostility was that *The New York Times* devoted considerably less space to the subject in 1949 and 1950 than it had in 1945 and 1946. Similarly, the anitsemitic tidbits that appeared in *The Chicago Jewish Forum*'s periodic columns of "Notes" or "Letters" from the West Coast, Washington, New York, and the Midwest showed up less frequently by the early 1950s. Jewish organizations documented the drop in antisemitic activity, and the ADL found a marked decrease in antisemitism in the 1948 political campaigns as compared to those earlier in the decade. It attributed "the comparative absence of antisemitic activity in the . . . [1948] campaigns . . . to the fact that there is economic prosperity and no national or international problems which are sharply and deeply dividing Americans."[45] And a search through the Nearprint Files at the American Jewish Archives reveals fewer materials on antisemitism for the years after 1952 compared to the previous twenty years.

Examples of friendlier attitudes were found across the American landscape. In Minneapolis, characterized by Carey McWilliams in 1946 as the nation's most antisemitic city, the Jewish Community Relations Council recorded fewer complaints about literature, rumors, and verbal attacks hostile to Jews. In Los Angeles, Screen Actors Guild President Ronald Reagan resigned from the Lakeside Country Club when it refused to accept a Jewish member.[46] The weight of the federal government's efforts, symbolized by President Truman's stands, no doubt also contributed to altering public attitudes. So, too, did the movies. Finally, and this is the most difficult aspect to measure or to quantify, the establishment of the state of Israel in 1948 gave many Jews renewed pride. Perhaps self-confidence and assurance in some intangible ways changed both their behavior and the reaction of non-Jews toward them?

Yet despite its decline, antisemitism remained a factor in American life. Fifty-seven antisemitic groups still existed in the United States in 1950 although thirty-five others had folded within the previous year. Moreover, even without organized associations to spur assaults, Jewish youths were set on by hoodlums, synagogues were desecrated, and high school youths wore swastika pins. During the early 1950s Boston teenagers engaged in physical attacks against Jewish adolescents reminiscent of violence in that city a decade earlier. And, as in the earlier period, policemen arrested the Jewish boys while allowing the others to flee.[47] How many of these incidents transpired is difficult to ascertain but they happened often enough for Jewish agencies to take cognizance of them.

Antisemitic vandalism was also a recurrent phenomenon. Such

behavior predated World War II but sociologists first began analyzing it after the war. They discovered that the synagogue desecrations and swastika daubings were almost always done by emotionally disturbed youths between the ages of 10 and 18 who had grown up in antisemitic homes without warmth or affection and who had been treated brutally by parents or the adults with whom they lived. The post-1950 perpetrators, often victims of discrimination themselves, generally came from lower income homes and had usually grown up with no father or father substitute.[48] Teenage violence against Jews was periodically exposed by the media, and whenever this happened some anguished Jews predicted a renewed wave of antisemitic fervor. After a particularly ugly period in 1959–1960, when more than thirty Jewish institutions in the New York area had been daubed with swastikas and racist slogans, where telephone threats had been made, and where the windows of Jewish businesses had been broken, Judge Justine Wise Polier opined that the "the acts might be a forerunner of another international or political conspiracy of neo-nazis." She thought it "fatuous and dangerous" to ignore such crimes as part of a passing craze, and, in a sense, Judge Polier was right.[49] Although the daubings and window-breaking activities did not lead to a growth in antisemitic fervor throughout the United States, this kind of recurrent vandalism, waxing and waning periodically without any particular rhyme or reason, serves as a constant reminder of the unpredictability of aggressive antisemitic incidents.

Instead of focusing on teenage vandalism, however, more commonly the media discussed the crackpots and extremists who tried to resuscitate the virulent antisemitism of the 1930s and the war years. The messages from the fanatics came in speeches, rallies, flyers, pamphlets, and periodical articles but they appealed to smaller audiences.[50] Gerald L. K. Smith continued to link Zionism with communism, sent a copy of *The Protocols of the Elders of Zion* to every member of Congress in 1947, sold copies of "The International Jew" in bulk in 1964, and then serialized it in *The Cross and the Flag* in 1966. Reverend Roy L. Laurin of the Church of the Open Door in Los Angeles asked in 1949, "What Shall We Do with the Jews?" If Jews were colonized in Africa, Laurin asserted, they would "soon be selling refrigerators and fur coats in every African Kraal." Throughout the nation one could read any number of antisemitic hate sheets like *The American Nationalist, The Cross and the Flag, Thunderbolt, The Defender,* and *Common Sense,* whose weekly circulation of 90,000 copies made it the leading antisemitic publication in the country in 1963. In *The Jewish Problem,* published by the Huron Church News in 1957, readers were told that "Jews are at the very root of all troubles, all conflicts, all revolts of the modern world," while in December 1959 and February 1960 two antisemitic articles appeared in *The American Mercury* calling Jews "the masters of the whiskey trade in the United States," linking Zionism with communism, and accusing the Anti-Defamation League of being the "secret policing organization operating on behalf of the Zionist conspiracy in the U. S. A. against the people." George Rockwell founded the American Nazi Party in 1959, and was perhaps the best known antisemite in the country until 1964, when he was shot by one of his own storm troopers.[51] Yet, unlike previous decades, the various accusations and charges against Jews attracted only a few malcontents and extremists and generated no significant antisemitic movement.

Even the association of Jews with communism failed to ignite popular imagination. When the House Un-American Activities Committee held hearings in a number of cities, and tried to tie Jewish names with communist activities, its inquiries sparked no antisemitic flames.[52] The nationally covered trial and appeals of Ethel and Julius Rosenberg, two Jews from New York's Lower East Side who were convicted of treason in 1950, frightened some Jewish defense agency professionals who thought that the publicity might once again trigger virulent tirades of antisemitism. They were wrong. Polls indicated that the Rosenberg case did not significantly alter a declining trend in antisemitic expressions.[53]

But people living through transition periods are not always aware of the significance of changing times and attitudes, especially when they are not the beneficiaries of altered views. While the staffs of defense agencies continually monitoring trends and opinions appreciated the transformations occurring, many well-to-do Jews still experienced the pain of exclusion. Those not hired for jobs for which they were well qualified or invited to become members of social clubs or sit on boards of museums and important civic institutions were keenly aware of the still existing antagonisms.

Moreover, scholarly studies and reminiscences of personal experiences of individuals in small and middle-sized towns in Ohio, Indiana, and Illinois during the 1950s reflected lingering apprehensions among upper-middle-class Jews. Speaking with these people, and reading their writings, reveals how uncomfortable many of them felt in their communities. A few showed inordinate concern with behaving properly and not provoking or offending Christians. Some even saw themselves as "ambassadors" to the Gentile world. In Muncie, Indiana, Jews "sensed" a tone of antisemitism, although one man acknowledged that "in every dealing I have ever had in this town, I have been treated by the *goyim* with utmost respect." Nevertheless, he concluded, "I know as well as you do what some of the bums are thinking." In a northern Ohio town, Jews who were promised anonymity acknowledged how uneasy they felt. A young dentist who had known prejudice during his childhood, professional training, and years in the Navy, candidly stated, "I don't act natural when I'm with non-Jews. . . . I am worried about whether they are going to say anything antagonistic against us Jews. This happens when they have a few drinks."[54]

A woman whose husband was on the faculty of a small college in

rural Ohio from 1956 to 1959 could not abide the attitudes and stereotypical thoughts of townspeople, which she interpreted as a form of
antisemitism. "Christian," she wrote, was used to describe every
human virtue. Although she made it clear that the family came from
Chicago, she was frequently asked, "Have you ever been west of New
York before coming here?" and "Are you going home to New York for
vacation?" It was also assumed that she and members of both her and
her husband's families were wealthy, intellectual, and against segregation. The woman found it galling to be told, as well, that other "foreign" families, like east European displaced persons, lived in the town,
while her seven year old was asked by a playmate, "Why do you call
yourself an American when you're a Jew?"[55]

Stereotypes die slowly. In an exercise on understanding different
religions at Fairfax High School in Los Angeles, three boys, Protestant,
Catholic, and Jewish, were on a panel talking to fellow students when
the Jew was asked, "Why do the Jews have all the money?" "Gee
whiz," he responded, "if that were only true! I'm a Jew and I've got
about a dollar and thirty cents and that's all I have to my name, so you
can't put me in that group." Then the Jewish student asked a question
of the questioner: "I'm curious, why do you think it's only the Jews
that have money?" The boy in the audience replied, "Well, look at
Rockefeller and Carnegie and Ford. They were all millionaires." The
Jewish panelist followed up with, "Why do you think those men you
have just named were Jews?" And the boy in the audience quickly
answered, " 'cause they were all millionaires.' "[56]

Such thoughts were not restricted to untutored high school students. In 1952 after a series of robberies occurred at several Jewish
homes in Kansas City, Missouri, a Gentile at a Rotary Club gathering
remarked that "it is strange that these robberies all occur in the homes
of Jews. The robbers would starve to death if they tried to rob Gentiles."
This statement, intended as a joke, was obviously based on deeply held
and possibly unconscious assumptions. When the commentator saw
the expression on the face of the man to whom he told the tale, however, he immediately apologized and claimed that he had meant no
harm.[57] The apology, and the desire of the joke teller not to be rude,
highlighted what so many Jews felt: while polls measured a subsidence
in antisemitic speech and behavior since the end of World War II, anti-
semitism had not been erased from people's hearts and minds. Anti-
semitism, they believed, had simply become unfashionable and unpopular, and had therefore gone underground.[58]

Will Herberg, who in 1955 published one of the seminal volumes
of the decade, *Protestant, Catholic and Jew,* captured the essence of the
latent prejudice when he wrote that large numbers of Americans were
still antisemitic but they did not act on it. Antisemitism had ceased
being respectable. Despite the "deep and far-reaching" changes that
had taken place in American society since the end of World War II,

Herberg felt obligated to acknowledge the reality of people's feelings. Judaism had been elevated to one of the three major American faiths in public utterances but many Gentiles were still uncomfortable with Jews in their midst. The junior executive suburbanite, Herberg acknowledged, could not understand the Jew who would not go along being bland and accepting the decisions of those in authority above him. Jewish restiveness, self-assertiveness, and independence made many other Americans uncomfortable in their presence.[59]

It was difficult for many Jews to accept the fact that people of different cultural backgrounds were generally more comfortable with others of similar values and behavior. Somehow, characteristics often associated with Jews of east European heritage, such as biting wit, intellectual accomplishment, independence of thought, outspoken women, and physically demonstrative behavior were not attributes that put other Americans at ease. Although such characteristics did not create antisemitism, they reinforced the views of parochial Christians who held negative preconceptions about Jews.

In fact, scholarly researchers in every section of the nation discovered a high correlation between antisemitism and devout Christianity. A 1940s study of students in the American West found that the more religious an individual was, the less likely he or she would have favorable attitudes toward African Americans and Jews. A similar test among New York University students in the early 1950s revealed that those belonging to Catholic and Protestant religious clubs "regardless of their religious affiliation, were found to be significantly more antisemitic than those not affiliated with religious groups."[60] And a major sociological study released in the 1960s again emphasized that the most religious people were also the most antisemitic. In their book *Christian Beliefs and Anti-Semitism* Charles Y. Glock, director of the University of California Survey Research Center, and his co-author Rodney Stark found that a majority of American church members were prejudiced against the Jews, that 60 percent of the Protestants and 40 percent of the Catholics thought that "the Jews can never be forgiven for what they did to Jesus until they accept Him as the True Savior," and that a majority of observant Christians still considered Jews materialistic, dishonest, and vulgar. In their conclusion the authors observed that they "were entirely unprepared to find these old religious traditions so potent and so widespread in modern society."[61]

Glock and Stark should not have been so surprised by their findings. After examining a large number of Protestant textbooks from the 1950s, Bernhard E. Olson, whose book *Faith and Prejudice* received widespread attention, concluded that in general Jews were "used as a convenient whipping boy for human ills and failings." Olson's findings did not lead to immediate changes. In 1968 an issue of The Church of the Nazarene's *Bible School Journal* called Judaism "a religion that is inadequate, based on ignorance and prejudice, non-satisfying, and providing no solutions to the problems of sin and death." The previous year an issue of *Gate*, published by the Lutheran Church—Missouri Synod, included the remark that "The Pharisees represented the one primary evil against which Jesus preached. He represents the one faith they could not tolerate and they conspired to kill him."[62]

Gerald Strober, of the American Jewish Committee, and Father John Pawlikowski of the Catholic Church, scrutinized hundreds of parochial school books in use during the 1960s and their conclusions echoed those of Olson. Both Strober and Pawlikowski argued in their works that Christian children learned a contempt for Jews from the reading materials provided to them. In these texts Jews were repeatedly chastised for the sufferings of Christ, for their rejection of Him, and for their "wicked ways." Father Pawlikowski noted that in the major religious texts in the Catholic schools Jews were not only the most frequently named non-Catholic group and but were also associated with negative connotations going back to the first century.[63] Statements in Catholic textbooks included phrases like, "the worst deed of the Jewish people [was] the murder of the Messiah," "in spite of the countless graces given to the chosen people, they voluntarily blinded themselves to Christ's teaching," "the Jew as a nation refused to accept Christ, and since his time they have been wanderers on the earth without a temple, or a sacrifice, and without the Messiah."[64] Not until the late 1960s, however, were serious attempts made to alter the perceptions children garnered from parochial school textbooks.

The major improvement in Catholic-Jewish relations started after the accession of Pope John XXIII in October 1958. This Pope tried to alter the harshness that Catholics fostered toward people of other faiths. In an attempt to build bridges with other religions, he called for a new Vatican Council to reexamine church doctrines, received representatives of the American Jewish Committee and the Anti-Defamation League, and eliminated antisemitic pejoratives from prayers. John XXIII recognized that Church teachings were partially responsible for antisemitism and he attempted to correct what he regarded as past sins and errors.[65] Admittedly it was difficult for Pope John to change almost two thousand years of church teachings and practices but he tried. In 1962 millions of Roman Catholics, especially in eastern Europe, still believed that hating Jews was "a legitimate expression of loyalty to the faith." In Poland, for instance, with few exceptions, "a 'good Catholic' is as a matter of course a Jew hater."[66] John endeavored to change this attitude.

One of Pope John's failures was an attempt to eliminate stories of Jewish blood libel that appeared on tapestries and stained glass windows in several European churches.[67] He also specifically ordered a cessation of the annual pilgrimages of children and adults to see the tapestry of the alleged ritual murder of Andrew of Rinn at Judenstein, Austria, but his edict was ignored. A 1950 protest had led the then

Cardinal Innitzer to state that it was wrong to suggest that "Jews had never done such things." Thus John's efforts in 1961 resulted in no more than a plaque being hung in the church at Rinn that stated "it is clear that the event [the murder of Andrew] had nothing to do with the Jewish people."[68] Folklorist Alan Dundes, commenting on the reluctance of the Austrians at Judenstein to give up their beliefs on this issue, stated, "It is not easy to legislate folklore out of existence."[69]

Despite his failure to eliminate the belief of some Catholics in the blood libel myth, Pope John XXIII inspired many people, both Catholics and non-Catholics alike, with the openness of his mind. Throughout his brief reign, which ended in June 1963, he caused the reexamination and alteration of many Catholic doctrines. A posthumous triumph was the 1965 schema "On the Relation of the Church to Non-Christian Religions," promulgated during the reign of Pope Paul VI. The text clearly stated that

> what happened to Christ in His Passion cannot be attributed to all Jews, without distinction, then alive, nor to the Jews of today. . . . Although the Church is the new people of God, the Jews should not be presented as rejected by God and accursed, as if this follows from Holy Scriptures.[70]

Newly enunciated church attitudes had great impact in the United States. Until 1963, opposition from the Catholic Church influenced members of the New York state legislature to oppose changes in state laws that would have allowed stores to remain open on Sundays.[71] A new Catholic mind set led to the withdrawal of such opposition as Catholics now sought closer relations with Jews. The American hierarchy created several permanent bodies to promote good will with Jews including the Bishops' Subcommission on Catholic-Jewish Relations and the Commission on Catholic Education and Ecumenism. Officials at the Catholic Theological Society of America invited Rabbi Mark Tanenbaum of the American Jewish Committee to deliver papers and give his interpretation of various Vatican Council declarations, and joint conferences were held with Jews at Catholic seminaries. In Cincinnati, the archdiocese cosponsored *Pacem in Terris* programs with the local Jewish Community Relations Council, the Hamilton Circle Daughters of Isabella held joint Christmas and Hannukah parties, and Jews were invited to luncheons to meet with Vatican representatives. Additionally, Father Victor J. Donovan was often referred to as "Rabbi Donovan" because of his strong involvement in trying to better Catholic-Jewish relations in the city. That the good will efforts succeeded was shown by the end of the 1960s when polls of Catholic college students revealed a lessening in antisemitic hostility.[72]

Success with Catholics inspired Jewish groups to meet also with Protestant fundamentalists to promote interfaith understanding. The first time that leaders representing major centers of evangelical Protestantism met with Jewish spokesmen for significant and sustained discussions occurred in the New York City headquarters of the American Jewish Committee in February 1966. Jews attending the meeting later called it a "truly historic occasion"; the gathering was a breakthrough on which future dialogues and exchanges would be based. No revolutionary changes in perceptions immediately occurred—Protestants in attendance still saw Jews "as abstractions and . . . as objects for conversion"—but fundamentalist representatives acknowledged their general ignorance of the Jewish people and Jewish communities in the United States.[73]

Continuing dialogues with both Catholics and Protestants inspired many Jews to believe that significant alterations might occur in Christians' perceptions of Jews. Jewish defense agencies recognized that Christian religious teachings had contributed to the hostile attitudes that Gentiles harbored toward Jews and saw all exchanges between groups as salutary. The new friendliness might permit joint attacks in those areas where antisemitic thoughts and feelings developed. For example, some Jewish community relations agencies had for decades been working against public school practices that allowed for the recital of prayers, continued celebrations of Christmas, and released time of students whose parents wished them to receive additional religious instruction outside of school during school hours. In these endeavors, Jewish agencies did not have the unanimous support of fellow Jews, many of whom either enjoyed the Christmas celebrations themselves, had no objection to released time or the saying of prayers, or who wanted to avoid controversies that were likely to be harmful to Jewish-Christian relations.[74] But with the new ecumenism it seemed possible that some Protestant and Catholic groups might work with them in eliminating practices that some Jews regarded as threatening or offensive.

That Jewish defense agencies even embarked on challenges to established school practices that favored the Christian faith suggests how much more secure Jews felt by the late 1940s and 1950s.[75] Instead of merely being content with the lack of attacks on, or challenges to, their own faith, they took a forceful position in opposition to long accepted American traditions like prayers in public schools. Ultimately their endeavors succeeded as the U.S. Supreme Court in *Engel v. Vitale* (1962) and *Abington Township School District v. Schempp* (1963) agreed with their contentions that prayers in public schools violated the concept of the separation of church and state.[76]

In sum, within two decades of the end of World War II near revolutionary changes occurred in public policies and attitudes toward Jews in the United States. The first big drop in antisemitic attitudes occurred in the five years after the end of World War II and, after continued gradual decline in the 1950s, another steep fall appeared in the early 1960s.[77] Commenting on the enormous transformation in Christian behavior toward American Jews, Benjamin R. Epstein, national director of the Anti-Defamation League, later observed:

> I think we will all agree that the two decades that followed World War II—the years 1945 to 1965—were a period of tremendous progress for the American Jewish community. It was a "golden age" in which our people took a vast leap forward, and achieved a greater degree of economic and political security, and a broader social acceptance than had ever been known by any Jewish community since the Dispersion.[78]

Scholars and laypersons alike agreed. Antisemitism seemed to be declining at a rapid pace while prejudices toward other groups also waned to varying degrees. Although still in its beginning stages, a greater tolerance toward, and acceptance of, different minorities could be noted throughout the country. John F. Kennedy, a Catholic, had been elected President of the United States in 1960 and in 1964 the Republican Party chose Barry Goldwater, whose father had been born Jewish, and William Miller, a Catholic, to head its national ticket. The Civil Rights movement also awakened Americans to the consequences of bigotry. Opportunities for minority groups had never been more abundant. Jews found less difficulty getting their sons into prep schools and their daughters onto debutante lists, charity boards invited some of the more distinguished Jews in the community to serve with them, and private parties of the movers and shakers in America included both Jewish hosts and Jewish guests. Antisemitism in the United States had so dwindled as an issue for the Jewish community relations agencies that in 1965 the plenary session report of NCRAC did not even include the topic in its table of contents.[79]

Statistics reinforced the impressions of those people who saw antisemitism as a vanishing problem. Formal policies of fraternity discrimination decreased from 61 national groups before World War II, to 25 in 1948, 10 in 1955, and 2 in 1960. Whereas 63 percent of Americans thought Jews had objectionable qualities in 1940, only 22 percent subscribed to that point of view in 1962. If 57 percent of the population stated that they would definitely not marry a Jew in 1950, only 37 percent clung to that opinion in 1962. To the question "Would you vote for a Jew for President?" 51 percent of southerners in 1958 responded "no" while only 30 percent of them felt the same way in 1965; 28 percent of non-southerners responded negatively in 1958 but that figure declined to 13 percent in 1965.[80]

Yet "the remarkable advance of tolerance, decency and justice," as John Higham put it 1966,[81] did not eliminate all American antisemitism. Overt antisemitism went out of fashion, accompanied in many cases by a sense of bad conscience. But antisemitism remained as a generally unexpressed undercurrent that surfaced at unexpected moments in different parts of the United States, a force that had to be dealt with when it erupted.[82] One such occurrence resulted from the strongly entrenched upper-class bigotry that appeared in preparation for an event in Scarsdale, a suburb of New York City, in December 1960. A debutante had selected two escorts for her coming-out party at the Scarsdale Golf Club. One of them, a recent convert to Episcopalianism, was considered Jewish by club officials and deemed unacceptable on that account. They informed the young woman of their decision which so upset her that she cancelled her debut. After the incident, Father George F. Kempell, Jr., rector of the Episcopal Church of St. James the Less denounced those individuals in his congregation who had participated in the decision as "morally reprehensible," and announced that "anyone who has in any way, by word or thought or deed, acquiesced with this position of the Scarsdale Golf Club is no longer welcome to receive holy communion at this altar." Three of the members of the parish had been on the golf club committee that made the decision and after Rector Kempell's admonition they and their friends forced his resignation. If any members of the church agreed with Kempell, their views received no publicity.[83]

Oscar Cohen, then research director at the Anti-Defamation League, was familiar with the Scarsdale incident and several others of that ilk throughout the country. He argued in 1966 that "anti-Semitism is a much more serious problem than we had previously thought. In particular, its potential has grave possibilities." Cohen believed that the NCRAC and local Jewish Community Relations Committees paid only "minimal attention" to the problem of antisemitism. "The big problems reside in the attitudes of the American people generally and in the discriminatory barriers which still exist against Jews, particularly in the economic and social fields."[84]

Several months after Cohen's pronouncement, and less than two years after *Time* reported that "anti-Semitism is at an all time low and publicly out of fashion,"[85] the single most shocking incident of post-World War II American antisemitism occurred on February 9, 1967, in Wayne, New Jersey, a town thirty miles west of New York City. Newt Miller, a member of the school board and a fiscal conservative concerned about school costs, opposed the proposed school budget and feared the consequences of an upcoming school board election where two of the candidates were Jewish. He telephoned a reporter and told her:

> Most Jewish people are liberals, especially when it comes to education. If [Richard] Kraus and [Jack] Mandell are elected . . . and Fred Lafer [a Jew already on the Board] is in for two more years, there's a three to six vote. It would take only two more votes for a majority and Wayne would be in real financial trouble.

Then, in a total nonsequitur, he continued,

Two more votes and we lose what is left of Christ in our Christmas celebrations in our schools. Think of it.[86]

The reporter with whom Miller spoke recognized the explosive nature of the comment and asked Miller to provide it in writing. The next day the story made headlines throughout the country, brought a mass of newspaper reporters and television cameramen to Wayne, and turned the 45,000-person community into a national goldfish bowl.[87]

Immediate attention focused on Miller. The day the story broke he called a press conference and apologized to any Jewish people whom he might have offended, but he did not retract the statement. That evening, February 9, the school board voted 8–1 to censure Miller and asked him to resign. He refused the request. As each member in the majority voted in favor of the censure applause broke out from the audience of townspeople who attended the open meeting. It seemed as if Miller had no public supporters. The Governor of New Jersey, the Mayor of Wayne, and several Wayne residents and clergymen deplored his comment. One community leader observed, "Open political anti-Semitism simply is no longer tolerated in American life." The people who spoke publicly were sure that Kraus and Mandell, both of whom had been favored to win earlier, would emerge victorious at the election the following week.[88]

The prognosticators were wrong. On February 15, a front-page headline in *The New York Times* read: "Two Jews Lose by Wide Margin." In an unprecedentedly large turnout for a school board election, the Wayne voters "delivered what was interpreted as a vote of confidence in Newt Miller."[89] Townspeople defeated the school budget by a count of 6,973 to 2,655. Yet in a five-man race for people to serve on the board, two of the non-Jewish budget supporters whom Miller also opposed won, and one of them received the largest number of votes. Another budget-supporting victor, Richard Davis, had been opposed by Miller because all his children attended parochial schools. The two Jews trailed badly. The votes for the five of them went:

David Celiri	7,488
James McLaughlin	6,741
Richard Davis	6,009
Richard Kraus	3,207
Jack Mandell	3,173[90]

After seeing the vote, the President of the school board commented: "Two fine men were defeated completely because they are Jewish. The vote is so obvious."[91] *The Christian Century* dubbed the incident a "patent case of anti-Semitism" and compared Newt Miller with Hitler: "Both stir up the beast slumbering in the gentile breast." *The New York Times* editorialized: "There is no reason to believe that what happened in Wayne could not happen in a thousand other communities."[92]

An election that most observers saw as a reflection of antisemitism in Wayne appeared differently to Rutgers University sociology professor Charles Herbert Stember who a year earlier had edited the widely acclaimed *Jews in the Mind of America.* Stember thought it "unwarranted" to conclude that bigotry affected the outcome. Aside from Miller's "gratuitous remarks about Christ and Christmas," the professor wrote that his other observations "do not strike one as an unfair or an untrue accusation and is not an attack on Jews for what they are but for their views on education." Stember blamed Jews for blowing up the affair by crying "anti-Semitism," thereby encouraging conservative voters who might otherwise have remained at home to come out and vote against the budget. Neither the professor's letter, nor his concluding remark, that "if the citizens of Wayne, or any other town, do not care about quality of education for their children, then they had better not elect Jews to the school board" led to any organized response from the Jewish community.[93]

Stember's observations seemed to ignore salient features about the election and the nature of the Wayne community. Four of the five candidates in the school board election, including two of the three victors, favored the proposed budget (McLaughlin opposed it). Also, the differences in votes between the three victorious Gentiles and the two defeated Jews was huge. Wayne counted about 25,000 Protestants and "15,000 Catholics who have not been affected by the spirit of ecumenism,"[94] among its 45,000 residents. In the 1920s the community had had a Ku Klux Klan chapter, and in the 1930s it housed a branch of the German American Bund whose members wore Nazi armbands. In the 1950s and 1960s there were also isolated and minor antisemitic outbreaks, and a small neo-Nazi Swastika club had existed in the high school.[95] The community also included two segregated country clubs and until 1965 two of the more exclusive residential areas in Wayne restricted purchases of homes in the area only to those applicants who won approval from residential screening committees. (The New Jersey Supreme Court outlawed that practice as soon as it was brought before them in a test case.[96]) "In short," as *The Economist* noted, "Wayne Township is a typical, small American town."[97] And while leaders and elected officials might deplore antisemitism, many denizens obviously harbored some doubts about Jews and alleged Jewish characteristics.

After the election Miller acknowledged that he had received over 200 letters about his statement with 98 percent of them favoring his position. The new school board also gave him a unanimous vote of confidence.[98] Community inhabitants who chose not to reveal their names seemed pleased about the outcome. One told a reporter, "No comment, but you can see I'm smiling." Another, like Professor Stember, seemingly confused as to who or what caused the commotion, said Miller was hard working but the press "terribly exaggerated" his statement.[99] A third woman also did not perceive Newt Miller's responsibility for what had happened. She commented:

It was a terrible shame that the Jews caused all this trouble and tried to make it a religious campaign. This is absolutely the first trouble they caused here in the eight years I've been here. It was a real shame because it hurt the town so. Why in the world did they do it? Can it really have been worth bringing in all those reporters and ruining the town just to get themselves on the school board? They probably would have been elected if they had just been decent. People really get sore about them trying to stir everything up.[100]

The events in Wayne did not mitigate the fact that antisemitism declined considerably in America after World War II. But they indicated that hostility toward Jews, no matter what pollsters found or laws prohibited, remained just beneath the surface in many typical American communities and could erupt without the slightest advance warning. And, as in Wayne, where an elected official deplored adding two Jews to the school board because it might lead to the elimination of "what is left of Christ in our Christmas celebrations," so too were there thousands of other American communities where people did not understand what constituted antisemitism and who was responsible for bringing on outbursts of bigotry. Thus, large contingents of Jews feared that the heralded postwar gains were illusions that could be destroyed by an episode like the one in Wayne. The real issue was whether American Jews had confidence in the new reality of greater tolerance or whether they considered it a temporary interlude before the noxious weed of antisemitism erupted anew in some unexpected place.

In the South, however, the roles and positions of minorities were more clearly delineated. In that region all things proceeded smoothly when everyone observed the rules, but a U.S. Supreme Court decision in 1954 led to chaotic changes over the next two decades. It is to the Civil Rights revolution, and its impact on southern Jewry, that we now turn our attention.

9

Antisemitism and
Jewish Anxieties
in the South
(1865–1980s)

While antisemitism declined throughout the North and West after World War II, manifestations of it increased dramatically in the South after the 1954 U.S. Supreme Court decision to outlaw segregation in public education. Despite the Supreme Court's decree in *Brown* v. *Board of Education*, a majority of southerners refused to comply with the ruling, remained steadfast in their opposition to integration, and attacked those who submitted to the new law of the land. Although most Jews opposed segregation, they were afraid to take a public stand that differed with the dominant regional values and sentiments of their communities.[1] Such reluctance was based on sound fears since poll data for the 1950s and 1960s showed the South to be the most antisemitic region in the country.[2] How southern Jews accommodated themselves to an evil system and how the larger community of whites and blacks in the South worked out these issues is a topic hardly discussed in American analyses of desegregation.

Since the Civil War the South sustained its institutions by warding off the views and opinions of people who questioned established val-

175

ues. Overwhelmingly committed to a Protestant fundamentalist faith, believing absolutely in the inferiority of African Americans and the need for racial segregation, southerners insulated themselves from new ideas. In general, people in the South expected everyone to conform to accepted thought and practices and, if they did not, they were subject not only to community wrath but ostracism. Physical attack itself was not unlikely since historically the South has been more violent than other regions, and people there have not hesitated to fight for their beliefs.[3]

An almost unique physical assault on Jews occurred when English-born James Joseph Sylvester arrived at the University of Virginia in November 1841 to begin teaching mathematics. Never before had a Jew been appointed to teach a non-Semitic subject at an American university. Local newspapers, led by the *Watchman of the South*, the voice of the Presbyterian Church, were appalled that the position had been offered to an English Jew. "Pure morality based on Christian principles may . . . be required of those who are to form the minds and morals of our young men," the *Watchman of the South* editorialized.[4] The University Board of Governors also hired a "Hungarian Papist" that same year, which also did not sit well with Virginians and complicated the reception that both new men received. Sylvester, only 27 years old when he assumed his teaching responsibilities, won no friends in Virginia with his strong antislavery position and short temper. At one point students physically assaulted the new professor. After an altercation with a student in the middle of the academic year Sylvester fled back to England in 1842 because he thought that he had killed an undergraduate. "It is indisputable," Philip Alexander Bruce, the historian of the University of Virginia, later wrote, "that there was a provincial prejudice against him as a Jew and a foreigner."[5]

What happened to Sylvester presaged future attacks on Jews who were thought to violate southern values. It was always possible, of course, for members of the local community elite to sidestep tradition and take an unpopular stand on a specific issue; such a stance by outsiders, however, was not tolerated. Eli Evans, a native North Carolinian, stated the point emphatically in 1973: no one crossed the Southerner in his own land.[6]

The minority of Jews in the region were always suspect no matter how long they had been there. In 1860 fewer than 30,000 Jews lived below the Mason-Dixon line with about 8,000 of them in Louisiana, 3,000 in South Carolina, and lesser numbers in the other southern states.[7] Even if their forebearers had arrived in colonial times, Jews could not escape the fact that most southerners believed their ancestors had killed Jesus; forgiveness was possible only if they accepted the truthfulness of Christian teachings. Southerners also regarded Jews, in the words of Mississippi-born journalist Jack Nelson, "as aliens in a land of uncompromising, militant, fundamentalist Protestantism." At times Jews were also seen as citizens of questionable loyalty and as "homeless" individuals who have always had a "dual allegiance" outside their countries of birth or residence.[8] A Duke University Professor of Education, without any basis in fact, wrote in 1972 that the Jew in America "has been considerably more homeless than other immigrants, for he dreams of an eventual return to a Jewish state, a promised land."[9]

Growing up in an environment where a majority considered them a people apart left Jews with a pervasive sense of anxiety. Nevertheless, they managed to survive, thrive, and interact pleasantly with their Christian neighbors. Regional customs dictated a surface cordiality between and among Caucasians. Therefore, by always watching themselves and never engaging in activities that might antagonize members of the dominant society, Jews hoped to avoid public discomfort. Most southern Jews were obsessed with the need to accommodate to community mores, and they never forgot that they were members of a minority group in an area where outsiders were considered deviant. Even Jewish attitudes toward religious devotion followed southern patterns. They were much more likely than coreligionists in other sections of the country to affiliate with a synagogue, their temples often resembled churches, they worshipped regularly, some ministers or reverends (almost never "rabbis") conducted services on Sunday mornings, and the overwhelming majority of Jewish youth received religious instruction.[10]

The almost total acculturation of Jews in the South allowed them to maintain a facile cordiality with Gentiles, even though just beneath the surface lay a bed of prejudice ever ready to label Jews as Christ-killers and Shylocks. After the Civil War virulent prejudice toward Jews declined and Jews participated in the rebuilding of southern society during the Gilded Age. Without any striking reasons for antipathy, however, individual Jews were accepted if they did nothing to disturb community tranquility and by the late nineteenth century they enjoyed prosperity and the respect of their neighbors in places like Selma and Montgomery, Alabama; Austin, Dallas, Houston, and Galveston, Texas; Little Rock, Arkansas, and Baton Rouge, Louisiana.[11]

But even when conditions seemed placid Jews could never feel completely comfortable.[12] Amazingly, it was a Christian, Zebulon Vance, the former Governor of North Carolina, who brought attention to southern antisemitism. Vance had several friends among the Jewish merchants in Statesville, North Carolina, and knew the community's attitudes toward them. He also knew that not until 1868 did the state constitution allow Jews to vote and that twice before, in 1835 and 1858, specific opportunities to give Jews the vote had failed to achieve majority support in the state legislature.[13] Inspired by what he believed to be unjustified bigotry, sometime between 1868 and 1874 Vance prepared a talk defending "the persecuted and despised Jew." Then, beginning with a speech in Baltimore in February 1874, he embarked on a fifteen-year crusade to make others more tolerant of Jews. Vance, an already accomplished speaker on the Chatauqua circuit, delivered this periodically altered address entitled "The Scattered Nation" in myriad communities throughout the country. It was also read from hundreds of pulpits in the South and printed in practically every southern journal and newspaper.[14] In the talk, Vance argued that the Jew should be judged

> as we judge other men—by his merits. And above all, let us cease the abominable injustice of holding the class responsible for the sins of the individual. We apply this test to no other people.

"All manner of crimes," he continued, "including perjury, cheating, and overreaching in trade, are unhesitatingly attributed to the Jews, generally by their rivals in trade. Yet somehow they are rarely proven to the satisfaction of even Gentile judges and juries."[15] That Vance prospered by repeating this speech indicates the strength of antisemitism in America, and especially in the South. Many Americans fretted over the existence of this feeling, and people who came to hear Vance at the Chatauqua speeches probably wanted to learn more about Jews and antisemitism even if they had no intention of acting on the knowledge thus acquired.

While Vance moved about the country advocating tolerance and understanding of Jews, hostile comments and vicious incidents in the South increased Jewish anxiety. Questions arose in 1873 about whether Jews qualified as witnesses in a Rome, Georgia, courtroom and periodically pieces appeared in the *South Georgia Times* (Valdosta) referring to Jews as people who "worshipped before the god of money," "engaged in under-handed business practices," and " 'wedded family business' in marriage more than men and women." Alabama's Senator John T. Morgan denigrated a political opponent as a "Jew dog" in 1878 without suffering any loss of status or position, and throughout the late 1880s and early 1890s invidious ethnic slurs about Jews as wife-beaters, bankers, and Shylocks cropped up in southern newspapers. During these same years roving bands of night riders, victimized by the severe economic depression of the era, attacked small Jewish stores in Louisiana, Mississippi, and Alabama.[16]

Southern traditions of violent actions encouraged brutality, and gangs of men, especially among the lower classes, often took justice into their own hands. During the agrarian crises of the 1880s and the Populist era in the early 1890s, many of the region's suffering farmers viewed themselves as victims of a Jewish conspiracy.[17] Making note of the conventional beliefs as if they were true, Mark Twain wrote that in the cotton states after the Civil War

> the simple and ignorant negroes made the crops for the white planter on shares. The Jew came down in force, set up shop on the plantation, sup-

plied all the negro's wants on credit, and at the end of the season was proprietor of the negro's share of the present crop and of part of his share of the next one. Before long, the whites detested the Jew, and it is doubtful if the negro loved him.[18]

Racial and religious prejudices had the greatest appeal to rural fundamentalists in the region and contributed to their violent behavior.[19] Most southerners were Baptists and Methodists, poorly educated, and wholly devoted to their Protestant faiths. Their ministers, equally ill educated, led their flocks both in religious devotions and bigoted attitudes. It was not unknown for the clergy to direct lynching parties and condemn those who refused to accept the accuracy of New Testament teachings. And it was not uncommon either for those who felt especially exploited economically to explode periodically and attack Jewish storekeepers.

Knowledgeable politicians across the South understood the beliefs and attitudes of their constituents. Men like Ben Tillman of South Carolina, James K. Vardaman of Mississippi, and Jeff Davis of Arkansas shaped their rhetoric and won popular support by exploiting existing prejudices. In Georgia, Tom Watson avoided racist attacks in the 1890s but achieved the greatest renown of his career two decades later when his newspaper, *The Jeffersonian*, published the vilest antisemitic diatribes ever seen in an American periodical to that point.[20]

In 1895, *The American Hebrew*, observing the recurrent savagery visited on Jewish storekeepers in the different southern states, thought antisemitism strongest in Mississippi and Louisiana. Jewish merchants there, and in other rural areas, had extended credit and held debtors in bondage, or so it seemed to many citizens of America's isolated hamlets. In some places Jews also owned a considerable amount of real property. One man in Lawrence County, Mississippi, complained in 1893 that "the accursed Jews and others own two thirds of our land." Coincidentally, Lawrence County was also a community in which Jews were frequently terrorized and attacked.[21]

Even in urban areas Jews could not escape antisemitism. Two men from Alabama, Joseph Proskauer of Mobile and Irving M. Engel, who hailed from Birmingham, later recalled the discomfort they felt growing up as Jews. Proskauer was often beaten up for being a Jew and a "Christ-killer," while Engel remembered being "constantly" teased for those reasons. Despite these occurrences Proskauer and Engel remembered pleasant childhoods although they were always conscious of being Jews in a land of Gentiles.[22]

Jewish adults in the turn-of-the-century urban South were often treated without respect. In 1897 solicitors for Atlanta's *Jewish Sentiment* found several merchants who refused to advertise in the paper because they wanted no Jewish patrons. Three years later a telling editorial acknowledged:

no one knows better than publishers of Jewish papers how wide-spread
is this prejudice; but these publishers do not and will not tell what they
know of the smooth talking Jew-haters, because it would but widen the
breech already existent. . . . But there is prejudice against the Jews everywhere.[23]

As the decade ended, the mayoralty campaign in Savannah witnessed
accusations from Christian pulpits that Jews were guilty of foisting
gambling and other vices on the city.[24] However, even when Jews in
Atlanta tried to support a citywide anti-vice crusade in 1906, leaders
of a Protestant-led reform group indicated that they could participate
only as long as they were willing to "follow the banner of Christ."[25]

To protect themselves, therefore, southern Jewry sought rabbis, or
reverends, who would represent them in Christian society and prevent
untoward events from impacting on Jewish survival in the region. During the first half of the twentieth century several men performed the
function well, including Edward N. Calisch of Richmond, Henry M.
Cohen of Galveston, and David A. Marx of Atlanta. A number of the
religious leaders wore clerical garb (black suits with white collars encir-
cling their necks), preferred being addressed as "Dr.," "Reverend," or
"Minister," and steered their flocks into appropriate conduct and
demeanor vis-à-vis members of the dominant community.[26]

Among those who stood out in this fashion, Calisch served as spiritual leader of Richmond's Beth Ahabah Temple from 1891 until his
death in 1946. Calisch, committed to eliminating foreign customs and
mannerisms from coreligionists, and dedicated to homogenizing and
Americanizing them, made patriotic speeches and saw to it that many
orthodox customs such as men wearing hats during services, keeping
the sabbaths strictly, and sitting at home on wooden boxes during periods of bereavement were discarded. Followers and other residents of
Richmond found his public posture beyond reproach and Christian
ministers regularly invited him to speak from their pulpits.[27] Thus to
protect, as well as to represent, his congregation, Calisch devoted his
life to developing the image of the Jew as thoroughly assimilated southerner, a "man completely unaware of any personal problem as a Jew,
at ease and unselfconscious, articulate but not argumentative, intelligent but not arrogant, worldly but not cynical."[28]

David Marx of Atlanta also won his congregants' respect by representing them well in the dominant community. A New Orleans
native, Marx replaced a much beloved orthodox rabbi who had been
in the United States for little more than a decade, but whose foreign
accent embarrassed some members of the Jewish community. He
served as Atlanta's premier reform leader from 1895 until 1946. Chosen
because of his appearance and heritage (his father was one of the best-
known men in New Orleans), Marx performed his job with aplomb.
As a cultured and articulate representative of the Jews, Christians
thought well of him. While Marx presided over Atlanta's Reform tem-

ple, he engaged in Americanizing activities similar to those introduced by Calisch in Richmond. Men wore neither skullcaps nor prayer shawls, traditional Jewish holidays that the Orthodox celebrated on two days were observed by Marx and his followers for only one, and religious services were conducted on Sundays rather than on Saturdays.[29]

Like Calisch, Marx behaved as he did to enhance the image of Jews in the dominant society. As spiritual leader to Atlanta's Jews he engaged in activities designed to win Gentile approval, and in his public speeches he always emphasized courage, loyalty, and good citizenship. His efforts won accolades from important Christians, the Atlanta *Journal* asked him to write a regular column, and the University of Georgia awarded him an honorary doctorate in divinity. The Gentile community regarded Marx as "their Jew"[30] and his congregants slept better because of that. Despite the approbation, Marx felt neither "secure nor accepted." Always aware of the general antipathy toward Jews, he had no misconceptions about the potency of antisemitism. "In isolated instances," he acknowledged in 1900, "there is no prejudice entertained for the individual Jew, but there exists wide-spread and deep-seated prejudice against Jews as an entire people."[31]

In pursuing his responsibilities to the Jews in Atlanta, Marx fought against prejudicial acts that he thought could and should be altered. He protested having Jewish students marked absent at school when they attended holiday religious services, he opposed required Bible reading in the public schools, and he objected to the use of *The Merchant of Venice* in Atlanta's high schools. In the 1920s he condemned articles on "The International Jew" in Henry Ford's *Dearborn Independent* and complained when the warden in the federal penitentiary in Atlanta refused Jewish prisoners permission to observe their religious holidays.[32]

The greatest challenge of Marx's career in Atlanta coincided with the most publicized event involving a Jew that ever occurred in the South. In that incident Leo Frank, President of the local B'nai B'rith chapter, became the innocent victim of a confluence of circumstances beyond his control. As the case evolved, it proved beyond any shadow of a doubt that no matter how hard they tried to acculturate, in a crisis the Jew would never be seen as a true southerner. In addition, the condemnation of Leo Frank developed into America's most horrifying example of antisemitism as all the nation's, and especially the South's, stereotypes of the Jew combined with age-old and deeply ingrained myths to lead to Frank's conviction for a murder that he did not commit.

As the Leo Frank case unfolded between 1913 and 1915 it aroused unprecedented concern in both Georgia and the American Jewish communities. Frank was accused, tried, and convicted for the murder of Mary Phagan, a thirteen-year-old girl who worked in the Atlanta pencil factory that he co-owned and managed. Mary's disfigured body had been found in the factory basement on the morning of April 27, 1913. Frank, after having been informed of the tragedy, indicated that he had paid her her wages the previous day; no one else ever admitted to seeing her alive afterwards. When questioned by the police a few hours after the body was discovered Frank seemed nervous and uncomfortable. The next day police arrested him. They had difficulty finding evidence to support their action but the community was so riled up by the murder of the innocent girl that the police had to jail someone for the crime.

A few days after the murder had been committed, Jim Conley, a black sweeper at the factory, was seen washing blood off of his shirt and the police took him into custody for that reason. No one bothered to test the blood on the shirt, and the garment eventually disappeared. Conley was forgotten about and left to rot in a cell until a few weeks later a chance remark by Frank led the district attorney to question him. The sweeper made an incriminating accusation indicating that Frank had committed the murder. Conley's statement was published, newspaper reporters questioned its accuracy, and the police reinterrogated the accuser. This occurred three more times. Conley altered his statement, it was published, reporters found flaws, the sweeper was reinterviewed, and a new affidavit was given out. Thereafter, the district attorney's permission was required before anyone could see Conley.

A careful reading of Conley's affidavits, as well as an examination of the evidence produced at the trial, indicated that Frank could not possibly have committed the crime for which he had been charged. Many of his supporters speculated that had Frank not been Jewish the district attorney would not have put him on trial nor would a white jury have found him guilty on the basis of a black man's testimony. Nonetheless, an Atlanta jury, and most of the population of the city, had no doubt that Frank murdered Mary Phagan, primarily because of the erroneous, but devastating, nature of the information presented in court and then published in the newspapers. Unfortunately, Frank's attorneys failed in their attempts to prove the testimony fraudulent.

The trial opened on July 28, 1913. Much of the prosecution's case focused on Frank's alleged inappropriate behavior with the girls and young women in his employ. Shortly after Frank was arrested on April 28, numerous rumors circulated in Atlanta to the effect that he had a previous history of pulling young girls off streetcars and using them for his pleasure, that he had had liaisons with a variety of young women, and that the Jewish faith encouraged males to exploit Gentile but not Jewish women.[33] On the witness stand Conley told a story that reinforced the views of those who believed Frank to be a "lascivious pervert." Under the gentle questioning of the prosecutor, the sweeper unfurled a tale in which he claimed that on the day of the murder he had served as lookout while Frank was alone with Mary Phagan. Then, Conley asserted, Frank had called him up to the office where he and Mary had been alone. Frank allegedly said that he had struck the girl, that she had hit her head against some heavy object, and that she had fallen and died. According to the sweeper's narrative, Frank asserted, "Of course you know I ain't built like other men," a statement never otherwise elaborated on or explained, and then the manager allegedly asked the sweeper to remove the dead girl's body.

When called on to testify in his own behalf Frank labeled Conley's tale a complete fabrication. He carefully accounted for all but ten minutes of his time during the period when the girl had been murdered. He added that he had had no contact with Mary except to pay her when she came for her wages at the factory on the day of her death, and that he knew nothing about her activities after she left his office. To discredit the state's main witness, Frank's attorneys tried a tactic that backfired. They asked Jim Conley whether he had ever served as lookout for Frank before and the man asserted that he had done so on many occasions. This opened up a line of questioning that alarmed the presiding judge. He declared the nature of the testimony unfit for innocent ears and ordered all women and children out of the courtroom. The next day newspapers indicated that the most "startling features of the negro's testimony are unprintable,"[34] thereby encouraging additional rumors among, and flights into fantasy by, those unable to hear the narrative.

Frank's attorneys' inability to expose the sweeper's story as erroneous, and their subsequent motion to have all of Conley's testimony about his employer's allegedly lascivious behavior struck from the record, undermined the defendant's claim of innocence. The prosecution agreed that the witness should not have been asked to talk about the factory manager's alleged previous activities with female employees, but once the defense counsel explored that area, one of the prosecutors argued, they could not expect to have the material expunged from the record. When word spread throughout Atlanta that Frank's attorneys could not break the story of an "ignorant negro," and that they then tried to remove his words from the court record, it all but convinced people that Frank had indeed committed the murder. The jury found Frank guilty, the judge sentenced him to hang, and Atlanta went wild with glee.[35]

What seemed like an appropriate verdict to most of the people in Atlanta struck the city's Jews as a blatant case of antisemitism. Shortly after Frank had been arrested, a Jewish woman in the community had written to the newspapers saying that this was the first time that "a Jew had ever been in any serious trouble in Atlanta, and see how ready is every one to believe the worst of him."[36] A few weeks after the trial ended some of Frank's friends in Atlanta wrote a letter to Louis Marshall, president of the American Jewish Committee in New York City, requesting assistance to combat an "American 'Dreyfus' case," in which "prejudice and perjury" determined the outcome.[37] Both Marshall and the American Jewish Committee shied away from involving themselves on behalf of Jews accused of crime and refused to take any official stance in regard to Frank. Nonetheless, members of the AJC recognized that he was a victim of a gross miscarriage of justice and, as individuals, decided to help him in his appeals.

Unfortunately, efforts that northern Jews made on Frank's behalf resulted in a vicious antisemitic response in Georgia. Citizens there claimed that the rich Jews of the North were undermining the state's judicial system to save a convicted coreligionist. "We cannot have one law for the Jew and another for the Gentile," wrote Tom Watson, a former Populist and then a Georgian journalist,[38] and no one could refute that logic. Moreover, it was generally—and accurately—assumed that the costs for the various judicial appeals were borne by Frank's coreligionists. But despite northern support, all of Frank's petitions for a new trial were turned down, thrice by the Georgia Supreme Court and twice by the U.S. Supreme Court. That background made it inevitable that thousands of Georgians were outraged when Governor John M. Slaton, in June 1915, commuted Frank's sentence to life imprisonment. Taking the law into their own hands two months after the governor's decree, a band of some of the state's "best citizens" stormed the prison farm, stole Frank away in the night, and hung him from a tree near Mary Phagan's birthplace in Marietta, Georgia.[39]

Not one shred of evidence linked Frank with the murder. The prosecutor had labeled a strand of hair as Mary Phagan's, but a scientific examination proved that it belonged to another person; drops of blood that Dorsey claimed had come from the victim's body were in fact chips of paint; and the location of where the body had been found and the description of how the victim reached that destination made it impossible for the crime to have been committed in the way described by the state's main witness. Nonetheless, for more than half a century afterwards many Georgians would claim with certainty that Frank had been the murderer; most Jews were equally convinced that antisemitism caused Frank's conviction and demise.

The impact of the Frank case in Atlanta was devastating to the assimilated southern Jews. What happened to Frank could have happened to any of them. For the next half century they lived in great apprehension, always concerned that another incident might arouse their Christian neighbors into another attack against Jews. After the lynching of Leo Frank southern Jews were so frightened that they might offend Gentiles that deviation from community values and standards would not be tolerated from any of their members or leaders.[40] In fact, Jews made extra special efforts to conform. An adult who later recalled his youth in Roanoke wrote how in restaurants the children ate their chocolate pudding a little more carefully, in buildings we shrieked a little less loudly, in streetcars we became a little less sick. There were aunts who were nervously polite to salespeople, and uncles who carefully referred to O.P.'s (our People) when talking in public places.[41]

Occasionally, an individual tried to challenge the timidity so carefully honed by fellow Jews but without success. In the early 1930s a new Jewish rabbi from the North spoke out against a particular lynching in the southern industrial town where he was employed. A majority in his congregation argued that since no one in the Christian churches had denounced the deed it "ill behooved a minister of a minority group like the Jews to denounce and excoriate those who were the powerful majority." The officers of the congregation then suspended the new rabbi and refused to renew his contract for the following year.[42]

There were other examples of where Jews, mostly from the North, tried to intervene on behalf of the civil rights of African Americans, but they were usually rebuffed by southern coreligionists. In the 1931 Scottsboro case, two young women of questionable virtue accused nine African-American youths of raping them. The accusation and subsequent prosecution of the boys attracted nationwide attention. Rabbi Benjamin Goldstein of Montgomery's Temple Beth Or planned to attend a rally in Birmingham to oppose their prosecution. His Temple board reluctantly agreed to allow him to go but shortly thereafter told the rabbi he must either cease all connection with the Scottsboro youths or resign. Goldstein resigned. The congregants did not disagree with Goldstein's belief in the innocence of the black adolescents but they feared his support of the boys threatened the welfare of the city's Jewish community.[43]

The Scottsboro case also heightened southern perceptions of the connection between communists and Jews. Communists throughout the nation backed the youths and denounced the nature of a judicial system that automatically condemned them because they were accused by two white women. Moreover, a New York Jew, Samuel Leibowitz, went to Alabama to serve as the boys' defense counsel. Although there was little doubt that a white jury of twelve men would accept the words of two southern white women over those of nine African Americans, a slim possibility existed that the flimsiness of the evidence might lead to a verdict of not guilty. That chance ended, however, when, during the summation, Wade Wright, the white prosecutor, pointed at Leibowitz and implored the jury: "Show them that Alabama justice cannot be bought and sold with Jew money from New York." Leibowitz immediately protested the remark and the judge admonished the prosecutor to be more careful in his assertions.[44] Yet Wright knew what he was doing. Reporters covering the trial thought that until the prosecutor made his antisemitic accusation the jurors had not completely decided how to vote. Thereafter the door slammed shut. Wright's "anti-Semitic summation," they concluded, "was the most effective single statement by the counsel for the prosecution."[45]

The charge that "Jew money" tried to influence Alabama justice impressed southerners and hurt the region's Jews more than all the fulminations of the Ku Klux Klan in the 1920s.[46] Southerners almost always had a "Black Beast" to fear, whether it be the African American, the Jew, or the Catholic. During World Wars I and II the Germans occupied that position. After World War I and the Russian Revolution southerners targeted the Communists. "By the 1930s," historian Joel Williamson has written, "the black beast of Radicalism had come to be, most often, the Commie-Jew-labor-organizer." Thus, in the southern mind, the involvement of the communists and Samuel Leibowitz, a non-communist Jewish lawyer from New York, "represented the nightmare come true."[47]

Obviously, Leibowitz's appearance in the Scottsboro case was an example of a northern Jew working to undermine southern values and inadvertently threatening the precarious position of southern Jews. The lynching of Leo Frank still hovered "like a black cloud over the region," and Leibowitz's appearance as a defender of African Americans sent additional chills down the spines of southern Jews. They had not yet recovered from either the Frank case or the waves of antisemitism that had swept several areas of the South in the early 1920s.[48] Here was yet another *cause célèbre* that threatened their status and sharpened the differences between the liberal thought of New York Jews and the more conservative mores of southern Christians and Jews.

Although the Scottsboro case also coincided with some of the worst years of the 1930s depression, during the rest of the decade no untoward events occurred even though Jewish apprehensiveness remained high. Occasional radio addresses, pulpit speeches, and fulminations at public meetings blamed Jews for the nation's economic problems; the charges, however, did not ignite further attacks in the South. One event that illustrated how reluctant Jews were to antagonize any Gentile, even in the mildest way, occurred during a 1938 High Holy Day service in Waycross, Georgia. The loud noises of carpenters working in another part of the building disturbed the proceedings. None of the worshippers, however, wished to approach the owner of the building and ask if the carpenters might be able to work more quietly. Finally, after much discussion, a representative of the group went to the building owner, carefully explained that an important religious service was being conducted, and timidly asked whether it might be possible for the employees to engage in some other activity so as not to disturb those engaged in prayers. The owner immediately agreed to assign his workers to some other task and thanked the individuals for their tactful presentation. The situation was handled appropriately by both sides—only the hesitancy of the Jews to even approach the building owner suggested their reluctance to do anything that might stir up trouble or reverberate negatively against the Jewish community.[49]

Outbursts of antisemitism were less frequent in the South during the 1940s than they were elsewhere in the country but once again southern Jews watched themselves carefully. For many white Christians "the Jewish Question" was closely related to "the Negro Question,"[50] and while both minorities lived in the South, they adhered to dominant behavioral codes. African Americans and Jews had their places and knew what was expected of them. Despite a continual uneasiness, however, few Jews would publicly acknowledge, as one Louisiana woman did, that "the average modern Jew is ashamed of his Jewishness."[51] Nonetheless they lived less than satisfactory lives. As the chronicler of several towns in southwestern Louisiana observed:

> the Jews in these small communities live out their lives, on the surface at least, as other people do, but they are never quite normal, never quite at ease, never completely secure, and seldom imbued with the sense of worthwhileness that members of other religious groups are. Outwardly they may be calm but inwardly they are worried.[52]

In some areas, in fact, like Richmond or Waycross, Georgia, Jewish and Gentile leaders often went to extremes to deny the existence of antisemitic feelings. Thus they inhibited any possibility of solving problems that emerged from the discomfort that most Jews felt. Another example of southern Jewish caution is seen in the refusal of the Richmond Jewish Community Council to take any stance on whether Jews should have a homeland in Palestine. "To be outspokenly Zionist before World War II in a Southern Reform congregation," we are told, "took courage on the part of a rabbi." Nor would southern Jews publicly condemn antisemitism. They believed the less said the better. "To many, this seemed a matter of survival." Richmond's Jews were also "more clandestine and negative," while in Waycross Jews often experienced "uneasiness." Similarly in El Paso, Jewish-Gentile relations appeared serene but there was "little ease, little real acceptance." When discussing conditions among themselves, however, southern Jews asserted that the solution to any problem with which they had to deal "lay in overcoming anti-Semitism."[53]

When in 1942 most of the nation's Jewish organizations voted to encourage the support of a Jewish state in Palestine, dissenters broke away from the majority and established the American Council for Judaism (ACJ), a distinctly anti-Zionist organization. Both Reverends Calisch and Marx stood in the forefront of the anti-Zionist movement and about one-third of the rabbis on the executive board of the new association hailed from the South. They believed that Judaism was a religion and not a nationality and they did not want to suggest otherwise to their neighbors. The elite Jews of Houston were so incensed by those

among them who favored a Zionist position that its Congregation Beth
Israel barred all Zionists from membership in the temple.[54] Fewer disputes erupted between American Jewish Zionists and southern non-
Zionists after the establishment of Israel in 1948. Southern Jews still
remained publicly quiet about the new Jewish state but the ACJ had
lost its battle, and strong positions no longer had to be taken over
already concluded events. It took many years, however, before a pro-
Israel stance became socially fashionable among Jews in the South.[55]

Southern Jews, however, did not need issues to temper their behavior or heighten their anxieties. Members of the Charlotte, North Carolina, Jewish community generally felt their security imperiled whenever the word "Jew" appeared in local newspapers in any context.
Then, in 1948, a letter circulated among members of the local chapter
of the Confederate Daughters of America stating that "nearly all the
Communists in America are Jews, and . . . most of the funds and agitators used in stirring up your Southern Negroes are Jewish in origin."
Charlotte Jews were besides themselves and feared all kinds of unpleasant consequences—none of which ever occurred.[56] Two years later
coreligionists in Durham opposed "Mutt" Evans' candidacy for Mayor
because they feared that if something went wrong during his administration Jews would be blamed. Similarly, they worried that some Jewish labor organizer from the North might come down and create trouble for them, or that some Jew might take a pro-integrationist stand.
When Rabbi Malcolm Stern organized an interracial religious service
in Norfolk in 1947 he aroused the fears of his executive board. "Only
the presence at the service of the Rev. Beverly Tucker White of the
socially elegant St. Andrew's Episcopal Church," he recalled, "rescued
me from more than a reprimand from the angry temple board." Despite
these various apprehensions, however, publicly Jews stated that everything was going well for them in the South.[57]

In his reminiscences Eli Evans tried to explain these paradoxes and
"to reconcile [as well] what I thought were unresolvable conflicts of
growing up Jewish in the South." In his part history/part autobiography he described how carefully Jews monitored themselves. At the
University of North Carolina, for example, one Jewish fraternity had a
rule that no more than five brothers could walk across campus together;
they did not want to be seen as people who travel in packs. Furthermore, Evans wrote, "Mr. Jew never disagrees fiercely, never looks for
a fist fight or an argument; he blends in well after he is accepted because
he works at friendship consciously, constantly, and sometimes devi-
ously."[58] While Jews in every region of the country, especially those
living in small towns or particularly anxious to ingratiate themselves
with members of the dominant community, always showed sensitivity
to their surroundings, the perceived necessity for such accommodation
in the South amounted to more than a difference in degree—it was a
difference in kind.

Despite their efforts to fit in, however, Jews could not control the turmoil that appeared in the South after the 1954 U.S. Supreme Court decree that called for the end of segregation in the nation's public schools. The ruling led to major upheavals across America and frightened southern Jews more than any single event had done since the lynching of Leo Frank. Most Jews knew that significant societal disruptions often resulted in their becoming one of the victims of the ensuing strife. And in this case they were absolutely right. *Brown* v. *Board of Education* ushered in more than a decade of chaos and violence as southerners strongly opposed the Supreme Court's decision. The memories of that era are not pleasant ones for many southern Jews.

One of the first results of the Supreme Court judgment was the establishment of the White Citizens Councils. On July 11, 1954, less than two months after the *Brown* decision, fourteen upper-middle-class white men, including the Mayor, a dentist, a lawyer, a druggist, two auto dealers, and others like them, met in Indianola, Mississippi, to form the new organization. They were "respectable" men who wanted a "respectable" movement.[59] In less than a year's time White Citizens Councils had spread to communities throughout the South. Their goal was to maintain segregation and prohibit any outsiders from telling southerners how to conduct their lives. Known nationally as white supremacists, which they were, they sought support from like-minded individuals. Many of the most fervent segregationists were also anti-semitic and associated Jews with racial integration. The overwhelming number of White Citizens Council members wanted to restrict membership in their organizations to those who believed in the divinity of Jesus Christ but in some small towns Jews were solicited and, once requested to enroll, did so.[60] "I did it with a heaviness I never did anything else with before," one Jew later admitted. But "I didn't feel I could stay here and not join."[61]

Only a small minority of Jews shared the views of the White Citizens Councils[62] but most other southerners were unaware of that fact. Journalist Hodding Carter, III, for example, thought that Jews had little to fear from the militant opponents of desegregation:

> one reason the Councils do not move against the Jewish citizens of Mississippi is that in many cases they do, in truth, share the Councils' views. Another is that while a latent anti-Semitism may exist in the white Christian community, it is not strong enough to make its positive expression worthwhile unless there should be some indication that the Jewish members of the community were going to question Southern racial attitudes. Few Southern Jews have openly done so. Accordingly, they are accepted as conforming members of the white community and not subjects for outright Council discipline.[63]

But northern Jews were not at all reticent in proclaiming themselves as foes of segregation and their southern brethren suffered because of it. Southern Jews got on as well as they did in the region because Gentiles thought they shared similar values. This was not the case but the southern Jews were not willing to publicly disassociate themselves from the norm; in fact they wanted Christians to believe that they fit in well. Therefore the active roles that the Anti-Defamation League, the American Jewish Committee, and the American Jewish Congress took in proclaiming the need to end racial barriers not only unnerved southern Jewry but led to counteroffensives from both respectable and lawless southerners. The Richmond *News-Leader*, for example, without understanding the discomfort that southern Jews successfully hid, editorialized on July 7, 1958:

> Relations between Jews and gentiles were excellent in the South before the ADL began setting up regional offices, as in Richmond, and stirring up clouds of prejudice and misunderstanding. What possible service the Jewish community finds in a Jewish organization that foments hostility to Jews, we have no idea. Perhaps some of the South's many esteemed and influential Jews will want to inquire into this matter.[64]

For a group like the southern Jews who had spent generations trying to avoid trouble with Christians the effects of this editorial were chilling.

So, too, were the temple bombings throughout the region in 1957–1958, and again in 1967. Although the perpetrators were not known, Jews attributed the deeds to staunch segregationists who associated them with the liberal views of their northern brethren. The bombings occurred in Charlotte and Gastonia, North Carolina, Birmingham and Gadsden, Alabama, Nashville, Tennessee, Miami and Jacksonville, Florida, and even in Atlanta, where the fear of repercussions from the Frank case finally ended when the bombings began. In November 1967, two months after the city's Temple Beth Israel had been bombed, another bomb shattered the residence of Jackson, Mississippi's integrationist rabbi, Perry Nussbaum, who had been born in Canada, educated in Cincinnati, Boulder, Colorado, and Australia, and had come South in 1954 after a stint as rabbi in Pittsfield, Massachusetts. Although only four or five of Jackson's 150 Jewish families had been even "moderately active" in the Civil Rights movement, the bomb planters sent the whole Jewish community a message. Thereafter the police provided twenty-four-hour-a-day protection to the rabbi while the city's Jews lived in constant terror of further attacks. According to members of his congregation, Rabbi Nussbaum's commitments more closely resembled those of a northern civil rights activist than those of a Jew who had been reared in the region. They felt that "he did not really understand what it meant to be a Jew in the South." He often spoke directly about things that were on his mind. Accordingly, it was not particularly startling when he angrily responded to the bombing of his house with these words: "It's the Sunday-school lessons from the New Testament in Baptist churches that lead people to commit such terrible acts."[65]

Although some southern Jews may have shared Nussbaum's views, they certainly would not speak as bluntly as he did on issues of concern to other southerners. Jews feared the anger and savage behavior of ardent segregationists in their midst; it was safer for southern Jews to vent their wrath on northern coreligionists. The southerners pleaded with the national Jewish organizations to desist from taking pro-integrationist stands in the South.[66] They argued that the Jewish agencies should focus on problems affecting Jews, not other groups. A southern Jewish women told an ADL staff member, "Everytime one of you makes a speech, I'm afraid my husband's store will be burned up."[67] Other southern Jews thought that if the northerners failed to heed their advice a Southern Jewish Committee should be formed to promote the interests of the region's Jews.[68] "You're like Hitler," an Alabama Jew charged a representative of a national Jewish organization, "you stir up anti-Semitism against us."[69]

The northern Jews had their own constituencies to deal with, however, and they refused to pull back. They also could not understand the reticence of their southern counterparts to protest antisemitism in the region. Northerners continued to advise their southern brethren to work together with Christians to improve race relations. Unity of this kind might have been possible in cities with liberal political structures such as Atlanta, Norfolk, New Orleans, and Nashville but almost perilous in rural areas and places like Birmingham, Montgomery, and Jackson.[70]

How perilous was illustrated by the tragedy that befell three civil rights workers outside Philadelphia, Mississippi, during the summer of 1964. It epitomized the savagery that many southern Jews feared possible from the rural, agrarian, and almost impoverished classes of fundamentalist whites. These groups lacked social and economic power but on the local and state level exercised a good deal of influence. Part of their need to maintain the status quo was to have minorities in their midst on whom they could vent their frustrations. Generally, they did not attack the white power structure because (1) they did not want to appear aligned with African Americans and (2) they also possessed what Lou Silberman, then a Vanderbilt University professor and internationally renowned scholar of Judaic studies, characterized as a "reverence for their betters" among the white Christians. "In looking for a more vulnerable group," Silberman continued, these lower-class whites "come upon the Jews of the South, with the resulting acts of violence that we know of." Silberman made these remarks after a number of temples had been bombed in 1957 and 1958 but his words were best illustrated by what occurred in Philadelphia, Mississippi, on June 21, 1964.[71]

During the summer of 1964 hundreds of northern civil rights workers had descended on the South to help local African Americans. They
participated in education, cooperated in giving instruction in methods
of peaceful protest, and tried to help some individuals register to vote.
Many white southerners hated these northerners who encouraged African Americans in their quest for equality. They were determined to
make their hostility to the outsiders known and found several ways of
doing so. Intimidation, petty harassment, verbal barbs, and menacing
body language were everyday events. Nonetheless, on June 21, 1964, a
band of venomous southerners engaged in an act that received worldwide notoriety.

That night, three men—two northern Jewish activists and a local
black civil rights worker—disappeared. One of the northerners, 24-
year-old Michael Schwerner, and his Mississippi associate, Richard
Chaney, had a month earlier encouraged an audience at the Negro
Methodist Church in the neighboring town of Meridian to continue
holding "Freedom Meetings." As the gathering ended angry whites in
the area beat up four black people and, referring to Michael Schwerner,
shouted, "Keep that Red Jew nigger-lover out of here or you'll all wind
up in the river." Then the white rowdies burned down the church.[72]

Shortly thereafter Schwerner returned to Ohio to train other civil
rights volunteers and came back to Mississippi on June 20 with 20-
year-old Andrew Goodman. The next day, as they were driving with
Richard Chaney, the police stopped the car, accused the driver of
speeding, and arrested all three. The "culprits" were taken to the Philadelphia, Mississippi prison, where a handbill tacked onto to the bulletin board outside of the sheriff's office read: "Jews Founded Communism."[73] Schwerner, Goodman, and Chaney remained in the
Philadelphia jail from 4:30 P.M. until 10:30 P.M., when they were
released. They were seen driving out of Philadelphia but exactly what
happened after that is speculation. Later on it came out that they had
been followed, driven off the road, captured, killed, and buried before
daybreak the next morning.

Thanks to a paid informant, their dead bodies were recovered six
weeks later after having been hidden in an earthen dam site that had
been under construction in June. Schwerner and Goodman had each
received a bullet in the heart; Chaney had been beaten and tortured
and then shot three times.[74] Their disappearance and discovery
received international attention, and observers wondered whether their
executioners would ever be brought to justice. (The following year,
several of the assailants were arrested. They were finally tried and convicted in 1967.[75]) At the time of the bodies' discoveries, however, an
anonymous resident confided to a reporter:

> This is a hate lynching. To the lynchers their victims were an agitating
> nigger and two Jew-atheist-beatnik-nigger-lovers. Hell, these killers

think they are patriots. They think they killed to protect the Mississippi way of life.[76]

The murders of the three civil rights workers confirmed the fears that existed among Mississippi's Jewish population. If they went "too far" they would certainly have to leave their communities. Speaking out for equality could mean death. "We have to work quietly, secretly. We have to play ball," one fourth-generation southern Jewish woman admitted. "Anti-Semitism is always right around the corner."[77] She continued:

> We don't want to have our Temple bombed. If we said out loud in Temple what most of us really think and believe, there just wouldn't be a Temple here anymore. They let it alone because it seems to them like just another Mississippi church. And if it ever stops seeming like that, we won't have a Temple. We have to at least pretend to go along with things as they are.[78]

Despite existing fears, a few southern rabbis spoke in favor of integration but almost all had northern or big city roots: Jacob Rothschild of Atlanta who was born and reared in Pittsburgh and educated in Cincinnati, Emmet Frank of Alexandria, Virginia, who grew up in New Orleans and attended college in Houston; Charles Mantinband of New York who spent his formative years in Norfolk and ministered to congregations in Alabama, Mississippi, and Texas, and Perry Nussbaum, a native of Canada who served the Jews in Jackson. Some had the support of their congregants; most did not.[79]

Rabbi Mantinband presided over the Hattiesburg, Mississippi, congregation during the height of the Civil Rights crisis in the early 1960s, and he had to contend with the opposition of his Temple board and other Jewish members of his community. On one occasion, after entertaining African Americans, a Jewish neighbor demanded to know who those people were who had just left his home. "Some of my Christian friends," Mantinband replied. In Atlanta, Rothschild had been an ardent supporter of the Civil Rights program proposed by Harry Truman in 1948 and generally had the backing of his more liberal (compared to other southern Jewish groups) Jewish community. Therefore his continuation of this position into the 1960s seemed natural and did not bestir congregants. Emmet Frank, who presided over a congregation in Alexandria, Virginia, in the late 1950s was not at odds with his temple members over his convictions but because of his tendency to showboat. On the eve of Yom Kippur in 1958 he denounced the most powerful Virginia politico, Senator Harry F. Byrd, for his segregationist views. The speech received national attention and most Americans congratulated Frank for his forthright stance. A local Jewish resident, however, scolded the rabbi. "As the years go on," this correspondent wrote, "you will reflect on the damage you have done those to whom you were supposed to be a religious leader."[80]

The fourth of the noted southern rabbis, Perry Nussbaum of Jackson, Mississippi, tried to be circumspect in his activities as he established a chaplaincy program for the arrested northern freedom riders of all denominations who had come south on integrated buses and who had been jailed for their activities. He engaged in other pro-integrationist activities and as word spread about his views both his temple and his home were bombed.[81] Following the second bombing several community residents, including a Baptist minister, came by to voice their concerns and support for the Nussbaum family. After the Baptist clergyman expressed his regret over the incident and denounced the unknown culprit who planted the bomb, he added, "But isn't it a shame that the rabbi doesn't know Jesus."[82]

In the 1970s both southern Christians and southern Jews tried to accommodate themselves to the inevitability of desegregation. As this process continued, violence ebbed and just about disappeared in regard to the Jews who then proceeded to go on with their lives. Tolerance among southerners increased somewhat and Eli Evans noted in 1973 that it was then "easier to be Jewish in the South" than it had been in earlier decades. In 1972 the Southern Baptist Convention, for the first time in its history, adopted a statement condemning antisemitism. And that same year Neil November, the new President of the Richmond Jewish Community Council, acknowledged a different atmosphere in his city. "We've at least begun to scratch the surface of changing the image of the fat, bald-headed, long-nosed, money-man Jew," he stated, "to that of the strong, wavy-haired, tall guy that you'd better watch out for."[83]

Also notable were changes in the nature of the southern Jewish population. About 75 to 85 percent of the 800,000 Jews in the South in the 1980s hailed from the North. At first it was only retirees from the colder northern climates who descended on Florida coastal towns but now senior citizens and younger Jewish families moved to Virginia, the Carolinas, Georgia, other parts of Florida, and Texas. Moreover, the changed nature of the national economy, the growth of university faculties, the impact of the research triangle in Raleigh-Durham-Chapel Hill, North Carolina, the national and international stature of cities like Miami, Atlanta, Houston, and Dallas, and the expansion of the federal government, resulting in the development of the northern Virginia suburbs, have all led to an influx of people, including Jews, to the South.

Most of these Jews are not cowed by southern traditions nor have they been swept up by southern Jewish anxieties. Furthermore, they are tied by neither regional ancestry nor economic necessity to southern viewpoints. It is also important to note that the South of the post Civil Rights era is much more enlightened than it had been in previous decades; there is a broader acceptance of diversity. Thus the evolution of southern Jewry is likely to be quite different from what it had been between the Civil War and the Vietnam war. There still may be pockets of Klansmen or those who think like them but the Jews who arrived in the South during the past generation have a different heritage and psychology and are more likely to behave like their coreligionists throughout the country than like Jews in the South of old.[84]

A gradual exodus of the children of rural Jewry has also contributed to the altered position of southern Jews. The narrow limitations of the small town combined with the economic opportunities and wider diversions of the larger cities to deplete the Jewish population of its younger generation.[85] No longer did sons think that they had to take over their fathers' retail establishments to earn a living. Nor were daughters constrained by the dictates of social conventions. Changes in American education, the economy, and society that began after World War II and accelerated in the 1960s offered possibilities to young Jewish adults that their parents and grandparents had neither perceived nor possessed. As a result, the numbers of Jews in southern hamlets waned while those who moved to urban centers found welcoming enclaves of other Jews. Living with large numbers of like-minded individuals of similar backgrounds and values contributed significantly toward changing the psychology of a small minority hovering in a community composed almost entirely of evangelical Protestants. Their defensiveness diminished, and they seemed less concerned than their parents had been about how their behavior would reflect on all Jews. The intermixture of southern and northern Jews and their consequent interaction modified generation-old patterns. During the past two decades one noticed southern Jews wearing Stars of David, occasionally using Yiddish expressions, and purchasing Jewish books and objects d'art for their homes. Before the 1970s, one scholar observed in 1986, "They did nothing that would mark them as significantly different."[86]

Nonetheless, while northern Jews who have moved to the South contributed to wider interactions for southern Jews, some regional practices have not altered, and southern Jews are uncomfortable with them. Carolyn Lipson-Walker, who studied Jews in the South in the early 1980s, found that close to one-third of her informants defined "their Southern Jewishness through the veil of real hatred. With almost no exceptions," she found that southern Jews spontaneously brought up antisemitism during the course of the interviews. In some communities Gentiles still see Jews in negative stereotypes. Expressions like "Jew down" and "to Jew" are still common phrases in the everyday conversations of lower-class whites in the region. Elites in New Orleans kept Jews out of some of the most prestigious krews during Mardi Gras season in 1969, and these barriers remained intact today.[87] Some fundamentalists still question whether "God Almighty ... [hears] the prayer of a Jew,"[88] and in parts of Houston people whose grandfathers had converted to Christianity were still referred to as Jews in the early 1970s. Thus many Jews, even though they are generally well-to-do, live in the best neighborhoods in the South, and enjoy a prosperity that most Americans covet, still "harbor deep-seated insecurities. Although many Southern Jews are able to move easily within the upper echelons of Southern society, they never drop their wariness."[89]

Another development, however, surprising to the general public but not to the cognoscenti, has became apparent since the 1960s. African Americans, whose views most other Americans rarely paid attention to before the desegregation crisis of the 1950s, emerged as the one group of southern background whose publicly expressed sentiments have aroused Jewish concerns. The blacks still living in the South have avoided accusations and confrontations; the most articulate African American opponents of Jews speak out in the North. During the Civil Rights era newspapers and periodicals assumed the existence of a Black-Jewish coalition since both groups battled together for Civil Rights legislation. Then, just as the Civil Rights Act of 1964 and the Voting Rights Act of 1965 passed and ostensibly accomplished what the coalition had been striving for, friction between blacks and Jews flared into the open. Why and how this happened will be subject of the next chapter.

10

African-American Attitudes (1830s–1990s)

It is ironic that just as African Americans benefitted from their greatest legislative accomplishments in a century—the passage of the Civil Rights Act of 1964 and the Voting Rights Act of 1965—many of them focused on Jews as a source of their difficult predicaments. To other Americans in the North who had followed the civil rights struggle it seemed as if Jews and blacks were firm allies in their quest for legislation to provide equality of status and opportunity to all Americans. And, in fact, their legislative goals in the early 1960s meshed. Nonetheless, several antisemitic outbursts in northern, urban areas—especially New York City—shocked contemporaries who had perceived a much stronger alliance than had ever existed between African Americans and Jews.[1]

If one considers African Americans as a group of people who spent several generations in the American South, it is much easier to understand the basis for black antisemitism. Living in the South they imbibed a white Protestant fundamentalist culture that saw Jews as people who had killed their Savior, who had never accepted the truthfulness of Christianity, and as cunning and exploitative individuals who had ruthlessly amassed fortunes as they allegedly acquired political and economic control of society. Such stereotypes, as discussed earlier, are part of both black and white cultures in the United States but seem to be more pronounced among religious conservatives and fundamental-

ists. For Christian Americans this heritage predates personal knowledge of, or interaction with, Jews.[2] As novelist Richard Wright wrote
about his youth in Arkansas and Tennessee circa World War I: "All of
us black people who lived in the neighborhood hated Jews, not because
they exploited us but because we had been taught at home and in
Sunday school that Jews were 'Christ killers.' To hold an attitude of
antagonism or distrust toward Jews," he continued, "was bred in us
from childhood; it was not merely racial prejudice, it was part of our
cultural heritage."[3] And in studying the history of African Americans
the Reverend Joseph A. Johnson, Jr., pointed out in 1971 that one must
be cognizant of "the influences and effects of the Christian faith" and
how it practically shaped and molded the black experience in America.[4]

Historically, black venom and distrust of Jews dates to the days of
slavery. Suggested catechisms for young "colored persons" reflected
the same views as those prepared for whites. Two typical ones read:

Q. Who killed Jesus?
A. The wicked Jews.

and

Q. The wicked Jews grew angry with our Savior and what did they do
to him?
A. They crucified him.[5]

Slaves sung songs like "Were you there when the Jews crucified my
Lord?," "De Jews done killed poor Jesus," and "Cry Holy," which had
a line about the Jews and Romans hanging Jesus.[6] In one spiritual a
verse ran:

Virgin Mary had one son
The cruel Jews had him hung.[7]

African Americans were continually instructed with the Christian
gospel, and within their culture the word "Jew" became synonymous
with the enemy, lacking humility or gentleness, always the antagonist
of Jesus.[8] In 1859 a female slave disappeared on the day she was to be
transferred to a new owner. When found and asked why she did not
want to go to her new mistress she offered the following explanation:

"I don't want to go to live with Miss Isaacs."

"Why don't you want to live with her? She is a good lady and will
make you a kind mistress, and besides, you won't have any hard work
to do."
"Ah! But Mass F . . . , they tell me Miss Isaacs is a Jew; an' if the Jews
kill the Lord and Master, what won't they do with a poor little nigger
like me!"[9]

Hostility toward Jews, emanating from Christian teachings in the early nineteenth century, carried over into the twentieth. Just before the United States entered World War I educator Horace Mann Bond responded to a 12-year-old boy who taunted him by shouting "Nigger, Nigger, Nigger, Nigger" with the expression "You Christ-killer." Bond explained the words that he chose to express his animosity toward the bigot by noting that he had grown up in a house where the family prayed before each meal, read Scripture every morning and evening, attended daily chapel and three services on Sunday. "Of course," he added, "the thought that Christ had been killed and by the Jews, and that this little boy was such a one, may have had a more ancient basis in my twelve-year-old mind than I can now bring myself to admit."[10] In 1948 novelist James Baldwin observed that among African Americans "the traditional Christian accusation that the Jews killed Christ is neither questioned nor doubted. . . . The preacher begins by accusing the Jews of having refused the light and proceeds from there to a catalog of their subsequent sins and the sufferings visited on them by a wrathful God."[11]

Secular stereotypes of the Jew reinforced prejudices already derived from religious teachings. Throughout much of the nineteenth and twentieth centuries Jews were seen as zealous and unprincipled in their quest for wealth and as people who used their cunning to amass large fortunes.[12] Stories exaggerating Jewish wealth also abounded in black newspapers in the late nineteenth and early twentieth centuries. Readers of these papers were told that Jewish riches stemmed from lending money at usurious rates to nobles and warrior in England after 1066,[13] that Jews controlled the southern money market thereby oppressing African laborers in the region,[14] that Jews not only had immense wealth but would shortly become the financial rulers of the world.[15] In 1903 a writer for the *Colored American* stated: "In an incredibly short time after the arrival of a Jew in any community he has nearly every family in his debt or under obligation to him."[16] In 1905 school teacher Jessie Fortune, writing about the immigrant Jews on New York City's Lower East Side, used phrases like "Jews will make money," "few of them are really destitute," and the Jew's "sole aim seems to be earning money."[17] Another article described the Jewish "race" as

> tribalistic rather than national in character, and parasitical and predatory rather than conservatory and constructive in tendencies—preying upon and devouring the substance of others, rather than creating and devouring the substance of itself. As a salesman, as a money-lender, the disposition of the Jew is to take the long end and let the other man take and hold the short end of every proposition . . . as a moneylender he holds the purse strings of the world and exacts his own terms of those, whether states or individuals, who need and must have money to finance their necessities.[18]

The great African American leaders at the turn of the century, Booker T. Washington and W. E. B. DuBois, shared the same negativism. Washington, who experienced a conventional southern upbringing, began his career, as his biographer Louis Harlan put it, "full of misunderstandings about Jews," and as late as the 1890s a friend cautioned him to keep his prejudices about Jews out of his speeches. Washington seems to have thought of Jews as exploitative shopkeepers and usurious creditors and from time to time differentiated between the Jew and the white man. However, disparagement of the Jews was impolitic since a number of them in the North made substantial financial contributions to black causes. As Harlan noted, "in [Washington's] effort to secure donations to his school . . . it was in his interest to drop his prejudice." Publicly he did so and many of his twentieth-century writings, in fact, praise Jews.[19]

As a young man DuBois also displayed hostility toward Jews that he later repudiated. While on an ocean liner crossing the Atlantic in 1895 DuBois confided to his diary that although he had met two congenial Jews, other Jews had "in them all that slyness, that lack of straight-forward openheartedness that goes straight against me."[20] In his major work, *Souls of Black Folk*, published in 1903, he relied on hearsay and folk tales to denigrate Jews, Yankees, and poor whites, but not "Southern gentlemen."[21] Several of his fallacious comments about Jews were corrected in a 1953 edition of this classic work but for half a century DuBois let stand observations like:

> I have seen, in the Black Belt of Georgia, an ignorant, honest Negro buy and pay for a farm in installments three separate times, and then in the face of law and decency the enterprising Russian Jew who sold it to him pocketed the money and deed and left the black man landless, to labor on his own land at thirty cents a day.[22]

and

> The rod of empire that passed from the hands of Southern gentlemen in 1865 . . . has never returned to them. Rather it has passed to those men who have come to take charge of the industrial exploitation of the New South—the sons of poor whites fired with a new thirst for wealth and power, thrifty and avaricious Yankees, shrewd and unscrupulous Jews. Into the hands of these men the Southern laborers, white and black, have fallen, and this to their sorrow.[23]

DuBois also wrote that in the Middle Ages Jews "used deception and flattery . . . cajoling and lying" and that these attributes left their stamp "on their character for centuries."[24]

Jewish leaders Jacob Schiff and Stephen Wise protested these prejudicial references and unsuccessfully tried to get DuBois to change them. "I . . . continued to let the words stand as I had written them," DuBois wrote in 1953, "and did not realize until the horrible massacre of German Jews, how even unconscious repetition of current folklore such as the concept of Jews as more guilty of exploitation than others, had helped the Hitlers of the world."[25] At that point he not only changed phrases like "the Jew is the heir of the slave-baron" to "immigrants are the heirs . . . ," "enterprising Jew" to "enterprising American" and "Jews of the Middle Age," to "peasants of the Middle Age,"[26] but he candidly admitted as well:

> I am not at all sure that the foreign exploiters to whom I referred in my study of the Black Belt, were in fact Jews. I took the word of my informants, and I am now wondering if in fact Russian Jews in any numbers were in Georgia at the time.[27]

Although their leaders and journals often criticized Jews, these same sources frequently advised fellow African Americans to emulate them. Beginning in the 1890s, and continuing into the twentieth century, blacks were constantly reminded of the Jews' group cohesiveness and economic accomplishments. In 1899 Booker T. Washington wrote:

> these people have clung together. They have a certain amount of unity, pride, and love of race; and as the years go on, they will be more and more influential in this country,—a country where they were once despised, and looked upon with scorn and derision. It is largely because the Jewish race has had faith in itself. Unless the Negro learns more and more to imitate the Jew in these matters, to have faith in himself, he cannot expect to have any high degree of success.[28]

The next year, a black attorney, practically echoing Washington, admonished his brethren, "If blacks were to be accepted, they had to become like the 'despised Jew, the representative of business and money.' "[29] A *New York Age* editorial in 1905 acknowledged that

> Prejudice against Jews is almost as general and persistent as it is against the Afro-American people; but it is displayed less, because the Jews are among the wealthy people of the country and know how to advance themselves by properly directing their wealth against those who offended them.[30]

James Weldon Johnson, one of the early leaders of the NAACP, also accepted a prevailing myth about Jews and wrote of "the two million Jews [who] have a controlling interest in the finances of the nation," and he urged fellow blacks to "draw encouragement and hope from the experiences of modern Jews."[31] Repeating this sentiment that blacks should emulate the Jews in standing together and making money, *The Norfolk Journal and Guide* also noted that "in many ways [the Jew] sympathizes with and helps us. He gets his pound of flesh for doing it."[32] This theme of Jewish wealth and what it could purchase has remained constant throughout the twentieth century. In the 1960s Martin Luther King, Jr., wrote that "Negroes nurture a persisting myth that the Jews of America attained social mobility and status solely because they had money."[33]

Such beliefs also reveal how folklore reinforces cultural stereotypes. "The persistence of folkloristic expressions of hostility and prejudice that have their origins in social conditions much different from those that presently prevail," folklorist Nathan Hurvitz suggested in 1974, "indicates that they serve a purpose in our society. This purpose is to maintain and create cleavages between groups" and thus promote intragroup solidarity. And in the case of black folklore, that point is certainly true. "The traditional stereotype of the Jew as a money-grubbing materialist," historian Lawrence W. Levine tells us, "was prominent in Negro humor." And folklorist Daryl Dance reported that the black jokes and stories she collected "usually depict the Jew as a dishonest, unscrupulous, but successful businessman." There are several humorous tales about a "Colored Man, a Jew, and a White Man" in which the Jew is distinguished from other Caucasians. The main thrust of almost all these jokes is the compulsive Jewish concern for wealth. There are many variations of black folk humor concerning Jews exchanging checks for cash in the coffins of deceased friends. One frequently told southern tale instructed other blacks in how to distinguish a "cracker" from a "Jew." "Some white people is crackers and some is all mean and stingy," the narrator relates to his listener. "If one of dem is more stingy than he is mean, he's a Jew; and if he's more mean that stingy, he's a cracker." A tale from Brooklyn in the 1930s also reflects black impressions of Jews. The folklorist, Richard Dorson, included it in one of his collections:

> The local Christian church had burned down and the Jewish congregation in the neighborhood, very interested in furthering relations between the groups, agreed to let them use their synagogue for their Sunday prayers. And so they had their service in there. And as they were walking out, the two Negroes were walking by after this service and they looked down to see these people coming out of this synagogue. And the one says to the other, "You know, dat dere's the poorest bunch of Jews I ever did see."[34]

African-American impressions of Jews, like the views of blacks on several other topics as well, reflected the attitudes and values of white American Protestants along with specific resentments felt particularly by the black community. Beginning in the 1920s observers noted African Americans berating Jewish merchants for overcharging in retail establishments, Jewish landlords for rent gouging, and Jews, in general, for desecrating the Sabbath and insulting Christians by keeping their places of amusement open on Sundays.[35] A number of blacks also internalized the ideology of Nordic supremacy and believed in the inferiority of Jews, Italians, and other southern and eastern Europeans.[36] An

August 1, 1925, editorial in *The Philadelphia Tribune* admonished readers about the inappropriateness of such beliefs:

> If over night, the twelve million Negroes of the United States turned white, there would be a decided boost in the membership of the Ku Klux Klan, judging from the manner in which many black men have absorbed the prejudices of white Americans. . . .
>
> There is for instance a group of Negroes, who will tell you that they don't like Chinese and can't stand Jews. As for a reason they haven't any.
>
> Many black servants, including bellmen, waiters, cooks and the like will talk of a summer resort as being over run with Jews as if that made the ocean less blue and the air less fresh and clean. . . .
>
> In a group of people so utterly the victims of prejudice, this tendency to assume the evil natures of their own oppressors, to dislike men in bulk because of race or nationality is the height of inconsistency, and the sooner Black Americans snap out of it, the better for their own cause and for humanity in general.[37]

But most African Americans could not "snap out of it." Many of their attitudes toward Jews were firmly in place by the 1920s, even before most members of either group had much contact with one another. After the worldwide economic depression began in 1929 their living conditions worsened, especially in comparison with the experiences of most white people. Thus, after Adolf Hitler came to power in Germany, his policies toward Jews allowed American blacks an opportunity not only to compare their plight with those who were discriminated against in Germany,[38] but to scapegoat American Jews. Black antisemitism was both a means of venting frustrations toward all whites and a complaint against the sorry predicament in which they found themselves. Perhaps half the black residents in Harlem in the 1930s were on relief.[39] A number of African Americans therefore gained psychic satisfaction in seeing another group—and one they disliked—being persecuted as well. W. E. B. DuBois, for example, revealed the sentiments of many African Americans when he wrote:

> Nothing has filled us with such unholy glee as Hitler and the Nordics. When the only "inferior" peoples were "niggers" it was hard to get the attention of the *New York Times* for little matters of race, lynchings and mobs. But now that the damned include the owner of the *Times*, moral indignation is perking up.[40]

Some African Americans even believed that the Jews could not be too bad off in Germany because "they have all the money."[41]

A significant number of blacks supported Hitler's policies. *The New York Age* observed, "If the Jewish merchants in Germany treated the German workers as Blumsteins [a department store] treat the people of Harlem, then Hitler is right."[42] And in Chicago, *Dynamite* declared, "What America needs is a Hitler and what the Chicago Black Belt needs is a purge of the exploiting Jew."[43] Lunabelle Wedlock, who surveyed the black press in the 1930s, concluded that most of the writers "are either indifferent to German anti-Semitism or view with evident pleasure the degradation of a minority group other than their own."[44]

Many African Americans believed that their own problems in the United States deserved greater attention than those facing Jews in Germany.[45] "To be a Jew in Germany is hell," *The Philadelphia Tribune* declared, "for one to be a Negro in America is twice as bad."[46] Nonetheless, several newspapers showed concern for the plight of the Jew in Germany, including *The Philadelphia Tribune*, which editorialized, "The Nazi treatment of the Jew is brutal and unjustified." Then, repeating a common American assumption, it added:

> it is necessary to remember, however, that perhaps most of what is told about Jewish treatment in Germany is propaganda since the Jews control to a great extent the international press.[47]

African-American newspapers in the 1930s and early 1940s noticed the increased evidence of antisemitism during the decade.[48] The policies and practices of white employers contributed to the growing animosity although Jews were singled out by blacks for special opprobrium. That merchants in the ghettos hired few African-American employees until community pressures forced a change rankled neighborhood residents. Jobs were hard to come by and many storekeepers barely scraped by employing only members of their families to help them. Frank's restaurant, allegedly the best steakhouse in Harlem, served all comers and was owned by a Greek immigrant who employed only those of Greek ancestry as waiters and waitresses. Yet the perception was that most employers were Jews, or, those who were not Jews were not to be attacked severely. In the 1930s Con Edison in New York employed only three African-American maids and two black inspectors; the telephone and transportation systems did not even interview blacks for skilled jobs, and the Fifth Avenue Coach Company, which ran many of the city's buses, hired only white drivers, preferentially those who had letters of recommendation from their parish priests, until a 1941 agreement altered that policy.[49] Banks, large supermarkets, and police, fire, and sanitation departments of the city were also overwhelmingly white but were not targets for open wrath. One survey of grocery stores in Harlem in the late 1930s found 241 of the shops owned by Jews and 151 owned by Greeks, but did not mention those owned by individuals who were of Irish, Italian, or other European backgrounds.[50] Nonetheless, a Harlem newspaper confidently reported in 1942 that Jews constituted 95 percent of Harlem's merchants and made no reference to those who employed the largest numbers of workers.[51]

Jewish stores were condemned for not hiring blacks, exploiting their customers, and driving African-American competitors out of business. In truth many Jewish-owned businesses fared well among neighborhood customers because they offered a wider variety of merchandise at lower prices, provided credit, and were especially courteous or solicitous of their patrons.[52] But that did not give sufficient reason for black leaders to respect them. Journalist Kelly Miller wrote in 1935 that although the Jewish businessman "treats the Negro more kindly and sympathetically than his white Gentile competitor," the Jew is "a born business man" and "looks upon the Negro as an easy field for exploitation." Miller also used the opportunity to reinforce ancient prejudices:

> the Jew seems to deem it his mission to cater to Christian needs and necessities. They [sic] wax fat and profiteer on Christian holidays. They violate their own Sabbath, gathering in shekels, to supply Christians with their requirements for Sunday. Christmas and Easter furnish their superlative opportunity.[53]

In one field, domestic labor, where an overwhelming number of African-American women employees encountered female Jewish employers, attempts were made by the former to avoid the latter. In several cities, including Chicago, Baltimore, Pittsburgh, and Cincinnati, domestics sought employment in Gentile households and placed ads in newspapers stipulating that fact.[54] One item in a situations wanted column in the *Cincinnati Enquirer* in 1941 read: "Colored woman wants week work; neat; with references; no Jewish people."[55] Paradoxically, even as African-American servants in middle-income Jewish households in Chicago repeated all the derogatory stereotypes of Jews common in the both the African-American and dominant cultures, two-thirds of these same women believed that their Jewish employers were less prejudiced and treated them "more like equals" than did non-Jewish whites.[56]

The plight of African-American domestics received the greatest amount of attention after a 1935 article entitled "The Bronx Slave Market" exposed the degrading hiring practices and exploitation of day laborers in one section of New York City. The article was not primarily an antisemitic piece. Nevertheless others perceived it as such and it confirmed existing black prejudices because the street locations where women congregated and waited to sell their services were in heavily Jewish neighborhoods.[57] Similar street markets existed in Chicago, Philadelphia, and other cities, and their origins could be traced to the Jacksonian era. They revived during the depression when desperate unemployed women converged in particular spots and bargained with middle-class housewives for day labor.[58] Women negotiated for pay in the range of 15 to 30 cents an hour, haggled over whether lunch and/or carfare would be added, and stipulated whether dangerous kinds of activities, such as window washing, would be included. But without any evidence provided in the article, the authors of "The Bronx Slave Market" wrote:

Fortunate indeed is she who gets the full hourly rate promised. Often, her day's slavery is rewarded with a single dollar bill or whatever her unscrupulous employer pleases to pay. More often the clock is set back for an hour or more. Too often she is sent away without any pay at all.[59]

The sensational nature of "The Bronx Slave Market," and the humiliation of women having to sell their labor to casual bidders in the street, gave legitimacy to the domestics' complaints and led to the establishment of state-run employment agencies where employee and employer could contact one another in a more professional atmosphere. But the animosity created by the original article, and subsequent embellishments by women who worked for Jewish employers in the Bronx, added grist to the black antisemitic mills.[60]

Housing constituted another area of tension between blacks and Jews during the depression. "A popular misconception in some urban ghettos where rent gouging was a constant irritant," historians Robert Weisbord and Arthur Stein wrote, "was that *all* white landlords and shopkeepers were Jews." The image of the "fiendish Jewish landlord" and the "Shylock landlord" predominated in the ghettos of Chicago, New York, Detroit, Philadelphia, Pittsburgh, and other urban areas even when someone else owned the property. One Jewish landlord in Harlem allegedly told his African-American tenants that they were lucky to get an apartment with hardwood floors, tiled bathrooms, and French doors at any price. That story spread throughout the neighborhood and was repeated over the years as an example of the patronizing and offensive attitudes of Jewish building owners.[61] Three decades later *Time* magazine noted, "A Negro will frequently refer to his 'Jew landlord' even though the man's name may be O'Reilly, Kawolski or Santangelo."[62]

These frequent and negative contacts between blacks and Jews during the 1930s made both groups wary of one another. Poor people found the Jews-as-scapegoat a satisfying target for blame. One woman asked, "Since the Jews controlled most of the money in the United States," why did they not use "that financial power to better the condition of the negro?"[63] A scholar concluded: "To Harlem it had become a way of life to blame the Jew for discrimination and abuse."[64] By the end of the decade, just as an upsurge in antisemitism began in white America, numerous observers noted a sharp increase in antisemitic feelings and expressions among African Americans in Baltimore, Chicago, Detroit, New York, and Philadelphia.[65]

The onset of World War II ended the depression but hardly altered black or white perceptions of Jews except, perhaps, to intensify the negativism and hostility.[66] In February 1942 *The Amsterdam News* observed, "There never has been such general anti-Semitic sentiment in Harlem as exists right now," while in April 1943 the newspaper saw signs of even greater animosity and predicted that Harlem was a "tinder box, ready to explode." [67] In 1943 the *Pittsburgh Courier* also spoke of "the dangerous and disastrous spread of anti-Semitism among Negroes,"[68] while the NAACP devoted part of its meager resources in an attempt to eradicate black antisemitism.[69] No particularly noteworthy incidents of antisemitism, however, occurred during the war years until the race riots broke out in Detroit and Harlem, in June and August 1943. In both cases, Jewish owners of small stores were often the targets of vandalism and looting. From that point on, historian Isabel Boiko tells us, "Jewish businessmen realized that there was no hope for the kind of peaceful co-existence they had envisaged because Negro bitterness toward the Jew appeared permanent."[70]

Contemporary observers underscored Boiko's analysis. Ralph Bunche, then head of Howard University's Political Science Department, wrote:

> In the home, the school, the church, and in Negro society at large, the Negro child is exposed to disparaging images of the Jew. . . . Negro parents, teachers, professors, preachers, and business men, who would be the first to deny that there is any such thing as "the Negro," or that there are "Negro traits," generalize loosely about "the Jew," his disagreeable "racial traits," his "sharp business practices," his "aggressiveness," "clannishness" and his prejudice against Negroes. . . . The Jew is not disliked by Negroes because he is "white," but because he is a "Jew," as the Negro conceives the Jew.[71]

Columnist Arthur Huff Fauset of *The Philadelphia Tribune* also deplored the fact that "too many Negroes" blamed Jews "for all the evils which the majority group perpetrate against Negroes. Most vicious and far-reaching kinds of exploitation and other evidences of racial prejudice may pass by unnoticed if the culprits are Gentile Americans, but let it be said that a Jew has taken the slightest advantage of a Negro and he becomes 'the dirty Jew', 'kike', 'sheeny', 'exploiter'." Fauset also noted how he had been on platforms where the expressions "man" or "white man" had been used to describe Protestant and Catholic individuals "but time after time I have noted, and on occasion reprimanded speakers publicly because they use the term 'Jew' when they want to describe some person of that group who has done something we do not like."[72] Even in the South, where African Americans voiced no public criticism of Jews during the previous two decades, similar feelings apparently existed. "The truth of the matter is," Tennessee's *National Baptist Voice* declared in May 1945, "Negroes are filled with Anti-Semitism. In any group of Negroes, if the white people are not around, the mention of the Jew calls forth bitter tirades."[73] One New York City poll in the mid 1940s also disclosed that 70 percent of the city's Protestants, most of whom were African Americans, held some negative attitudes toward Jews.[74]

Despite Jewish knowledge of black animosities, no white group in America provided as much enthusiasm, organizational and financial assistance, and sincere involvement in the Civil Rights movement as they did.[75] Jews had been among the founders and promoters of the NAACP in 1909, they had provided three presidents of the organization: Arthur and Joel Spingarn and Kivie Kaplan, and Jack Greenberg worked with Thurgood Marshall before succeeding him as counsel of the NAACP Legal Defense and Educational Fund. Jews also joined and helped finance groups like the United Negro College Fund, the National Urban League, CORE (Congress of Racial Equality), SNCC (Student Non-Violent Coordinating Committee), and SCLC (Southern Christian Leadership Conference), while wholeheartedly devoting themselves to the cause. Jews constituted a plurality of the white civil rights attorneys, more than half of the "freedom riders" of the early 1960s, and approximately two-thirds of the college students who volunteered for Mississippi's "Freedom Summer" in 1964. Black civil rights leaders like Roy Wilkins of the NAACP, Bayard Rustin of the A. Philip Randolph Institute, and James Farmer of CORE recognized and appreciated Jewish involvement and assistance in their battle to achieve eqality in America. Martin Luther King, Jr., spoke for many knowledgeable activists when he said: "It would be impossible to record the contribution that Jewish people have made toward the Negro's struggle for freedom, it has been so great."[76]

A major area of contention developed, however, in Jewish endeavors at assistance. Jewish leaders assumed the role of senior partner because they believed that they had the necessary know-how to take charge. Jews thought in terms of doing things "*for* Negroes, rather than *with* them."[77] "It was kind and benevolent," Jewish community activist and scholar Albert Vorspan later recalled, "but it was also colonial."[78] At the time, though, Jews did not realize how such behavior could intensify existing resentment and animosity. "There was an alliance," Philadelphia AJC director Murray Friedman later wrote, "but side by side much tension which Jews tended to ignore but Blacks clearly recognized."[79]

Nonetheless both Jews and blacks wanted civil rights legislation and frequently their interests meshed. Not only did Jewish community relations leaders support the Civil Rights movement but so too did most ordinary Jewish businessmen who sometimes may have been less than honest in their dealings with customers in the ghetto, and Jewish housewives, who occasionally tried to pay their domestics with old clothes instead of cash. Even Jewish bigots, who in private conversations used the derogatory Yiddish term "*schwartzes*" (black people) when speaking about African Americans, enthusiastically backed civil rights legislation. There is an ethical component in the Jewish faith and a tradition of social justice in the Jewish culture that is strong even though the rhetoric and behavior of some Jews belie the existence of these values. However, none of this minimizes the fact that Jewish self-interest also played a part in their quest for civil rights legislation.

Similarly, black leaders had their eyes on the target and were not about to jettison a partner who shared their goals and bankrolled their cause. Like most white people who may not have felt great affection for Jews, African Americans carefully refrained from outbursts of antisemitism during the late 1940s and 1950s. That did not mean that either black or white hostility toward Jews had disappeared; yet while it was definitely on the wane among whites, one might say it was relatively quiescent but stirring among blacks. Thus, if most African Americans did not embrace Jews in the immediate post-World War II decades, they rarely excoriated them publicly.

Despite the appreciation that black leaders heaped on Jews for their promotion of civil rights legislation a growing number of the rank-and-file resented their take-charge attitudes; some even considered Jewish participation as little more than enlightened self-interest, which for some Jews it no doubt was. Several black professional staff members who worked with Jews in the Civil Rights movement considered them hypocritical and suspected their motives. In the early 1950s a citizen's committee in Chicago formed a Council Against Discrimination (CAD) but African Americans were its weakest supporters. Several participants even raised charges of Jewish "dictatorship." Many local NAACP groups thought they should control the movement's direction and did not want to share the helm with Jews. In 1950s Chicago, therefore, although Jews were the most well organized and well financed of the civil rights coalition they found it difficult to get members of the NAACP or other African American groups to engage in activities for the cause.[80]

As in previous decades, major African-American journalists tried to mend fences and pointed out the inappropriateness of black antisemitism. In 1958, Louis Martin of the *Chicago Defender* wrote, "No other minority in American life, including ourselves, has fought more vigorously or more effectively against prejudice and bigotry than the Jews." Then, in 1960, the *Pittsburgh Courier* reminded readers that there probably would not have been a lasting NAACP if it had not been for Jewish support. "Not only have Jews stuck their necks out for us," the editorial continued, "they have fought gallantly and intelligently for social justice for everybody. The Jews are a people to be emulated, not despised."[81] Yet, despite these exhortations a significant percentage of African Americans questioned the sincerity of Jews who involved themselves in the fight for civil rights.[82] Polls taken from the 1960s through the 1990s consistently found African Americans more antisemitic than American whites. In a 1964 analysis 47 percent of blacks compared to 35 percent of whites scored high on antisemitic beliefs; in 1981, one of five whites and two of five blacks were found to be antisemitic;

and in 1992 Henry Lewis Gates, Jr., director of the Afro-American Center at Harvard University, wrote that African Americans, especially the younger and better educated adults, were twice as likely to be antisemitic as their white counterparts.[83] One 1960s study, in fact, found that black migrants from the South to the North harbored "an amazingly high percentage of anti-Semitism, reflecting some of the white attitudes of that region."[84]

In the 1960s angry and rising younger leaders like Cecil Moore of Philadelphia, who regarded every Jew in the civil rights movement as "a goddam phoney,"[85] and Malcolm X, who thought that the six million Jews annihilated by Hitler "brought it on themselves,"[86] won enthusiastic followings. During that same decade the depth of black rage also became evident. One African American in Philadelphia asserted:

> You know all those Jews in the civil rights marches and going down South—you know why they do it? They do it to take the heat off themselves. They've got a bad conscience because they live on black dollars.[87]

Both knowledgeable blacks and Jews in the major defense and community relations organizations like the NAACP, the Urban League, CORE, ADL, AJC, and the American Jewish Congress knew of the rising antisemitism among blacks. Most of the leaders of both Jewish and African American groups, however, did not make too much of that hostility lest they split the fragile coalition working for passage of the 1964 Civil Rights Act.[88]

What many Jews then discovered, and what the younger and more combative blacks spoke out about fiercely, was that the goals of the two groups could not be the same because their conditions differed. Unequal treatment of blacks and whites in this country, as well as different economic circumstances, cultural values, historical experiences, and levels of skills, made shared strategies after passage of the Civil Rights Act difficult to obtain. Sociologist Nathan Glazer captured African American sentiments exactly when he wrote in 1964, "The Negro anger is based on the fact that the system of formal equality produces so little for them."[89] Jews had been well educated, ranked as the wealthiest ethnic group in America, and were positioned to benefit from equal opportunities in a wide variety of areas once they became available.

But equality of opportunity for blacks did not significantly advance the group's position. African-American spokesmen recognized that their people had to make up for centuries of slavery, racism, and discrimination that had caused significant domestic and social problems. They needed additional education and training before most of them could fully participate in the elusive American dream.

As the disappointments mounted from the failure of both the 1954 *Brown* v. *Board of Education* decision and the Civil Rights and Voting Rights Acts of 1964 and 1965 to produce significant improvement in their lives, many African Americans rallied behind younger and more aggressive leaders. Black rhetoric turned increasingly nationalistic, and their new militancy included a greater identification with dark-skinned peoples in the Third World, Marxism, opposition to Israel, and mounting antisemitism. The escalation of black nationalism cemented the younger leaders' beliefs that African Americans must control the direction of their cause.[90]

But "black power" was not the reason that the United States witnessed 400 urban riots, some of which broke out every summer between 1964 and 1969. The underlying causes of these outbursts related more to unfulfilled promises in the United States than to antisemitic hostility. Nonetheless, once the barriers of public civility were broken, outspoken antisemitism from African Americans received much media attention. Although Jewish leaders, and especially Jewish liberals of all ages, tried to be "understanding" about the manifestations of black fury some racist attacks could not be ignored.[91]

One of these occurred on the night of February 3, 1966, at a meeting of the Mt. Vernon, New York, school board. Desegregation of local school facilities dominated the increasingly bitter discussion. During the course of a heated exchange Clifford A. Brown, an official of CORE, became so riled up that he shouted out at the audience, "Hitler made a mistake when he didn't kill enough of you." Brown later apologized for this remark but the damage had been done. No black officials made any immediate comment about Brown's infelicitous expression. The following week former baseball player Jackie Robinson chastised this "man, who, if he is not vicious, is combat-weary in the struggle and ought to be retired from the front lines as unfit to speak for anyone but his fevered self."[92] Then, a month later, Martin Luther King, Jr., indicated that he did "not view this horrible outburst as anti-Jewish. I see it as anti-man and anti-God. It would be a statement to harshly condemn, coming from anyone."[93] Jews were not only shocked by Brown's remark, but disappointed by the small number of African-American spokesmen who distanced themselves from Brown or his outrageous comment. Brown's words, however, resulted in the downfall of CORE. Financial contributions to the organization, most of which had been coming from Jews, had totalled $44,500 in January 1966, and then declined to $19,900 in February and $7,500 in March. In February 1969 Roy Innis, president of CORE, stated, "A black man would be crazy to publicly repudiate anti-Semitism." A year later CORE ceased functioning.[94]

It was obvious by 1966 that African-American antisemitism was no longer a topic to be shoved under the rug and whispered about by those still trying to maintain a weak coalition of blacks and Jews.[95] A year after the Mount Vernon incident, in April 1967, *The New York Times Magazine* highlighted the undeniable rift. In one article James Baldwin argued that "Negroes Are Anti-Semitic Because They're Anti-White"; in a counterpoint Robert Gordis responded, "Negroes Are Anti-Semitic

Because They Want a Scapegoat."[96] Never before had black animosity toward Jews been proclaimed so dramatically by a major American publication. But in the next two years the pace of the incidents accelerated and practically shattered the remnants of solidarity between African-American and Jewish organizations.

The most noteworthy controversy centered in the New York City public schools and received nationwide attention. Because passage of the Civil Rights Act of 1964 and Voting Rights Act of 1965 failed to provide the hoped-for changes in black lives, other avenues for improvement were explored. Militants focused sharply on economic and political goals, particularly on gaining influence over institutions that limited their opportunities, such as schools. The idea of small community-controlled school districts appealed to many civil rights supporters because the public schools in the city had not been successful in educating black children. In the autumn of 1967 one African American school teacher, John Hatchett, wrote:

> We are witnessing today in New York City a phenomenon that spells death for the minds and souls of our black children. It is the systematic coming of age of the Jews who dominate and control the educational bureaucracy of the New York Public School system and their power starved imitators, the Black Anglo-Saxons.[97]

The opening of a few experimental schools in New York City promised greater control by African Americans of their destinies. The one in Ocean Hill-Brownsville, established in 1967, received the most publicity because of some financial assistance from the Ford Foundation and the personal attention given to the project by then Senator Robert F. Kennedy.

Almost immediately the Ocean Hill-Brownsville school board found itself locked in with a set of tenured teachers who belonged to the American Federation of Teachers (AFT) union. In an attempt to reorganize the district, Rhody McCoy, the chief school board administrator, dismissed several of the old-timers, most of whom were Jewish, and replaced them with younger people, 40 percent of whom were also Jewish. Those fired naturally sought union protection. The head of the teachers' union, Albert Shanker, accused the Ocean Hill-Brownsville school board of antisemitism, and New York City's teachers went out on strike in the fall of 1968. What was an attempt by blacks to control the local school district escalated into an employer-employee conflict that then erupted into the most vicious and visible black-Jewish confrontation in the history of the United States. The strike polarized the city and made national headlines. No provocative statement made by either side remained unpublicized. The teachers' union took the remarks of perhaps half a dozen vicious antisemites and circulated thousands of copies of the most defamatory comments. A few black parents and others in the district paraded with signs reading, "Jew

Pigs," and calling Hitler the "Messiah." Jews were accused of practicing "genocide" on black children, and Jewish teachers received notes reading "Watch yourself, Jew, crossing streets, drinking tea, etc. You have been marked for elimination." The numerous insults and glares exchanged by both sides intensified mutual hostility.[98]

McCoy and the governing board of the Ocean Hill-Brownsville district went on record as opposing antisemitism, but the actions by the teachers' union and several black militants exacerbated the conflict. Although the school strike finally ended on November 18, 1968, with places found elsewhere in the system for the fired teachers, its repercussions lasted for decades. Many teachers who experienced that incident still have vivid memories of it. For many blacks, the behavior of the Jewish teachers and their union confirmed long-held prejudices. The New York Civil Liberties Union later condemned Shanker for "proving" accusations of antisemitism during the strike with "half-truths, innuendoes, and outright lies," while the Anti-Defamation League observed in its report of the same controversy:

> anyone familiar with the events of the last few years would know that this [conflict] has been building up and has been waiting for an incident to release the mounting wrath, and . . . the planned and calculated incitement of Blacks against Jews.[99]

Both the New York Civil Liberties Union and the ADL were correct. Albert Shanker and the teachers union had misjudged the nature of what McCoy and others in the experimental school district were trying to do and steadfastly adhered to a literal interpretation of the union contract with the Board of Education. In a situation demanding heightened sensitivity, members of the union kept referring to contractual clauses. On the other hand, many of the Jewish teachers had been victimized by sniping African-American colleagues in the 1960s and they felt that in the rush to promote civil rights for African Americans their concerns had been overlooked. Furthermore, had black wrath been directed solely at the union's intransigence, American Jewish leaders would not have been unduly alarmed. But the use of antisemitic slurs and slogans in promoting an otherwise worthy cause suggested that some blacks welcomed an opportunity to vilify Jews rather than just seek improved schools for their children.[100]

Three other incidents in New York City in 1968–1969 convinced Jews that blacks were becoming more antisemitic. Surprisingly, blacks interpreted the same incidents as proving that Jews controlled the media and other power points in American society. The first concerned John Hatchett's appointment as director of the New York University's Afro-American Center in 1968. Hatchett, a militant black teacher, had accused Jews of controlling and dominating the New York City school system. Thoughtful Jews felt uncomfortable with NYU's decision and tried, unsuccessfully, to get it reversed. But after only a few months on the job Hatchett accused the respective Democratic and Republican presidential nominees, Hubert H. Humphrey and Richard M. Nixon, of being, along with Albert Shanker, "racist bastards." The use of such language in describing these men led NYU to terminate his contract. According to a *New York Times* reporter, however, Hatchett's supporters viewed him "as a victim of racism," and they questioned "whether the fear of losing contributions by wealthy Jews was behind N.Y.U.'s action in dismissing him."[101]

After the teachers' strike ended in November, an inflammatory poem read over New York City's WBAI radio station in late December eclipsed the Hatchett affair in people's minds. An African-American school teacher, Leslie Campbell, hesitated to quote the verse of one of his students while being interviewed on a radio talk show because he feared the public relations consequences of doing so. The program's host, Julius Lester, nonetheless prevailed upon him to go ahead. Dedicated to Albert Shanker, it began:

> Hey, Jew boy, with the yarmulke on your head
> You pale faced Jew boy—I wish you were dead.[102]

Campbell's reading evoked no immediate response but when *The New York Times* carried the story on January 16, 1969, a few Jewish spokespersons called for the revocation of the radio station's license.[103]

Two days later, on January 18, 1969, an exhibit entitled "Harlem on My Mind" opened at New York City's Metropolitan Museum of Art. Intended as a tribute to blacks and their cultural accomplishments, it created a furor because of some of the comments that a fifteen-year-old girl wrote in the introduction to the exhibit's catalog. Critics zeroed in on these sentences:

> Behind every hurdle that the Afro-American has yet to jump stands the Jew who had already cleared it. Jewish shopkeepers are the only remaining survivors in the expanding black ghettos. The lack of competition allows the already exploited black to be further exploited by Jews.

Outraged Jews, and Jewish organizations, demanded that these remarks be expunged from the catalog and any other suggestion of antisemitism be immediately eliminated from the exhibit. The catalog, however, had already been published and exhibit organizers argued that the girl's remarks reflected cultural values and beliefs and were not designed as attacks on anyone. Nevertheless, the protests were so strong that the museum withdrew the catalog from public circulation before the month ended.[104]

The events of the late 1960s frightened American Jews. The attacks by African Americans, along with the 1967 school board incident in Wayne, New Jersey, combined as the most overt manifestations of bigotry toward Jews in the United States since the end of World War II. And the hostility seemed to spread from New York to several other cities. The editor of the Los Angeles *Herald-Dispatch* wrote that "what the Negro hates is the awful deceit by the Jews. The whole civil rights movement is a branch of Zionism."[105] An African-American leader of Chicago's West Side Organization told listeners, "the Negro hates the Jew, at least around here"[106]; Philadelphia blacks regarded Jews as parasites; black antisemitism spread "like wildfire" in Atlanta,[107] while a speaker at a meeting of the Angry Black Young Men in New Haven, attended by over 600 people (75 percent black, 25 percent white, 90 percent under thirty), received enthusiastic responses from the audience for his "anti-Jewish diatribes."[108] On the night of April 4–5, 1968, after word spread of Martin Luther King, Jr.'s assassination, angry blacks broke into Cincinnati's Rochdale Temple and tore it apart. Two members of the congregation recalled the devastation:

> they came in and overnight the whole temple was destroyed. They came in and pulled out every fixture, every piece of brass, all the beautiful lights were pulled down. . . . And it was just unbelievable, the destruction. All the pews were broken and turned over, and the place was just ramshackle. . . . Most of the destruction took place overnight. They just came in and they just ripped the place apart.[109]

A Boston congregation only barely avoided that kind of destruction. Members of the local SNCC chapter wanted the Jewish Temple, Mishkan Tefila, turned over to them because it was in a changing neighborhood. The synagogue board of directors was told, "We get the temple mortgage free or else we burn, baby, burn." While the board deliberated, another message arrived: "Put the temple in the hands of the black community or we'll burn it down with Jews in it." The board capitulated and gave the temple to the local African-American community.[110]

These aggressive and hostile actions of African Americans left most Jews bitter,[111] and they reacted by reducing or eliminating their financial support of civil rights causes. Other Jews still believed that Jewish welfare was tied to the improvement of all minorities, however, and they tried to rationalize the viciousness and the assaults because basically they felt deeply for black people who had suffered so much over the centuries.

In the 1970s, leading African-American and Jewish organizations attempted to heal the raw wounds of the 1960s. To some extent they did, but one overriding issue—affirmative action—prevented a complete rapprochement. Both groups wanted to bring more blacks into industry at skilled levels and into professional positions. Both favored training and education for those formerly denied quality instruction. But blacks also favored specific numerical goals while historically minded Jews interpreted that stipulation as "quotas"—something that white Christians had used against them. Jews had been working for equal opportunity based on individual merit; they were not willing to negate their advances in this sphere by acceding to the demands of group rights.[112] Moreover, many non-Jewish whites also favored some kind of "affirmative action" that did not stipulate specific numerical goals.

This issue did not have universal agreement from Jewish organizations, however. When in 1974 the University of Washington passed over an individual, Marco de Funis, for admission to law school and accepted another candidate with lower test scores but from a different cultural background, de Funis sued on the grounds of reverse discrimination. Both the National Council of Jewish Women and the Union of American Hebrew Congregations stood with the National Urban League in filling *amicus curiae* briefs with the courts, supporting the university's contention that test scores alone were an inadequate criteria for admission. The case went into federal courts but the University of Washington subsequently admitted de Funis to law school before the U.S. Supreme Court had the opportunity of rendering a judgment on the matter.[113]

Another dispute, however, that of Allan Bakke, a white Christian of Norwegian ancestry, did get to the Supreme Court and in 1978, by a slender 5–4 verdict, the majority ruled that race alone was not an acceptable criterion for determining school admissions.[114] Although polls showed that 77 percent of white people approved of the Court's decision and only 12 percent disapproved, among blacks the figures were practically reversed. "By 74–15 percent," a Lou Harris survey concluded, "a majority of blacks tend to feel that 'unless quotas are used, blacks and other minorities just won't get a fair shake.' "[115] This real conflict over affirmative action and quotas would certainly have strained black-Jewish relations but would not have aggravated antisemitism per se if underlying hostility did not exist.

The same might be said for another incident: Andrew Young's forced resignation as U.S. Ambassador to the United Nations in 1979. The Carter administration decided that Ambassador Young had to be replaced because he had violated American policy by meeting secretly with representatives of the Palestinian Liberation Organization (PLO) and then misinformed officials in Washington about its occurrence.[116] Young's indiscretion was more than a minor mistake since two Presidents had promised Israel that American officials would not meet with representatives of the PLO until that organization recognized Israel's right to exist. When word of Young's meeting leaked, both Israeli and American Jews wondered whether it meant a reversal of policy. American Jews were furious about Young's clandestine encounter and many of them spoke out in anger. But Jewish *leaders* did not want Young to be sacrificed because of it, and they feared, correctly, that if he did go they would be blamed for his release. Almost all African Americans, however, were convinced that American Jews had forced Young's resignation. From the black perspective Young's departure highlighted both their own powerlessness and the apparent clout of American Jewry.[117]

It is impossible to exaggerate the impact of Young's resignation. Not since the New York City school strike of 1968 had any incident drawn such a strong response from African Americans, their spokespersons, and their newspapers. Aside from Supreme Court Justice Thurgood Marshall, Young was the highest level African American official in the government and served as a source of pride to the entire black community. His alleged misdeed did not even appear to be particularly inappropriate. After all, he was using his office and influence to try to obtain peace in the Middle East. Therefore, in the minds of African Americans, he was punished not for what he did but for what Jews did not like him to have done.[118] One cab driver in Chicago exclaimed: "It's them [the Jews]. They hate us. They rob us. They block every move we try to make to get a piece of the action. Now they got Andy."[119] A minister in a New York suburb complained: "Our people had been amazed that the Jews had such political clout with the President and that they cause [sic] the resignation of our Ambassador."[120]

African-American spokespersons and columnists forcefully articulated their concerns in responding to Young's exodus from his post at the United Nations. William Rasberry of the *Washington Post* wrote that "the Young affair served to legitimize real anti-Semitism" among blacks. Julian Bond of Georgia observed that "Young has finally angered the one lobby that cannot be angered." Esther Edwards of the National Black Human Rights Caucus called Young "a scapegoat to appease Jewish ethnics here and in Israel," while the *Pittsburgh Courier* quoted one black leader who attributed Young's removal to "the power, influence and money of the American Jews. To tie it with any other thing is ludicrous." Jesse Jackson, who would soon emerge as the most adored and respected African-American leader since Martin Luther King, Jr., called the resignation a "capitulation" to the Jews. "The real resistance to black progress," he continued, "has not been coming from the Ku Klux Klan but from our former allies in the American Jewish Community."[121] Andrew Young later admitted privately, but would not do so publicly, that Jews were not responsible for his resignation.[122] Nevertheless, five years later a "black scholar" with no additional evidence beyond that available at the time of the resignation, wrote that Young had been "pressured from his post as U.N. Ambassador for the U.S. by the Jewish establishment."[123]

Young's ouster shattered any possible reconciliation between Jewish and black leaders. The helplessness that African Americans felt would have resulted in the destruction of any black person's credibility in the community if he or she made overtures for cooperation with Jewish organizations at that time. On the other hand, there was a sense of disaffection and general deprivation among blacks and the field was

wide open to anyone who would pick up where Young left off and try
to assume the mantle of leadership.[124] And that is exactly what Jesse
Jackson did.

Jackson had been part of Martin Luther King Jr.'s entourage and
then directed an anti-poverty program, People United to Save Human-
ity (PUSH), in Chicago. In the early 1970s he told followers that four
out of five of Nixon's top advisors in the White House were Jewish,
and he attributed the administration's insensitivity to poor people to
the President's chief associates, Robert Haldeman and John D. Ehrlich-
man, two non-Jews whom he referred to as "German Jews."[125] Appearing on the CBS television program "60 Minutes" in 1979, Jackson
opined that Jewish-controlled industry kept blacks out of positions of
authority. The next year he was quoted as saying that "Zionism is a
kind of poisonous weed that is choking Judaism."[126] Later, when
informed of CBS White House correspondent Leslie Stahl's ethnic background, he seemed surprised: "She doesn't look like she's Jewish. She
doesn't sound like she's Jewish."[127]

Jackson always seemed to have had an eye for the limelight and
wasted no time enhancing his stature in the African-American community after Young resigned as Ambassador to the United Nations.
Within weeks he had flown to the Middle East, embraced PLO leader
Yasser Arafat, and seen his picture displayed on front pages of major
newspapers throughout America. Claiming he was making every effort
to promote peace in the Middle East, Jackson returned to this country
a hero in his community.[128] In September 1979 he told a 400-person
audience of African Americans that support for the PLO marked "Black
America's finest hour."[129] Aside from the Black Muslims, African
Americans had not been noticeably concerned about the plight of the
Palestinians before or after Jackson's visit, and until the ouster of Young
there had been no indication that Jackson had had an affinity for the
PLO cause in the Middle East.[130] His impromptu efforts at negotiation
therefore startled many Jews and impressed most other Americans.
Officials of the federal government, however, did not approve of private citizens engaging in diplomatic affairs without the sanction of the
President or the State Department. Jackson's perceived diplomatic triumph, however, and the attention it brought him, resulted in the development of an enthusiastic following in the African-American community and suspicion in Washington, D.C.

Jackson's Middle East tour resulted in more than national attention
and community respect. On his return he called Arab-American businessmen together at his Chicago headquarters and told them that if
they wanted the support of African Americans they would have to
show them that it was mutual. The Arab League understood the message and donated over $200,000 to PUSH. When critics denounced Jack-
son's pandering to Arabs he responded, "Arab-Americans have been
treated like pariahs. I'm working to change that." Julian Bond

explained Jackson's solicitation of assistance somewhat differently. He observed that previous sources of financial contributions had fallen off and added, "Gee, these people have all this money. We need it, and we are potential allies of theirs."[131]

Jackson's rise in stature in the early 1980s led to his decision to seek the Democratic presidential nomination in 1984. Never before had there been a black candidate with so many followers for the nation's highest office. African Americans backed him in unprecedented numbers as he pledged new assistance to the downtrodden and promised to reverse the fortunes of those who had been ignored by the federal government for so long. His message rang true to many liberals and those anxious for a significant change of direction in national policies.

Unfortunately, however, an indiscretion of his, made in a conversation with a reporter, concerned nicknames he claimed to have had for merchants and clothiers in a downtown section off of Maxwell Street in Chicago: "Jewtown is where Hymie gets you if you can't negotiate them suits down."[132] He then stated how he had referred to New York City in the past as "Hymietown."[133] The remark further alienated millions of American Jews who had already had concerns about his commitment to issues they considered important, and put another nail in the coffin of Jewish-black amicability. Jackson apologized for his gaffe but columnists and most Jews refused to forgive or forget.

Jackson made other remarks in his 1984 quest for the presidency that Jewish Americans considered insensitive to their contributions to the civil rights crusade decades earlier. His recollection of the involvement of Jews in the civil rights causes minimized their dedication and contributions. He stated that in the South at that time

there were signs that said, "No dogs, blacks or Jews." The Klan was lynching blacks and using violence against Jews as well. So in many ways, neither was doing anybody a benevolent favor. Our interests converged.[134]

Jackson's attitudes toward, and remarks about, Jews did not affect his standing with African-American supporters. In assessing Jackson's candidacy historian John Hope Franklin observed:

For the first time white America saw a black Presidential candidate who spoke out on all the issues, not just black issues. Jesse showed that he could discuss any subject. He held his own with other candidates and, in many instances, bested them. It was a marvelous experience for whites.[135]

Although some young and idealistic Jews perceived Jackson's candidacy in the same vein as Franklin did and worked in his campaign, the overwhelming majority in the American Jewish community neither liked nor trusted Jackson and withheld its support. When queried afterwards, twelve African-American congressmen said their constituents believed that Jews opposed Jackson because of his race.[136]

Jackson's campaign also helped propel the career of Louis Farrakhan, a Black Muslim minister. Some of Farrakhan's followers, strong young men known as the "Fruit of Islam," served as bodyguards for Jackson. During the furor over the "Hymie" remark, Farrkhan warned, "I say to the Jewish people . . . if you harm this brother, I warn you in the name of Allah, this will be the last one you do harm."[137] Farrakhan, who had been preaching self-help within the black community during the 1970s and receiving little public notice, also attracted attention from Jews in March 1984 when he stated, "Hitler was a very great man."[138] Then, in a summer address at the National Press Club, he informed listeners that Israel had had no peace in the previous

> 40 years and she will never have any peace because there can be no peace structured on injustice, lying, thievery, and deceit using God's name to shield your dirty religion or practices under His Holy and Righteous name.[139]

As a result of Farrakhan's growing national recognition, his stature rose among African Americans and he became one of the most sought-after speakers on college campuses throughout the country. The Black Muslim minister's comments reached the ears and eyes of millions of Americans. He shared his views with college youths and more mature adults in every part of the United States. At some of his forums he invited others to sit on the platform. In Los Angeles, he had a former leader of the Ku Klux Klan; elsewhere he featured Arthur Butz, a man who claimed that the Holocaust had never occurred. Psychiatrist Alvin Poussaint attributed Farrakhan's popularity to his skill at tapping into the frustrations and anger that African Americans felt about the unfulfilled promises of the Civil Rights movement.[140]

The Farrakhan speech that received the greatest attention took place at New York City's Madison Square Garden on October 7, 1985, before an overflow crowd that exceeded 20,000 enthusiastic supporters. Farrakhan used Biblical idioms to emphasize themes familiar to his audience, ones almost guaranteed to evoke emotional responses. His words enthralled listeners and periodically brought them to their feet cheering.[141] His speech was supposed to discuss economic matters but reporter Michael Kramer wrote, "Anti-Semitism was the very centerpiece of his ramble. Economics was an aside."[142] The minister's statement that "the Jewish lobby has a stranglehold on the government of the United States"[143] elicited responses of "Yes!" and " 'Tell 'em Brother!' "[144] When Farrakhan asked, "Who were the enemies of Jesus?" the audience responded: "Jews! Jews! Jews!"[145] And another time, "Are the Jews that are angry with me righteous?" "No," came the passionate response.[146] Comparing himself to the Messiah, Farrahkan observed:

Jesus had a controversy with the Jews. Farrakhan had a controversy with the Jews. Jesus was hated by the Jews. Farrakhan is hated by the Jews. Jesus was scourged by Jews in their temple. Farrakhan is scourged by Jews in their synagogues.[147]

Then, in remarks directed at Jews, he said: "The scriptures charge your people with killing the prophets of God,"[148] and "I am your last chance, Jews," it will be too late "when God puts you in the oven."[149]

Assessments of the speech afterwards suggested that Farrakhan had told listeners what they wanted to hear. Julius Lester, then director of the African American Studies program at the University of Massachusetts, Amherst, was among those present in Madison Square Garden. He wrote:

> the audience greeted each anti-Semitic thrust by rising to its feet, cheering, arms outstretched at 45-degree angles, fist clenched. As this scene repeated itself throughout the evening, I wondered, Is this what it was like to be at the Nuremburg rallies in Nazi Germany?[150]

Also commenting on the presentation, Roger Rosenblatt of *Time* noted that people in the audience were already "predisposed to Farrakhan, they seemed to be ahead of him, rising to his message so eagerly it was hard to tell if the incitement preceded the response."[151] Commenting on both Farrakhan's speech and general positions, Reverend Lawrence Lucas, pastor of Resurrection Roman Catholic Church in Harlem, told newspaper reporters: "I see 99 percent of Louis Farrkhan's message as positive."[152] In 1988 Chicago Alderman William Henry echoed that stance: "You don't have to agree with everything one says, but most of the things Farrakhan says are true."[153] In May 1988 one of Farrakhan's disciples, Steve Cokely, an aide to Chicago's acting mayor, Eugene Sawyer, made the front pages of the Chicago *Tribune* when it became known that he had earlier recorded a message stating that Jewish doctors injected the AIDS virus into black babies and that Jews were involved in a conspiracy to take over the world.[154] Cokely apologized for his remarks amid cries for the Mayor to dismiss him. It then took a week before Sawyer asked for Cokely's resignation.

Cokely's words were both outrageous and without foundation but within Chicago's African-American community few condemned him publicly and many supported him openly. Black columnists Clarence Page and Vernon Jarrett rebuked him for what he said but other members of the African-American community and their elected officials responded differently. "When I walk down the street," Cokely acknowledged, "people want to touch my clothes, they want my autograph. It's like I'm Muhammad Ali." Farrakhan praised him and said Jews were offended by the remarks because "the truth hurts."[155] Only three of the city's eighteen African-American aldermen called for Cokely's resignation.[156] Listeners to the black radio phone-in talk show on WVON heard defenses of Cokely and "vicious" antisemitic comments as people vented their wrath. A typical caller, one Chicago reporter wrote, "quotes from the New Testament a Christian denunciation of the Jews as Christ-killers and urges that ministers teach this doctrine to their flocks."[157]

There is little that Jews can do to eliminate the anger that African Americans feel toward them. They are continually being held responsible for a variety of frustrations that blacks still experience. In July 1991, for example, Leonard Jeffries, who received his Ph.D. from Columbia University and who directed the African American Studies Department at CCNY, told an audience that Jews controlled the African slave trade centuries ago and that Hollywood's Jewish moguls have engaged in a "system of destruction of black people." Henry Lewis Gates, Jr., an African-American scholar who deplores anyone's "unscrupulous distortion of the historic record" countermanded Jeffries' assertions by pointing out that only about 2 percent of those involved in the slave trade were Jewish and that all the Jewish slave trading combined involved fewer slaves than those bought and sold by a single American Gentile firm, Franklin and Armfield. Gates also censured "the tacit conviction that culpability is heritable." Should everyone be condemned for the alleged crimes of their ancestors?[158] Many African Americans, however, see Jeffries as a man of great wisdom and a target of inappropriate attack. "He's telling the truth—he's got the documentation," a listener averred after one of Jeffries' talks.[159]

The opinions and accusations of Jeffries, Cokely, and Farrakhan have little basis in reality but the fact that they arouse Jews and win recognition from other African Americans are sufficient reasons for them to continue making inflammatory statements.[160] "That's a manifestation of the widespread rage, frustration and despair we have in the black community," columnist Clarence Page acknowledged.[161] In a similar vein, journalist Taylor Branch noted that any black leader who stands up to the white establishment or makes "white America leap on a chair in fright or revulsion will win the generous admiration of suffering black America."[162]

African Americans believe that their concerns receive little positive response from society at large. Some of their problems are overwhelming and apparently insurmountable. People in these circumstances need a scapegoat to whom they can attribute their plight. Cornel West, Professor of Religion and Director of the Afro-American Studies Program at Princeton, has written that

> the rhetoric of Farrakhan and Jeffries feeds on an undeniable history of black denigration at the hands of Americans of every ethnic and religious group. The delicate issues of black self-love and self-contempt are then viewed in terms of white put-down and Jewish conspiracy. The precious quest for black self-esteem is reduced to immature and cathartic gestures that bespeak an excessive obsession with whites and Jews.[163]

The question that must be asked, of course, is "Why the Jews?" And the answer to that must be that among African Americans anti-semitism serves several functions. It not only allows them to identify with the white majority but it provides as well a socially acceptable outgroup on whom they might vent their frustrations. (Until 1992 blacks were rarely criticized for their antisemitic comments except by Jews and occasionally by elected political officials who valued the support of their Jewish constituents.) Antisemitism is also an expression of their deeply felt religious beliefs.[164] The fact that a Jew "doesn't believe in Jesus Christ," a Columbia College student athlete told his audience in May 1990, "offends many, including me."[165]

Black newspapers have also played a role in promoting prejudice. They may have praised Jews and encouraged African Americans to emulate their achievements and cohesiveness but editors generally also warned of the Jews' inability to be trusted and their desire always to obtain their pound of flesh. Ideas about Jewish wealth and cunning were so conspicuous in society that African Americans absorbed them simply by living in America.[166] Feminist writer Barbara Smith put it in exactly those words: "I am anti-Semitic. . . . I have swallowed anti-Sem-itism by living here, whether I wanted to or not."[167]

Psychological, and even unconscious, factors also exacerbated anti-semitism among African Americans. In the early 1970s psychiatrists Sidney Furst and Curtis Kendrick tried to explain why the Jews serve the function that they do among African Americans. Furst observed that "Jews present a convenient [scapegoat] on which troubled people can project their particular difficulties." Kendrick indicated that among blacks antisemitism was, in the main, "an instance of the projection onto someone else of the disowned, negative aspects of the self." He added that "the unconscious feelings that one is bad as bigots say join with other unconscious, reprehensible feelings the person wants to displace on to someone else, and add a special power to the resulting prejudice."[168]

Harold Orlansky in 1945, and sociologists B. and M. Sobel in almost the same language twenty years later, also interpreted the psychological factors underlying black hostility toward Jews. For the African Americans, they concluded, "Attacking Jews serves through indirection a dual purpose denied expression by the overwhelming power of white society—revenge on the white man (the Jew is white) and alliance with him (against the Jews)."[169] By being antisemitic, therefore, blacks both identified with, and expressed their rage against, members of the dominant culture. And they could do this without really attacking whites or distancing themselves from Christians.[170]

Another explanation for black antisemitism that surfaced in the 1960s is that contact between blacks and Jews in the northern urban ghettos became common after World War I. In these ghettos, they had many interchanges with small Jewish merchants, landlords, and rent collectors, some kindly, others less so. Many black women did day work for Jewish housewives and felt uncomfortable by the way in which they were hired and how they had to negotiate the terms and wages of their labor. And in the 1930s the United States witnessed the worst depression in its history. Much of white America was virulently antisemitic in the depression decade and so, too, were blacks. This continued in both the dominant society and in the black community through World War II.

After the war ended the civil rights crusade began and blacks and Jews had difficulty in adjusting to one another in a new kind of relationship. Jewish organizations provided a good deal of the money and leadership for a cherished goal but along with the assistance there seemed to be a patronizing and superior attitude. Black resentment of this joined rage against the injustices of society, and turned toward the Jews because they were conditioned to do so in earlier years and because Jews had already become the traditional scapegoat for community woes.

But black interactions with Jews no more caused antisemitism than white contact with Jews caused them to be prejudiced. There is no doubt that black interactions with Jews exacerbated and brought out latent or previously less articulated animosity, but such interchanges did not create the antagonism. As sociologist Richard Simpson argued, "Contact between members of different groups, where one is subordinate to the other, is more likely to intensify any stereotypes and hostilities which exist."[171] And since for the better part of the twentieth century Jewish contacts with African Americans have almost always been as philanthropist to recipient, shopkeeper to customer, landlord to tenant, employer to employee, teacher to student, welfare worker to client, and so forth, it is easy to see how Simpson's analysis might be confirmed.

In the mid-1960s, after passage of the Civil Rights and Voting Rights Acts, younger and more militant leaders discovered that they received a great deal of attention by attacking Jews. They also had a concrete example of what they perceived as Jewish intransigence when the predominantly Jewish teachers union in New York City demanded that the labor union contract be adhered to before innovative changes could be made in the public schools. Much of the pent-up wrath African Americans felt was again discharged against the Jews. One aspect of the hostility that took on new meaning was the resentment that younger blacks felt about Jews participating in causes that they recognized had to be done by themselves. Also contributing to negativity was the rise of the Muslim faith among African Americans which, like Christianity, held the Jew up to ridicule and contempt.

By the 1970s there were both real and perceived differences between Jews and blacks that prevented an easy rapprochement. Most

Jews and most blacks stood on different sides of the issue of quotas to obtain affirmative action. Most Jews would not support quotas; most blacks felt them necessary for group progress. At the end of the decade, the resignation of Andrew Young once again kindled flickering embers into a roaring fire. It was no secret that Jews did not trust Young or his attempts to bring the leaders of the PLO to peace talks with Israel, but Jewish leaders did not demand his resignation. Nevertheless, when Young left his post African Americans were convinced that "the Jews" forced his departure.

During the 1980s, Jesse Jackson towered amongst fellow African Americans as a leader of stature and won recognition from white America. Others who had their moments in the sun included Louis Farrakhan, Steve Cokely, and then, in 1991–1992, Leonard Jeffries. All received a great deal of publicity because of their antisemitic comments but that helped, rather than hurt, them among other African Americans.

Jews are not responsible for the plight of African Americans. If anything, they have done more to promote civil rights for all groups than any other identifiable American people.[172] Nonetheless, as non-Christians, as "oppressive allies" who even the white establishment seems wary of sometimes, and as the traditional scapegoat in the Christian world, many blacks have internalized the thought that Jews are the sources of their woes.

Jews have tried in many ways to avoid that role but have not yet been able to solve that dilemma. Jewish defense organizations, and most Jewish individuals, support liberal and progressive domestic programs that would benefit economically disadvantaged people. They are zealous in their defense of equal rights and opportunities for all Americans, and they use every opportunity to speak out against prejudice and discrimination. Leading Jewish organizations employ people on the basis of their skills and qualifications and thus have numbers of African American employees. There are no specific policies that most Jews and their representatives pursue that are specifically designed to hurt or retard the welfare and progress of any other groups of Americans even though their strong opposition to quotas is often interpreted in that fashion.

As with white antisemitism, there is little that Jews can do or refrain from doing to combat black antisemitism. It is possible that some African Americans proclaim what millions of whites believe but are afraid to say publicly because of the social stigma attached to the utterances of bigoted remarks. African Americans suffer no such constraints. Their antisemitic verbiage seems to make them more, rather than less, popular among their peers. The opposite is the case among whites, especially on campuses where members of the dominant culture are continually admonished to be respectful and tolerant of the values and backgrounds of different groups.

Still another explanation for the more recent bouts of African American antisemitism is a battle for political turf. Gates explained it as the

> bid of one black elite to supplant another. . . . The strategy of these apostles of hate . . . is best understood as ethnic isolationism—they know that the more isolated black America becomes, the greater their power. And what's the most efficient way to begin to sever black America from its allies? Bash the Jews, these demagogues apparently calculate, and you're halfway there.[173]

Among African Americans, however, Gates notes, "college speakers and publications have played a disturbing role in legitimating" bigotry.[174] *NOMMO*, the newspaper published by the African American Studies Center at University of California's Los Angeles campus, for example, has defended the importance of *The Protocols of the Elders of Zion* as a means of understanding the ways of Jews.[175] University of California officials have tried to avoid interfering in *NOMMO*'s right to publish whatever it chooses. When the June 1991 issue, however, contained a piece by entertainment editor Darlene Webb indicating that most Jews were "typical cave-dwelling (Khaza mountains, to be exact), white, zionist f———!"[176] the college withdrew its advertising from the paper. At that point Winston Doby, the school's highest ranking African American official and vice chancellor for student affairs, wrote a letter expressing his "deep sadness and dismay over several highly offensive and blatantly anti-Semitic statements in the June issue of *NOMMO*."[177] Although the language used by Webb may not be typical of other African American campus writers, the ADL reported that "many black student leaders and representatives repeatedly and enthusiastically support speakers who are well known for their Jew baiting."[178]

The ferocious hostility that some African Americans harbor toward Jews, combined with the Christian and dominant culture's heritage of animosity toward Jews, makes it unlikely that significant changes will occur in the near future. Existing anti-Jewish attitudes among African Americans seem resistant to change. Well-educated and middle-class blacks appear to be leading the charge.[179] And, as one scholar has written,

> stereotypes tend to reinforce themselves because people are likely to expose themselves only to messages they want to hear; they interpret messages to coincide with their own preconceptions; and tend to remember only what they want to remember. Logic, reason, understanding, education have little impact. We are moved by those facts that bolster our stereotypes.[180]

Jewish actions are often viewed as either paternalistic or colonial. Any help provided, as Rabbi Robert Gordis pointed out in 1967, "inevitably breeds ill will, ingratitude and a sense of inferiority."[181] Rational policies and programs will not combat deep-seated prejudices. Beliefs that a people hold dear are almost impossible to eradicate. The challenge is to discover new ways of assailing old myths.

Some African-American leaders have shown signs, however, that they want to stop Jew bashing and focus on more positive ways in which members of the community can improve their lives in the United States. In the 1990s Gates, West, and even Jesse Jackson have strongly condemned expressions and feelings of black antisemitism and have called for greater tolerance and understanding among all peoples. They are particularly concerned about the hostility that stems "from the top down, engineered and promoted by leaders who affect to be speaking for a larger resentment."[182] With men of such stature in the African-American community—Gates, West, and Jackson—assuming this stance of tolerance, it is possible that black antisemitism might weaken in the distant future. It may take several generations of promoting respect for, and acceptance of, Jews, however, before centuries-old beliefs are eradicated.

11

At Home in America
(1969–1992)

Despite well-publicized outbursts of black antisemitism in New York in the summer and fall of 1992, American Jews have never been more prosperous, more secure, and more "at home in America" than they are today. Not only has antisemitism, according to historian Edward Shapiro, "diminished almost to the point of insignificance," but Jews have been increasingly accepted into the American mainstream. This is equally true for the Orthodox, Conservative, Reform, and non-observant Jews. Unlike previous eras, articles are no longer written advising parents how to explain the pitfalls of being Jewish to their children, printed debates no longer assess the relative merits of changing one's name, and discussions of "the Jewish problem" are neither in vogue nor in need. Contemporary Christian theologians are questioning church teachings about Jews, no mainstream politician campaigns against them, and there are no antisemitic demagogues of national stature. Moreover, the circulation of the most popular antisemitic weekly, *The Spotlight*, declined from 315,000 in 1981 to 112,000 in 1987. The critical issues of yesteryear—expanded immigration, limited educational and employment opportunities, residential and resort segregation—have ceased being concerns of the community. More significantly, Jews no longer rely exclusively on behind-the-scenes diplomacy to obtain political objectives. Their organizational spokespersons are dynamic and above board as they clearly enunciate and vigorously promote issues of concern to Jews. Most important for Jewish thinking, it is no longer assumed that the community as a whole must take a unified stand on every issue. Public dissent among Jewish organizations and publications is almost as common as it is among individual Jews. As a result, unlike the past when Jews were virtually impotent but were perceived to have a great deal of power, they now wield real political influence but only a minority of other Americans think that they have "too much power."[1]

Professionals in the field of Jewish community relations who follow the waxing and waning of American hostility toward Jews generally agree that antisemitism has been on the decline in the United States since the end of World War II. A changed atmosphere in the nation after World War II and Supreme Court decisions requiring desegregation, congressional legislation mandating equal opportunities for all Americans, and a resulting call for tolerance in the rhetoric of community, national, and church leaders all contributed to the decline of bigotry in the United States. In 1969, Earl Raab, then executive director of the Jewish Community Relations Council in San Francisco, wrote:

> for the past quarter of a century, there has been no serious trace of political anti-Semitism in America. Any suggestion today that "it could happen here" has an antique flavor and would be widely branded as phobic, paranoid, and even amusing.[2]

ADL research director Oscar Cohen, only six years after worrying about how much antisemitism still existed in the United States, confided in a 1972 memo, "my own feeling is that the Jewish position in this country has probably never been as secure as it is nor has there ever been less prejudice." Two years later the *Jewish Post and Opinion* noted that "the position of the Jew is stronger and more less insecure today than at almost any time since the dispersion." A decade later, in 1982, Milton Himmelfarb of the American Jewish Committee observed, "When one looks at the history of modern anti-Semitism here and now, one is struck precisely by the minor position it now has." In 1989 sociology professor Steven M. Cohen of Queens College in New York wrote that "Fortunately, American anti-Semitism is largely confined to the political periphery rather than the institutional center." The following January Jerome Chanes of NCRAC reiterated what other knowledgeable people had been stating for the past twenty years: "The long-term downward trend of antisemitism, well-documented since the 1960's, probably did not reverse during the past two years." A 1991 AJC study concluded: "On most indicators anti-Jewish attitudes are at historic lows," and in 1992 another sociology professor, William Helmreich of the City College of New York, wrote that polls show the decline of American antisemitism, fewer people in the United States harbor negative stereotypes of Jews than they had in the past, and only a small percentage of Gentiles object to their children marrying Jews. Finally Jews, in thought or deed, are not central to the agenda of most other Americans.[3]

Can we conclude, therefore, that antisemitism is dead in the United States? Absolutely not. It has declined in strength from its high points in the mid-1940s but we see it openly among African Americans, periodically among die-hard hate groups, and sporadically on college campuses. A 1992 Intergroup Relations survey conducted for the AJC indicated that much animosity exists between and among the various groups of people living in New York City. More than 40 percent of those polled perceived discrimination against Jews and more than 60 percent of the African Americans and Hispanics in that survey expressed the opinion that Jews possessed "too much" influence in the city's "life and politics." Many of the respondents who hold such opinions probably are antisemitic. The hundreds of incidents that the ADL has recorded every year since 1979 might possibly, although not probably, spark a widespread antisemitic campaign in the United States, but public constraints against defamation are so thoroughly institutionalized that even bigots are aware of them. Nevertheless Jews are still excluded from some prestigious social clubs, a number of Jewish feminists have seen bigotry in the women's movement, and antisemitism has been observed in Philadelphia's cultural establishment, in German sections of St. Louis, among Democratic pages in the U.S. Senate, in high schools throughout Connecticut, in the outspoken expressions of literary figures like Gore Vidal and the late Truman Capote, and among lunatics who deny that the Holocaust ever occurred. United States Senators and Congressmen have received antisemitic communications in their mail bags, and random assaults, swastika daubings, and fire bombings have occurred in places as different as West Hartford, Connecticut, New York City, and Los Angeles. One couple who had retired to northern Idaho in the 1980s left that area because they felt so much antisemitic fervor about them, perhaps inspired by the tone set there by a small group of antisemitic Arayan Nation members. Then, in the fall of 1992, Marge Schott, owner of the Cincinnati Reds baseball club, received negative publicity because she allegedly referred to some people of African-American heritage as "niggers." When interviewed by a reporter for *The New York Times* she acknowledged receiving a gift of a Nazi arm band with the swastika on it. "I keep it in a drawer with Christmas decorations," she stated. Ms. Schott also opined that "Hitler was good in the beginning but he went too far."[4]

That antisemitism still exists in an unmeasurable degree among millions of mainstream Americans is not debatable. In 1974 *The Christian Century*, which in the 1930s and 1940s had tried to induce Jews to convert to Christianity, observed that the virulent disease of anti-Semitism is a reality in our culture—one that the Christian in particular must acknowledge as the product of Christian history. The tendency to characterize the Jew as "Christ-killer" haunts us too deeply to be ignored as an occasional individual aberration.[5]

Franklin Littell, a scholar of Christian theology, observed in 1985 that the great body of American Protestantism still retained its bedrock of theological and critical hostility toward the Jews while in 1990 another student of Christian teachings noted that feminist theologians have depicted Judaism as the "antithesis of feminist values." Classroom observers know that many parochial and public school teachers do not have the ability to distinguish between secular and Christian cultures or to teach church history and philosophy in a neutral fashion.[6] As one critic noted, "With no background material given for teachers, one can only conclude that the cumulative influence on the students, over the years of Christian education, will in the final analysis be little different than that of the sixteenth-century catechisms."[7] Polls of high schoolers in the early 1980s confirmed this observation, as students stated, "I believe the Jews do drink the blood of Christians" and "Jews are Christ-killers." Even in the 1990s nursery school children in Tucson hear from their friends that Jews killed Christ.[8]

The Yankelovich firm conducted a poll about American attitudes toward Jews in 1981. Its findings indicated that one in five white Americans and two of five African Americans harbored some degree of hostility toward Jews. But a decade later an ADL survey found less animosity. The 1992 ADL study, based on a relatively small sample, concluded that only 17 percent of white Americans and 37 percent of African Americans might be classified as "hard-core" antisemites.[9] But if these percentages are to be believed—and I am not suggesting that they should not be—it means that over forty million Americans are bigots. But do these people act on their bigotry in the same way as their ancestors did? For at least the past decade the answer is no for the whites. Institutional bias is practically a thing of the past, physical violence toward Jews rarely occurs, and if one is not Jewish one might hear antisemitic expressions from a wide variety of individuals who, despite their unfriendly feelings, constitute little threat to Jews. Yet, as Steven Cohen pointed out in 1989, antisemitic stereotypes and sentiments remain "far more widespread than anti-Semitic behavior."[10]

One issue many Jews use to gauge American antisemitism is criticism of the government of Israel and its policies. To be sure, anyone who follows the byzantine political happenings in the Middle East has found many events difficult to understand. Complete sympathy with the positions taken by any side is rare, and criticism of any government's actions is often a thoughtful, rather than a bigoted, response to a particular situation. Thus it is sometimes difficult to separate criticism of Israel that is "just" or inappropriate, depending on one's political views, from that which is merely a disguise for antisemitism. Former *Jerusalem Post* reporter, Wolf Blitzer, however, noted in 1985 that in the United States antisemites "come out of the closet when relations with Israel are strained." Many Jews throughout the world believe that most criticism of Israeli policies by Christians is merely a pretense for manifesting antisemitic views. Historian Henry Feingold, also in 1985, observed that antisemites, by shifting their focus to anti-Zionism, continued "to demonstrate an unerring instinct for what lies at the center of Jewish sensibility"; to him antisemitism was "still a dynamic force."[11] This aspect of American antisemitism remains a gray and fuzzy area, more easily "sensed" by those who see it than proven to the satisfaction of those who do not.

Many other clearly enunciated prejudices, however, make it unnecessary to focus on political attitudes toward Israel. When tired shibboleths consistently reappear, even the most complacent of Jews understands that entrenched ideas of yesteryear can materialize into public consciousness. Notorious examples of this tendency are the thoughts and remarks of former President Richard M. Nixon who set an extremely nasty tone for his administration. Throughout his tenure in the White House he made scurrilous references to "kikes" and "left-wing" Jews, attributed leaks of news stories to "our Jewish friends," and blamed "a Jewish crowd in Baltimore" when federal attorneys investigated accusations that Vice President Spiro Agnew took bribes while in office. In 1971 the President reportedly accused a "Jewish cabal" in the Bureau of Labor Statistics of trying to undermine him.[12]

Then, in the summer of 1974, an interview appeared in an Egyptian newspaper that probably would have been heavily publicized in the United States if Nixon was not already in a difficult public position. The President had refused to provide some taped conversations that he had had with subordinates concerning the attempted burglary of Democratic headquarters in the Watergate complex in Washington two years earlier and a congressional committee had gone to the Supreme Court to adjudicate the matter. On July 12, 1974, two weeks before Nixon's attorney heard the tapes and indicated that the President had not been truthful in his remarks about his participation in the Watergate affair, the transcript of part of an interview with Nixon appeared in the Egyptian newspaper, *Al Mussawar*. The reporter, Fikri Abbaza, had asked Nixon, "Do you believe . . . that the Jews in the U. S. have a part in the campaign [concerning your participation in the Watergate affair] that is being waged against you?" According to the published interview the President responded:

> there may be some truth in that if the Arabs have some complaints about my policy towards Israel, they have to realize that the Jews in the U. S. control the entire information and propaganda machine, the large news-

papers, the motion pictures, radio and television, and the big companies, and there is a force that we have to take into consideration.[13]

Leonard Garment, one of the President's many assistants in the White House, doubted whether the interview had ever taken place, but there was never an official denial of its essence.[14]

The comments attributed to the President were hardly unusual for his administration. While revising the government's list of subversive organizations in April 1974, Attorney-General William Saxbe, a former Republican Senator from Ohio, attributed the decline of communist groups in the United States to the changed attitudes of "the Jewish intellectual, who in those days [1940s-1950s] was very enamored of the Communist party." *The Christian Century* called Saxbe's observations "an outrageous slander on American Jews," while *The New York Times*, in a rare departure from its usually measured tone and language, charged the Attorney-General with "a harebrained obtuseness, mixed with ignorance. . . . his unthinking statements suggest strongly the limitations of mind that Mr. Saxbe brings to a once-revered post that demands judgment, balance, discernment—and common sense."[15] Further loose thought was evident in October 1974, six months after Saxbe's observation, three months after the alleged interview with the President appeared in the Egyptian newspaper, and two months after Nixon resigned. The chairman of the Joint Chiefs of Staff, General George S. Brown, casually mentioned that the "Jewish influence" in Congress was "so strong you wouldn't believe. . . . They own, you know, the banks in this country, the newspapers. Just look at where the Jewish money is."[16]

Prompt publication of the general's fallacious beliefs led to a minor brouhaha. Brown quickly apologized for his "unfortunate and ill-considered" remarks, but editorials were written condemning his bigotry and insensitivity, and more knowledgeable people produced data exposing the General's misconceptions. *Newsweek*'s Meg Greenfield dismissed Brown's observation and wrote, "Jewish control of the banks is almost as real a phenomenon as Jewish control of the Archdiocese of New York."[17] An American Jewish Committee survey of the 25 largest banks in the United States found only one Jew among 377 senior management positions, 38 Jews among 3,027 middle level managers, and 26 Jews among 491 officers. The chairman of the U.S. Senate's Banking Committee, William Proxmire of Wisconsin, noted, "No industry has more consistently and cruelly rejected Jews from positions of power and influence than commercial banking." Then he provided his committee's statistics on newspaper ownership: Jews owned only 3.1 percent of the nation's 1,748 newspapers accounting for 8 percent of the daily circulation.[18] On November 21, 1974, Stephen Isaacs, author of *Jews in American Politics*, wrote, "The notion that Jews control the banks and newspapers of this country is among the most enduring of myths."

Issacs' observation was confirmed by John Coleman, president of the McConnel Foundation, who told Oscar Cohen in December 1978, "General Brown was honestly saying what a very large number of people believed but were too smart to say, too polished to say." There is a fair amount of feeling, Coleman continued, that the Jews run this country and that the "New York Times is probably the most influential voice in this country and one equates the New York Times with Jews."[19]

Repercussions from the Brown incident were still reverberating when he again spoke out in a manner that alarmed Jews. In June 1976 he told the Senate Armed Services Committee that Israel was a "burden" to the United States and that Jews had too much influence with members of Congress. That remark prompted New York City Congresswoman Bella Abzug to inform President Gerald Ford that "General Brown's inappropriate and unacceptable anti-Jewish remarks made yesterday during Senate Armed Services Committee hearings should totally disqualify him from assuming the highest leadership post in our nation's military."[20] But Abzug's words did not result in Brown's dismissal, and he remained chairman of the Joint Chiefs of Staff until Jimmy Carter became president in 1977.

An incident that attracted great press coverage from March 20, 1977, through June 1978 concerned the National Socialist White Workers Party of America's (Nazis) request for a permit to parade in full regalia down the main street of Skokie, a Chicago suburb, home to 40,000 Jews, many of whom had survived the Holocaust. This local Nazi group consisted of its 32-year-old leader, Frank Collin, and perhaps 25 to 30 of his youthful supporters. Had the permit been granted and the parade held, it would have attracted almost no press notice. But when the Skokie village council denied the request, Collin sought and received the support of the American Civil Liberties Union, which interpreted the hate group's request for a parade permit as a classic case of First Amendment rights of freedom of speech. For fifteen months legal efforts to secure the Nazis' right to march whipped up community anxiety. Ultimately, a federal court of appeals granted the right of the Nazis to march, and they immediately announced that the location would be switched from the main street in Skokie to a public park in Chicago. Collin's small band of Jew haters had gotten more for their efforts than they had originally hoped: fifteen months of national publicity and the support of the major civil liberties organization in the United States. The fact that the Nazis wanted to parade rekindled the most frightening memories of Jews throughout the country and led to a series of thoughtful articles by and about Jewish Americans and whether the principle of free speech should have been supported in this situation. (It should also be noted that in March 1978, while the Illinois group awaited the court's decision, a small band of Nazis paraded in St. Louis in virtual isolation.[21])

Many of the Illinois Nazis were adolescents, an age group in the forefront of the antisemitic vandalism and harassment that has occurred in the United States in every decade since the 1930s. The activities of these angry teenagers are almost always done on the sly, are horrifying and disturbing when noticed, and rarely involve physical violence. Since the end of World War II they have never led to, or represented, any broad based antisemitic movement.

A more troubling indication of adolescent antisemitic aggression occurs regularly in colleges and universities throughout the United States suggesting the attitudes of the nation's elite-to-be. Although the number of publicized incidents occurred on fewer than 10 percent of American campuses from the late 1970s through the early 1990s they often received media attention and careful scrutiny from Jewish defense organizations. Most of the non-African-American hostility documented since 1978 has arisen in connection with athletic events or fraternity activities and none of the individuals involved is known to have engaged in antisemitic actions afterwards. Indeed, the antisemitic behavior observed regarding fraternities and athletic events seems quite different from campus bigotry involving other minorities, women, or gays and lesbians, who have been subject to much broader assaults in a wider variety of settings. Smaller groups of campus antisemites support the Palestinian Liberation Organization and its goals in the Middle East, and almost always equate Zionism with racism.[22]

The most publicized antisemitic campus activities consisted of fraternity vulgarities and athletic excesses. On November 9, 1978, the fortieth anniversary of *Kristallnacht* in Germany, about 150 students from the Kappa Alpha and Sigma Phi Epsilon fraternities at the University of Florida campus gathered in front of the Tau Epsilon Phi fraternity house and shouted expressions like "F——— the Jews," and "Your mother was bright but she was a lampshade." Institutional memory of such incidents is fleeting, and when I spoke at the University of Florida in October 1990 no one in the audience had any knowledge or recollection of the incident. In 1979, a short-lived group at the University of South Florida in Tampa called itself "Sons of Hitler." The sporadic nature of antisemitica on southern campuses was reflected in a February 9, 1982, analysis of four southeast states by the ADL regional office in Atlanta; the report found "the kind of anti-Semitism that existed in years past is almost absent on the campuses today."[23]

During the same period a few New England schools also had minor antisemitic disturbances. On November 3, 1978, Babson College (Massachusetts) soccer players, preparing for a match with Brandeis, wore shirts with swastikas underneath their uniforms, hung a sign reading "Brandei$" in the dining room, and rallied one another during practice by shouting, "Kill the Jews." That same Brandeis soccer team also encountered antisemitic slurs and jeers when they played in North Adams, Massachusetts. The University of Massachusetts in Amherst in 1981 had such a wave of bigotry involving racism, sexism, and anti-semitism, that its chancellor, Henry Koffler, proclaimed a "Year Toward Civility" on campus. Workshop coordinators giving sessions designed to promote tolerance on campus kept hearing that Jews were loud, pushy, clannish, rich, cheap, materialistic, sly, spoiled, and ambitious. A Jewish student sadly responded at one of these sessions, "People have stereotypes. They were brought up in a Christian society."[24]

Throughout the decade of the 1980s sporadic antisemitic episodes involving undergraduates occurred across the nation. A succoth built in 1982 by Dartmouth students was destroyed, the *Protocols of the Elders of Zion* was distributed at the University of California in Berkeley, and Jewish students at the University of Michigan accused the student newspaper, *The Michigan Daily*, of Jew baiting, especially after a Pan American flight crashed over Lockerbie, Scotland, on December 21, 1988. *The Daily's* February 14, 1989, editorial suggested that Israel might have been responsible for blowing up the plane because a group of Hassidic Jews who had been scheduled to fly on it had changed their reservations. In 1989 the Phi Gamma Delta fraternity at Penn State held a Rosh Hashona theme party that mocked Jews. In the early 1990s observers noted hostility toward Jews on the campuses of California State University in San Francisco, and at the branch campuses of the University of California at Los Angeles and at Davis. A Phoenix student at Tucson's University of Arizona, referred to in the state as the U of A, recalled in 1993 being asked by friends and acquaintances on several occasions when he returned home for vacation, "Do you still go to the Jew of A?"[25]

Perhaps no antisemitic campus activity of the 1980s received as much attention as the teasing of "JAPs" ("Jewish American Princesses"). Unlike other forms of antisemitism, this one may have had its origins among young Jewish men on campus who used the term to describe young Jewish women whom they regarded as crass, materialistic, or vulgar. Originally employed within the in-group, the term eventually spread to those who had neither fondness nor sympathy for Jewish men or women, but who were also knowledgeable enough to know that outspoken bigotry against ethnic minorities was not socially acceptable. The "JAP" attacks were also a way of obliquely harassing and denigrating women. Ultimately, the "joke" got out of hand, the term "JAP" was thrown around carelessly, and even was chanted during athletic contests when some women got up to get a coke or leave the arena. The phenomenon seemed centered in the Northeast and JAP baiting received its greatest attention after Syracuse University sociology Professor Gary Spencer wrote about it. He then appeared on national television and campuses throughout the United States to discuss the subject.[26]

Focusing on college students as sources of prejudice, however, distorts reality. At a November 1, 1989, ADL conference that addressed the issue of campus bigotry, Stanley Levy, then Vice Chancellor for

Student Affairs at the University of Illinois in Urbana, approached the subject from a different perspective. The University of Illinois, he explained, was in "the plains of Central Illinois, a hundred forty-five miles from any place. If you extend the Mason-Dixon Line westward, my campus is below the Mason-Dixon Line. It is Bible Belt America." Antisemitism, he went on,

> is endemic in my community. It is there, it has always been there, and it is just beneath the surface. So is racism. Second, it has been taught to our students long before they get to campus. A thousand years of religious teachings and behavioral characteristics simply don't disappear in the four hundred years of this country or the three years since the Pope's statement that Jews aren't to blame for Christ's death.[27]

Occasionally Christian mythology also reared its ugly head in the workplace. Wallace H. Weiss, the only Jewish employee at the federal Defense Logistics Agency, endured years of abuse (1980–1983) from fellow workers who called him, among other things, "Jew faggot," "Christ-killer," and "rich Jew." One even told him, "You killed Christ, Wally, so you'll have to hang from the cross." Since supervisors made little attempt to end their underlings' denigration of Weiss, and at least one manager even shared the attackers' views, Weiss sued. He won a decision in the federal courts under Title VII of the 1964 Civil Rights Act that specifically prohibited religious harassment on the job.[28] Weiss' case proved that old discriminatory attitudes lingered into the 1980s. In industries like banks, utilities, and chemical, auto, and oil companies, covert policies, perhaps made only by individuals, maintained the status quo. Recruiters for some firms in these industries visited campuses like Notre Dame but avoided the colleges of the City University of New York which have large numbers of minority students.[29]

At the same time, however, there were official indications that these same companies recognized that corporate bigotry had to end. The most spectacular sign of change occurred when Irving Shapiro became the first Jewish chief executive officer of the world's largest chemical company. At the time he received the job his son exclaimed, "It was an absolute bombshell. A Jew does not become chairman of Du Pont."[30] By 1978 John Coleman of the McConnel Foundation explained that the upper echelons in American business were not so much antisemitic as they were ill-at-ease with Jews in their midst; they found them "too intellectual." As companies began hiring Jews, Coleman pointed out instances of Christian discomfort. He noted how his firm, the Provident Insurance Company (Rhode Island), had added several attorneys in the mid-1970s and at one gathering of newcomers each of the WASP males was introduced with a few brief remarks. On the other hand, the Jewish woman attorney was given an elaborate introduction detailing her outstanding qualifications. "I think this is anti-Semitism," Coleman noted, "and some anti-woman stuff."[31] But it is noteworthy that the Jewish woman was hired and not passed over as she would have been only a decade earlier.

No industry has been more antisemitic than commercial banking. David Rockefeller, of New York City's Chase Manhattan bank, stood in the forefront of those in his profession who wanted to end bigotry. Chase not only hired more Jews but also installed a kosher cafeteria for its workers in 1971. Later Bankers Trust, another major New York City firm, raised eyebrows when it began issuing Jewish calendars. The number of new Jewish bank appointees in the 1970s, however, was not significant. Only in the 1980s, when banks discovered that they had to compete in international markets, did they alter their recruitment policies. *The New York Times* heralded the shift in an article entitled: "No Longer a WASP Preserve," and explained that "searching for talent, banks have begun to open key jobs to Jews, Italians, and others. But at the very top tradition dies hard."[32] The article explained how more Jewish names were seen in some of the highest banking circles. In 1983 Boris S. Berhovitch became vice-chairman of Morgan Guaranty Trust's banking department; by 1986 there were two Jewish vice presidents at Manufacturers Hanover Trust, and prominent Jews near the top of Citibank and Chemical Bank. *The New York Times* explained that

> deregulation has forced large commercial banks to engage in new activities that require skilled, aggressive executives. To fill the posts, the banks are tossing aside old barriers and converting their upper managements increasingly into meritocracies.[33]

A study of 4,340 senior executives in the nation's largest businesses in 1979 and 1986 compared the percentages of people of different religions in executive positions. The statistics quickly revealed the kinds of shifts that were occurring:

	1979	*1986*
Protestants	68.4%	58.3%
Catholics	21.5	27.1
Jews	5.6	7.4

In 1986, however, Jews, who constituted less than 3 percent of the nation's population, accounted for 13 percent of executives under the age of 40. Throughout America major corporations like Disney, McDonalds, and United Airlines brought Jews aboard and allowed them to compete for top-echelon positions.[34] The American Jewish Committee, which two decades earlier had complained about discrimination at the highest levels of the corporate world, noted in 1988, "There is no evidence of widespread discrimination against Jews in the executive suite."[35]

The academic world also reflected new perspectives. By the early 1970s religious discrimination for admission to prestigious colleges had just about disappeared, Princeton opened a kosher kitchen for students in 1971, and the 40 or so Jewish studies programs of 1966 expanded to over 300 by the 1980s. And, in 1977, Davidson College in North Carolina ended its "Christians only" policy when it engaged a Jewish political science professor.[36] Changed hiring policies were particularly noticeable at the top, and whereas in the 1960s the American Jewish Committee lamented the small number of Jewish deans, provosts, and university presidents, by the 1980s it had no reason to be upset. Jews had been appointed to the presidencies of some of the most prominent schools in America. They included, among others, Bard, Barnard, California Institute of Technology, Columbia, Dartmouth, Indiana University, Massachusetts Institute of Technology, Princeton, Rutgers, and the Universities of Chicago, Cincinnati, Michigan, New Hampshire, Pennsylvania, and Wisconsin. In addition, Harvard Dean Henry Rosovsky chose to remain in Cambridge and therefore turned down offers to preside over Yale and the University of Chicago.[37]

Jewish students also perceived the change. Even at the schools where antisemitic incidents had recently occurred they enjoyed a sense of ease and well-being. Not a single Jewish student in a 1988 survey at Dartmouth thought that being Jewish made any difference in his or her future opportunities in America. A 1989 poll of 100 Jewish leaders on college campuses showed that most Jews were quite comfortable with their surroundings, had lost the self-consciousness of previous generations of Jews, and believed that antisemitism was weaker among their college and university peers than in society at large. In 1992 a junior at the University of Michigan, when asked about past charges of antisemitism at the school or in the student newspaper, indicated that he had not noticed it in the two years that he had been there.[38]

One of the most far-reaching breakthroughs, and one with potential for enormous change in Christian beliefs and behavior, has been the reexamination of Christian perspectives and teaching about Jews. Many of the nation's most prominent academic theologians, including among others, Rosemary Radford Ruether, Franklin Littell, Paul M. Van Buren, Krister Stendahl, and Clark Williamson, have reexamined and questioned the accuracy of the role Christian churches have ascribed to Jews for almost 2,000 years. Several of these scholars began to question the careless rhetoric and the unrestrained diatribes of the *New Testament* which, in the words of theologican Carl D. Evans, "makes us see the worst in Judaism rather than the best."[39]

Ruether may have been the first who noted that to end Christian antisemitism required "a dramatic shift in the spirituality which it teaches," but others have joined her position. Both Ruether and Walter Burghardt, S. J., agreed that "there is no such thing as a collective responsibility of the Jews, then or now, in the crucifixion of Christ. Such responsibility is historically and theologically untenable." Father Burghardt also stated without equivocation that "the history of Christian attitudes and actions [toward Jews] is so cruel, so constant, so unchristian."[40]

Others like Krister Stendahl, professor of divinity and former dean of Harvard Divinity School, became convinced after teaching undergraduates for eight years at the University of South Carolina that "false witness against our Jewish neighbors is commonplace, even habitual, in the routine life of the church. . . . If this were not so, students entering my classes would not hold so many erroneous ideas about Jews and their religion."[41] Carl Evans also joined the chorus of those condemning past Christian behavior and attitudes toward Jews.

> From the New Testament times to the present, it is difficult to find a single period when the Church has not acted shamefully toward the Jews. I'm convinced that anti-Semitism has been such a powerful and persistent nemesis largely because of the church's false witness against the Jews.[42]

In 1983 theologian Robert Andrew Everett also recognized a problem and argued that there is a "need to find a way to create a Christian identity free of anti-Judaic teachings. Obviously, this will not be an easy task."[43]

The Catholic Church has been in the vanguard of trying to bring on significant changes in Christian perspectives about Jews. Not only did Pope John XXIII inaugurate a new emphasis on interfaith dialogue but the 2nd Vatican Council specifically "exonerated" Jews for Christ's death. Catholics also led Christian efforts to revamp textbooks used in their parochial schools.[44] In 1984 the Bishop of Pittsburgh specifically acknowledged that "the study of the Catholic [textbook] material revealed many unfortunate treatments of Jews and Judaism." He also found it

> regrettable that we Catholics have to confess that the relationship of trust and friendship [between Catholics and Jews] . . . is a development. For it was not always thus. The road to Jewish-Christian brotherhood has been, at times, a rough one. There have been literally centuries of persecution, harassment and hostility.[45]

There are also other examples of Catholic attempts to rectify past "misunderstandings." Catholic-Jewish relations workshops began in 1973 and have met annually ever since. In them adherents of the two faiths share successes and problems, try to encourage formation of Catholic-Jewish dialogue groups where they do not exist, and "explore directions and motivations for continuance of the dialogue." Pope John Paul II has endorsed the reconciliation efforts, and in 1985 he received a Jewish delegation to the Vatican. The following year he became the first Pope to visit a synagogue when he entered the main orthodox center in Rome. And on May 11, 1992, Msgr. Alex J. Brunett stated at a gathering of the Ecumenical Institute for Jewish Christian Studies at the Shrine of the Little Flower, Father Coughlin's church in Royal Oak, Michigan, "I would change history if I could. We need to find forgiveness in our lives whenever possible." Across the street from this church where thousands once gathered to hear or catch a glimpse of Father Coughlin, a lone protester marched. He carried a sign reading: "Father Coughlin was on target concerning the Jewish Communist conspiracy."[46]

With the Catholic church obviously making friendly gestures and so many scholars reexamining Christian attitudes toward Jews, one is hopeful that eventually their conclusions will filter through and that church teachings will eliminate the words and thoughts that in the past have stimulated negative responses toward Jews. One major work designed to accomplish that goal is currently in progress. James F. Moore of Valpariso University's Department of Theology is preparing a book to "push the new way of Christian thinking to a third stage—the stage of dialogue in which dramatically new ways of thinking about Christian identity and Christian scripture become an antidote to the history of the teaching of contempt."[47]

In human terms, the best indication of the decline of American antisemitism is the number of intermarriages that have occurred between Jews and Gentiles.[48] When there exists great hostility, individual group members are afraid to give up the security of their own environment for the perils of the unknown or association with a partner of another status. Since the mid-1960s, when Jewish intermarriage rates had not yet approached 10 percent, there has been an enormous breakthrough. For the period since 1965, the statistics on the religious heritage of the life partners that Jews have chosen are as follows:

Spouses	Jewish	Gentile	Converts to Judaism
Before 1965	89%	9%	2%
1965–1974	69	26	5
1975–1984	49	44	7
1985–1990	43	52	5[49]

Obviously, what was once taboo in Jewish and much of Gentile society is now a matter of much less private, and little public, concern. People of various religious backgrounds intermarry and they do not have to separate from family and friends. Moreover, about 28 percent of the children of interfaith marriages are brought up as Jews.[50] It is no longer a stigma in most parts of the United States to choose a marital partner of another religion, and those who do intermarry obviously see no negative consequences. They are not afraid that their children will be without playmates or scorned as pariahs. And that very fact—the American indifference to, or acceptance of, intermarriage—documents the decline of antisemitism. People marrying Jews do not feel that they are sacrificing secure positions in their communities by doing so.[51]

No matter how much antisemitism has declined in the United

States, however, some people refuse to accept the facts at face value. In addition, any major national or international problem or controversy in which Jews may be involved—even tangentially—creates waves of anxiety among Jews with fine tuned historical memories. In 1971 sociologist Melvin Tumin, who had analyzed the literature on research dealing with antisemitism, observed, "There is more anti-Semitic prejudice and discrimination of all kinds than the polls, surveys, and institutional studies currently reveal."[52] There is no way that scholars or anyone else could refute that statement. Four years later another sociologist, Nathan Glazer, wrote: "In 1971 it seemed to me that there could not fail to be serious domestic consequences from the loss of the Vietnam war, and it seemed likely that Jews would suffer from those consequences. It hasn't happened yet—and it probably won't."[53]

Another common problem in interpreting contemporary events is that observers are often unable to distinguish between momentary setbacks and long-term trends. Sparks are sometimes regarded as fires out of control and incidents of little concern to non-Jews heighten the fear of anxious Jews that new waves of antisemitism are imminent. In 1969 and in 1974, respectively, *Commentary* magazine published articles entitled "Is American Jewry in Crisis?" and "Is There a New Anti-Semitism?" Also in 1974, Benjamin Epstein, who had co-authored several books on antisemitism in the United States during the previous two decades, declared that the "golden age" of progressively less antisemitism in the United States had ended and he cited the radical left, the radical right, violence as a political weapon, Black Power, and anti-Zionist sentiments as evidence to support his contention that "a change for the worse" had taken place. The Middle Eastern oil embargoes in 1973 and again in 1978 also unsettled Jews who worried that foreign events would have antisemitic repercussions in the United States. After a number of antisemitic flurries in the early 1980s, a 1982 poll found that 77 percent of American Jews believed that antisemitism was a very important issue in this country and another 13 percent saw it as "somewhat important." Then in 1985 came the Pollard affair, in which two American Jews were convicted of spying for Israel, and the Wall Street inside-trader scandals in which Ivan Boesky, and eventually Michael Milken, were found guilty of illegal activities. Some Jews believed, erroneously, that these criminal convictions would lead to increased antisemitic fervor in the United States. Contributing to Jewish fears, of course, were also tidbits like the one President George Bush's staff provided when they told reporters in 1991 that "a flood of supportive mail . . . much of it anti-Semitic" reached the White House backing Bush's stance on curtailing loan guarantees to Israel.[54] And then on January 11, 1993, *NewYork* magazine showed a burning Jewish star on its cover with its lead story in boldface print: **"The New Anti-Semitism."** Inside, Craig Horowitz informed readers that "anti-Jewish attitudes have . . .

become more insidious, resembling the anti-Semitism that was prevalent in Europe in the first part of this century."[55] Obviously, anyone who would make a comment like that is not well versed in the history of twentieth-century European antisemitism.

Today antisemitism in the United States is neither virulent nor growing. It is not a powerful social or political force. Moreover, prejudicial comments are now beyond the bounds of respectable discourse and existing societal restraints prevent any overt antisemitic conduct except among small groups of disturbed adolescents, extremists, and powerless African Americans. By emphasizing the hostility of some Americans toward Jews, one attributes too much power and/or influence to fringe people and overlooks the whole new network of positive and diversified interactions between Jews and Gentiles, black and white. By comparing antisemitism in the United States today with that sentiment in other countries, or with its expression in previous eras of American history, one sees fewer alarming episodes. Much less prejudice exists in our own time than in any other period in the history of this nation.[56]

On the other hand, the paradox of declining antisemitism paralleling Jews' fears about its increase is not necessarily paranoia. Jews have an acute sensitivity to oppression, and their two thousand year history in Christian lands, the Holocaust, and the annual ADL figures listing antisemitic incidents remind Jews that they are still vulnerable.[57] Perhaps only the passage of generations can really end the "shadow of a doubt" that remains alive in the Jewish psyche.

Antisemitism, however, is not currently one of the major social concerns in the nation. It is upsetting to Jews that most Gentiles cannot understand Jewish anxiety, but there are so many problems with race, gender, and ethnicity in the United States that the apprehensions of Jews are not foremost in the minds of Christian Americans. For Jews that is both good and bad. Good, because most Christians are not thinking about them and are not out to get them; and bad, because it means that most non-Jews are not giving enough thought to a prejudice that has lasted for almost two thousand years and which erupts at unexpected moments. Jews never again want to read an editorial, such as the one that appeared in the April 24, 1974, issue of *The Christian Century:* "In retrospect we confess now that we have been too slow to recognize the latent anti-Semitism that pervades our Christian society."[58] Jews want all antisemitism eradicated; even with every effort of the ADL and concerned others, that is unlikely to happen.

As John Coleman told Oscar Cohen in 1978, "Anti-Semitism is there and anybody who denies it is crazy. More subtle with the better educated people but it's there. But it is not a big thing. It is not something that somebody wants to do something about."[59] Although most Jews would not dispute the substance of Coleman's remarks, many still worry that some spark or episode might turn what is "not a big thing" and "not something that somebody wants to do something about" into a "big thing" that somebody *wants* to do something about. And this fear cannot be obliterated easily. History has taught Jews never to let down their guard.

Summary and Conclusion

Antisemitism is unlikely to vanish in the United States. It is too much a part of the Christian culture to disappear. It must also be emphasized, however, that in no Christian country has antisemitism been weaker than it has been in the United States. Although always present—and at times virulently so—conflicting American attitudes and traditions of tolerance have for the most part minimized its impact. There have never been pogroms in America; there have never been respectable antisemitic political parties in America; and there have never been any federal laws curtailing Jewish opportunities in America. However, the Jew has always been an outsider in Christian lands.

The stereotypic impressions of the Jew as Christ-killer and Shylock are too deeply imbedded in Christian societies for them to be ignored or passed over. Even when individual Christians have had pleasant relationships with individual Jews, group myths never disappeared. Jews still serve as the personification of evil, the quintessential alien, the perpetrators of whatever problems members or groups in society feel threatened by but cannot understand or explain. In different historical periods Jews were regarded as having caused plagues, killed children, brought on unwelcome events, sponsored policies or philosophies uncongenial to members of the dominant culture, and acting as underminers determined to destroy Christian societies. Jews have had no responsibility for any of these things, but that did not matter to those who had learned of their rejection of the Savior, who society had traditionally mistrusted, and who could never really be part of the dominant Christian culture.

These beliefs carried over into American society where countervailing traditions of tolerance, democracy, acceptance of pluralism, legal equality for white people, some sort of religious toleration, and compassion for oppressed groups became part of the nation's self-image. Christians from other lands brought their cultural baggage with them to the New World where, after a generation or two, Old World notions began to blend with the values espoused in the United States. Yet much as the United States proclaimed itself as a haven for others, it still emphasized its Christian traditions and most people in the country have always considered it to be a "Christian nation."[1] The Jews, therefore, would always be a people apart no matter how many other attributes of the nation they incorporated into their personas. And people judged outsiders are always vulnerable in times of crisis.

Jews were welcomed in colonial America because their numbers were small and white people were needed to help build new communities. Other minorities—such as the Indians and the blacks—occupied the lowest rungs in society and there were more fears about these groups than about Jews. During the nineteenth century many devout Christians either scorned Jews or tried to convert them to Christianity but never regarded them as equals of a different faith.

The first major crisis occurred during the Civil War. In a time of great national stress Jews were excoriated in both North and South and accused of disloyalty and profiteering. It was generally assumed in the North that Jews favored the southern cause and in the South that Jews were loyal to the Union. That was the first time in American history when Jews became scapegoats for the ills in society; that tradition continued afterwards. Jews were held responsible for economic woes, for threats to undermine society, and for disloyalty during engagements with foreign foes.

Social discrimination also appeared and spread in the last third of the nineteenth century. As American parvenus sought status and world respectability, they aped European aristocrats and values. Too insecure to be comfortable with their own accomplishments and ways of life, upper-crust Americans felt the need to distance themselves from everyone perceived of as deviant and allegedly lower-class standing or breeding. Jews, by definition, were considered inferior because they were not Christians. At that time, and through the 1920s, not only were there intellectual rationalizations that justified the sense of WASP superiority but institutional barriers were erected to keep Jews, and to a lesser extent other white minority groups, from "infiltrating" upper-class society.

The onslaught of tens of millions of immigrants in the late nineteenth and early twentieth centuries once again seemed threatening to "Americans" of every social station who feared that most of the newcomers would undermine existing values and traditions. Some Americans even believed that these lower-class hordes of people ranked barely above the status of barbarians. Among the new immigrants were about two million Jews whom Christians "knew" could not possibly fit in with members of the dominant society. Immigration laws in the 1920s successfully limited the entry of southern and eastern Europeans American Protestants found offensive.

And then came the Great Depression. Jews did not cause the depression and were not responsible for its continuance but for desperate people who needed some explanation for the seeming disintegration of society the answers provided by Hitler and by some of America's fundamentalist and other religious leaders rang true. During World War II the hostility honed earlier continued almost unabated.

Not until after the war did American public attitudes toward Jews and other minorities begin to change. Certainly the economic growth that accompanied the end of the war and a renewed optimism in American society contributed to the change. But more important, I think, were the activities of the various ethnic defense agencies, including the Jewish ones. Never before had these groups been so active in pursuit of forcing the American states and federal government to give legal support to the ideas of freedom and equality that have been so much a part of the nation's ideology and rhetoric. The agencies involved have never been sufficiently recognized by the dominant society for their accomplishments. Within less than a generation Americans not only witnessed legislation prohibiting discrimination in various states and localities but they saw the federal government pass a comprehensive Civil Rights Act in 1964 that just about eliminated every facet of legal discrimination against religious and ethnic groups. Moreover, since the 1960s the federal government has, to a degree, sparked the forces needed to eliminate bigotry in this country. This goal, however desirable, is still elusive but the various ethnic defense agencies remain active in pursuit of it.

Prejudice against Jews and other minorities used to be taken as a matter of course until after the end of World War II. Then through the activities of defense agencies, the courts, and ultimately the state and federal legislatures, expressions of bigotry became socially and publicly anathema—even among those who still harbored bigoted thoughts. With the changes in societal mores prejudicial attitudes gradually began to decline—especially in negative feelings toward Jews. Just as prejudice toward perceived outgroups will always be part of our society, so, too, will negative feelings toward Jews. But they have lessened and most American Jews have to read about antisemitism because they so rarely experience it in their daily lives. (In November 1990 an adult who witnessed two African Americans harass a Jew on a New York City subway car wrote, "I had never before seen a Jew abused for being a Jew."[2]) In matters that at one time counted the most—such as educational and employment opportunities, housing and resort segregation, and to a lesser extent membership in the various components that make up what is defined as the American "elite" or members of the "in-group"—Jews have few, if any, complaints. In the most visible areas of society, antisemitism is simply a non-factor. As writer Paul Berman observed in October 1992,

> Anti-Semitism is nearly impossible to debate. All responsible people, and even most irresponsible ones, agree that anti-Semitism is a bad thing and should be beaten down every time it rears its head.[3]

This conclusion has to be modified, however, in terms of one well-defined American group: African Americans. Until the 1960s most Americans, if they thought of the subject at all, assumed that African Americans and Jews were allied in common cause. That was never quite true, and by the late 1960s expressions of public antipathy, especially on the part of blacks toward Jews, made headlines throughout the country. In subsequent decades respected black leaders have often targeted Jews for opprobrium in a manner that would have ended the public careers of white persons who articulated identical words or sentiments.

African-American antagonism toward Jews, however, is no more a phenomenon of recent decades than white antisemitism is. Of all American groups, few are as tied to their fundamentalist Protestant religious heritage as are those of southern ancestry—blacks and whites. The genesis of their hostility toward Jews has more to do with religious teachings than with personal interactions. Certainly contact exacerbated any negative attitudes that had already been nurtured and harbored but Jews are not responsible for the difficulties encountered by other groups.

During the past half century African-American attitudes toward Jews have been particularly virulent. As a minority group that has suffered grievously in American society, African Americans have arguably had much greater hurdles to leap than members of other ethnic backgrounds. One means of identifying with the majority, and therefore gaining greater self confidence and esteem, has been to take on the prejudices of the majority group. To some extent, black antisemitism has been a psychological, perhaps an unconscious one at that, response to their positions in American society. By adopting a majority prejudice, members of minorities can feel superior to the group that they too despise. Moreover, like members of the dominant culture, a number of African Americans have been comfortable scapegoating Jews for their difficulties. Having the strong religious undergirding of Christian theology, it can easily be seen how African Americans have personified the Jews as the sources of their woes. And, added to that underlying belief, since the 1930s there have been enough specific examples of negative interactions between African Americans and Jews for African

Americans to assume that they are responding to realities in their lives rather than building on myths from the past. Moreover, most non-Jewish white Americans have been loathe to "get involved" in black tirades against Jews. Whites are either indifferent to what the African Americans are saying or secretly share some of the same attitudes but are hesitant to express them publicly. For blacks there is little censure in the dominant society for attacking Jews and more praise than condemnation in their own group for giving "whitey" a reason to stand up and pay attention to what African American people are saying.

As blacks become more integrated in American society and as they participate to a greater extent in the power points, they, too, will recognize that outspoken bigotry provides few rewards. In the American political arena, the most successful African Americans of the 1990s are those who have acknowledged the necessity of respect for cultural pluralism in the United States. And as long as the nation as a whole grapples with the issue of integrating minorities into the mainstream ethnic and religious prejudices will wane.

Unlike most immigrants to the United States, however, Jews have always been willing to acculturate yet somewhat more reluctant to assimilate. In recent times the trend seems to have changed and a majority of young Jews have chosen marital partners who have been reared in another faith. That such a phenomenon is occurring suggests a lessening of prejudice toward Jews in this country. Rarely do members of a more prestigious social group seek life partners among those who are less favored in society.

On the other hand, what appears to be a trend of increasing proportions may be a result of diminishing animosity and hostility between and among Christians and Jews. Recent surveys show that antisemitism in the United States is declining—it is not disappearing. Jewish defense agencies are still vigilant in their attempt to alert society to the bigots in our midst and, to a certain extent, they themselves are victimized by the diminution of prejudice. Rationally, they seek to end all evidences of bigotry. But with the ostensible achievement of most of their goals they have had to reshape their agendas. To a certain extent this dictates the kind of work they do, the kind of surveys they commission, and the kind of information they disseminate to the public.

Because of much greater publicity given by both defense agencies and the media to antisemitic acts engaged in since the late 1970s, larger numbers of Jews are worried and sense an increase in hostility. This is especially true in and around the New York area where a plurality of American Jews live. New York City is different in many ways from the rest of the nation, especially since a majority of its population is made up of African Americans, Hispanics, and Asians. In day-to-day encounters many Jews perceive intense hostility from some other city residents, especially African Americans. As a result, 37 percent of New

York Jews questioned in one 1992 survey believed that antisemitism was a "very serious problem" in the city while another 54 percent regarded it as "somewhat of a problem."[4] The yearly audit of antisemitic incidents, conducted and published by the ADL, also heightens public and Jewish awareness of antisemitism in the United States although the numbers of vile acts counted rarely correlate with the personal experiences of the overwhelming majority of American Jews. Negative stereotypes about Jews are still greater in the United States than are negative activities toward Jews.

Thus, while antisemitism has always been a problem for Jews in a Christian society it has always been weaker in the United States than in European nations. Moreover, since 1945, and especially since the mid-1960s, the American government has conducted serious campaigns designed to lessen and eradicate (a goal if not a realistic possibility) prejudice in American society. The cessation of hostility toward Jews falls into this greater American goal. There is no reason to suspect that antisemitism will not continue to decline in the United States even though there will always be sporadic outbursts and temporary flare-ups. Protestant religious fundamentalists may give lip service to stated national policies and rhetoric but generally find it difficult to accept the fact that Jews are the equals of, or deserve the same respect as, believing Christians. The same may also be true of some Catholics convinced of the truthfulness and superiority of their own faith. Also, supporters of the PLO on college campuses throughout the nation periodically play on latent antisemitic feelings among students to create tense situations between Jews and Gentiles. However, the nation is too strongly committed to cultural pluralism for long-term trends toward greater tolerance and acceptance of diversity in the United States to reverse themselves sharply. By comparing the strength of antisemitism in the United States today with what it had been in previous decades or centuries, the obvious conclusion is that it has declined in potency and will continue to do so for the foreseeable future.

Acronyms and Abbreviations

ACJ	American Council for Judaism
ADL	Anti-Defamation League of B'nai B'rith
AH	*American Hebrew*
AI	*Israelite* (to June, 1874), *American Israelite* thereafter
AJA	American Jewish Archives
AJA	*American Jewish Archives*
AJC	American Jewish Committee
AJH	*American Jewish History*
AJHQ	*American Jewish Historical Quarterly*
AJHS	American Jewish Historical Society
AJYB	*American Jewish Year Book*
Blaustein	Blaustein Library, AJC, New York City
CC	*The Christian Century*
CCNY	City College of New York
COHC	Columbia University Oral History Collection
Cohen Mss.	Oscar Cohen mss., AJA
CORE	Congress of Racial Equality
FDR	Franklin Delano Roosevelt

FDR Mss.	FDR manuscripts, FDR Library, Hyde Park, N.Y.
IRL	Immigration Restriction League
JAP	Jewish American Princess
JCRC	Jewish Community Relations Committee, Cincinnati, Ohio; mss. are housed in AJA
JPS	Jewish Publication Society of America
JSS	Jewish Social Studies
KKK	Ku Klux Klan
LC	Library of Congress
Minutes	Minutes of the AJC Executive Committee, housed in Blaustein Library, AJC
NAACP	National Association for the Advancement of Colored People
NCRAC	National Community Relations Advisory Committee
NYU	New York University
PAJHS	*Publications of the American Jewish Historical Society*
PLO	Palestine Liberation Organization
PPF	President's Personal File
PSF	President's Secretary's File
PUSH	People United to Save Humanity
T	*The New York Times*
WASPs	White, Anglo-Saxon, Protestants

Notes

Prologue

1. Jacob Katz, *From Prejudice to Destruction* (Cambridge: Harvard University Press, 1980), 322; Jacob R. Marcus, *The Colonial American Jew, 1492–1776* (3 volumes; Detroit: Wayne State University Press, 1970), III, 1114; M. Ginsberg, "Anti-Semitism," *The Sociological Review*, 35 (January–April, 1943), 4; Joshua Trachtenberg, *The Devil and the Jews* (New Haven: Yale University Press, 1943), p. 6; Glenn T. Miller, *Religious Liberty in America* (Philadelphia: The Westminster Prees, 1976), p. 92.

2. James Parkes, *The Conflict of the Church and the Synagogue* (New York: Hermon Press, 1934), p. 373; Hyam Maccoby, *The Sacred Executioner* (New York: Thames and Hudson, 1982), pp. 147, 151.

3. Rosemary Radford Ruether, "Anti-Semitism and Christian Theology," in Eva Fleischner, ed., *Auschwitz: Beginning of a New Era? Reflections on the Holocaust* (New York: KTAV Publishing House, Inc., 1977), pp. 79, 85; Marcus, *Colonial American Jew*, III, 1114–1115; Trachtenberg, *The Devil*, p. 159; Esther Yolles Feldblum, *The American Catholic Press and the Jewish State, 1917–1959* (New York: KTAV Publishing House, Inc., 1977), p. 12; Katz, *From Prejudice*, p. 323; Sergio I. Minerbi, *The Vatican and Zionism* (New York: Oxford University Press, 1990), p. 93; "Spiritually, We Are All Semites," *America*, 152 (March 9, 1985), 185; Cecil Roth, "The Feast of Purim and the Origins of the Blood Accusation," in Alan Dundes, ed., *The Blood Libel Legend: A Casebook in Anti-Semitic Folklore* (Madison: University of Wisconsin Press, 1991), p. 270.

4. Parkes, *The Conflict*, p. 158; Feldblum, *The American Catholic Press*, p. 12.

5. Quoted in Clark M. Williamson, *Has God Rejected His People? Anti-Judaism in the Christian Church* (Nashville: Abingdon, 1982), p. 99; see also Edward

H. Flannery, *The Anguish of the Jews: Twenty-three Centuries of Antisemitism* (Revised and Updated. New York: Paulist Press, 1985), p. 62.

6. Quoted in Claire Huchet Bishop, *How Catholics Look at Jews* (New York: Paulist Press, 1974), p. 53.

7. G. Peter Fleck, "Jesus in the Post-Holocaust Jewish-Christian Dialogue," CC, 100 (October 12, 1983), 904.

8. Maccoby, *The Sacred*, pp. 150–151; Paul Johnson, *A History of the Jews* (New York: Harper and Row, 1987), p. 165.

9. Jeremy Cohen, *The Friars and the Jews* (Ithaca: Cornell University Press, 1982), p. 19ff.; Flannery, *The Anguish*, pp. 47, 62; quote from Marc Saperstein, *Moments of Crisis in Jewish-Christian Relations* (London: SCM Press, 1989), pp. 10, 51.

10. Joseph L. Lichten, "Polish Americans and American Jews: Some Issues Which Unite and Divide," *The Polish Review*, 18 (1973), 57; Flannery, *The Anguish*, p. 47.

11. *New Testament*. Matthew, chapter 27, verses 15, 24, 27.

12. Jules Isaac, *The Teaching of Contempt: Christian Roots of Anti-Semitism* (New York: Holt, Rinehart and Winston, 1964); p. 109; "The Virulent Disease of Anti-Semitism," CC, 91 (April 24, 1974), 443; Claire Huchet Bishop, "Learning Bigotry," *Commonweal*, 80 (May 22, 1964) 264ff.; Bruno Lasker, *Race Attitudes in Children* (New York: Greenwood Press, 1968; originally published in 1929), pp. 179, 181–182, 252, 279; John B. Sheerin, "Catholic Anti-Semites," *Catholic World*, 203 (July, 1966), 201; James Brown, "Christian Teaching and Anti-Semitism," *Commentary*, 24 (December, 1957), 495; Feldblum, *The American Catholic Press*, p. 14; Parkes, *The Conflict*, p. 376; Rodney Stark, Bruce D. Foster, Charles Y. Glock, and Harold E. Quinley, *Wayward Shepherds: Prejudice and the Protestant Clergy* (New York: Harper and Row, 1976), p. 39; John T. Pawlikowski, *Cathechetics and Prejudice: How Catholic Teaching Materials View Jews, Protestants and Racial Minorities* (New York: Paulist Press, 1973), p. 8; Flannery, *The Anguish*, p. 62; Hannah Adams, *The History of the Jews from the Destruction of Jerusalem to the Present Times* (London: n. p., 1818), p. 53.

13. Isaacs, *The Teaching*, p. 43; Adams, *The History*, p. 53; Minerbi, *The Vatican*, p. 93; Katz, *From Prejudice*, p. 323; Maccoby, *The Sacred*, p. 175; Bishop, *How Catholics Look at Jews*, p. 53.

14. Thomas F. Gossett, *Race: The History of an Idea in America* (New York: Schocken Books, 1965), p. 10; Flannery, *The Anguish*, pp. 67, 73; Trachtenbereg, *The Devil*, p. 159; Sidney Ahlstrom, *A Religious History of the American People* (New Haven: Yale University Press, 1972), p. 574; Saperstein, *Moments of Crisis in Jewish-Christian Relations*, p. 16.

15. Ellen Schiff, *From Stereotype to Metaphor: The Jew in Contemporary Drama* (Albany: State University of New York Press, 1982), p. 5; J. Von Dollinger, "The Jews in Europe," *Popular Science*, 21 (July, 1882), 301.

16. Poliakov, *History*, I, 41; Gavin I. Langmuir, *From Ambrose of Milan to Emicho of Leningen: The Transformation of Hostility Against Jews in Northern Christendom* (Spoleto: n.p., 1980), p. 21; Flannery, *The Anguish*, p. 91.

17. Poliakov, *History*, p. 49.

18. Rosemary Radford Ruether, *Faith and Fratricide: The Theological Roots of Anti-Semitism* (New York: The Seabury Press, 1974), p. 206.

19. "Jews and Christians in the Middle Ages," *Saturday Review* (London),

56 (July 14, 1883), 41; Florence H. Ridley, "A Tale Told Too Often," *Western Folklore*, 26 (1967), 153.

20. Roth, "Feast of Purim," p. 155.

21. Maccoby, *The Sacred*, p. 155; Gavin I. Langmuir, "The Knight's Tale of Young Hugh of Lincoln," *Speculum*, 47 (July, 1972), 462; Ridley, "A Tale," p. 155; Flannery, *The Anguish*, p. 99; Edward J. Bristow, *Prostitution and Prejudice: The Jewish Fight Against White Slavery, 1870–1939* (New York: Schocken Books, 1983), p. 46; *The Los Angeles Times*, April 20, 1965, part 2, p. 6.

22. See Langmuir, "The Knight's Tale."

23. Flannery, *The Anguish*, p. 99.

24. "Jews and Christians," p. 41; Abraham G. Duker, "Twentieth-Century Blood Libels in the United States," in Leo Landman, ed., *Rabbi Joseph H. Lookstein Memorial Volume* (New York: KTAV Publishing House, Inc., 1980), p. 85; Roth, "Feast of Purim," p. 270; Maccoby, *The Sacred*, p. 153; Flannery, *The Anguish*, p. 99; Trachtenberg, *The Devil*, p. 124 ff.; Robert S. Wistrich, *Antisemitism: The Longest Hatred* (London: Metheun London, 1991), p. 31.

25. "Bias Study Blasts Complacency," *B'nai B'rith Messenger* (Los Angeles), November 5, 1965, pp. 1, 4; Alan Dundes, "The Ritual Murder or Blood Libel Legend: A Study of the Anti-Semitic Victimization Through Projective Inversion," in Dundes, ed., *The Blood Libel Legend*, p. 343; Flannery, *The Anguish, p. 100*; R. Po-Chia Hsia, *The Myth of Ritual Murder: Jews and Magic in Reformation Germany* (New Haven: Yale University Press, 1988), pp. 1–2.

26. Eunice Cooper and Marie Jahoda, "The Evasion of Propaganda: How Prejudiced People Respond to Anti-Prejudice Propaganda," *Journal of Psychology*, 23 (1947), 15.

27. Langmuir, "The Knight's Tale," pp. 459–460, 467; Ridley, "A Tale," p. 154; "The Jew's Daughter," New York *Tribune*, August 17, 1902, II, 2; Christopher Magee Leighton, "On Sacrifice and Forgiveness: Toward the Containment of Violence" (unpublished E. D. D., Teachers College, Columbia University, 1990), p. 48 ff.

28. "The Jew's Daughter," New York *Tribune*, August 17, 1902, II, 2.

29. Leslie A. Fiedler, "What Can We Do About Fagin?: The Jew Villain in Western Tradition," *Commentary*, 7 (May, 1949), 413, 417; Marcus, *Colonial American Jew*, III, 1117; Foster B. Gresham, "The Jew's Daughter: An Example of Ballad Variation," *Journal of American Folklore*, 47 (1934), 358; "The Jew's Daughter," New York *Tribune*, August 17, 1902, II, 2; Martha W. Beckwith, "The Jew's Garden . . . ," *Journal of American Folklore*, 64 (1951), 224–225.

30. Dundes, "Ritual Murder," pp. 350, 350, 352–353.

31. Bishop, *How Catholics*, p. 53; Flannery, *The Anguish*, p. 96; Maccoby, *The Sacred*, p. 166.

32. My colleague, Alan Bernstein, has pointed out to me that in the Middle Ages the word "usury" was a synonym for "interest." In our own day, however, "usury" means excessive interest.

33. Maccoby, *The Sacred*, p. 166.

34. Trachtenberg, *The Devil*, p. 188.

35. Poliakov, *History*, p. 79.

36. Quoted in Cohen, *The Friars*, p. 243; Poliakov, *History*, I, 64.

37. Quoted in Flannery, *The Anguish*, p. 102.

38. Cohen, *The Friars*, p. 242.

39. Cohen, *The Friars*, p. 244; Ruether, "Anti-Semitism," p. 88; Poliakov,

History, pp. 104, 109–110, 113; Paul E. Grosser and Edwin G. Halperin, *Anti-Semitism: Causes and Effects* (New York: Philosophical Library, 1983), p. 132; Jacobson, *The Affairs*, p. 159; Trachtenberg, *The Devil*, pp. 100–101; "The Inquisition, the Reformation, and the Jews," p. 715; Flannery, *The Anguish*, pp. 108, 109.

40. Flannery, *The Anguish*, p. 111; Poliakov, *History*, p. 165; Arthur A. Goren, *The American Jews* (Cambridge: The Belknap Press of Harvard University Press, 1982), pp. 7–8.

41. Egal Feldman, *Dual Destinies: The Jewish Encounter with Protestant America* (Urbana: University of Illinois Press, 1990), p. 4; Marcus, *Colonial American Jew*, III, 1116; "The Modern Jews," *North American Review*, 60 (1845), 348; Jonathan I. Israel, *European Jewry in the Age of Mercantilism, 1550–1750* (Oxford: Clarendon Press, 1985), p. 1.

42. Poliakov, *History*, pp. 216–217, 222–223.

43. Franklin H. Littell, "American Protestantism and Antisemitism," in Naomi Cohen, ed., *Essential Papers on Jewish-Christian Relations in the United States* (New York: New York University Press, 1990), p. 176; Heiko Oberman, *Roots of Antisemitism in the Age of Reaissance and Reformation* (Philadelphia: Fortress Press, 1984), pp. 84ff., 140.

44. Feldman, *Dual Destinies*, p. 4.

45. Quoted in Poliakov, *History*, p. 123.

46. Joseph Krauskopf, "Is the Jew Getting a Square Deal?" *The Ladies Home Journal*, 27 (November 1, 1910), 12; Hubert Howe Bancroft, *California Inter Pocula* (New York: McGraw-Hill, 1967; originally published, 1888), pp. 373–374; James Samuel Stemons, *As Victims to Victims* (New York: Fortuny's, 1941), p. 235; Nathan Hurvitz, "Blacks and Jews in American Folklore," *Western Folklore*, 33 (October, 1974), 302; Jacob Katz, "Misreadings of Anti-Semitism," *Commentary*, 76 (July, 1983), 43; Heywood Broun and George Britt, *Christians Only* (New York: The Vanguard Press, 1931), p. 277; Robert Michael, "America and the Holocaust," *Midstream*, 31 (February, 1985), 14; Trachtenberg, *The Devil*, p. 12; Naomi W. Cohen, "Antisemitic Imagery: The Nineteenth-Century Background," JSS, 47 (Summer/Fall, 1985), 309; *A Course of Lectures on the News* (New York: Arno Press, 1977; originally published, Philadelphia: Presbyterian Board of Publication: 1840), p. 193. Writing in 1980, Julius Lester recalled that although his mother was not antisemitic, she occasionally disparaged an individual with the remark, "oh, he's like an ol' Jew. You can't trust him," "All God's Children," *Moment* (April, 1980), 11.

47. Lucy S. Dawidowicz, "Can Anti-Semitism Be Measured?" *Commentary*, 50 (July, 1970), 36–43; Ginsberg, "Anti-Semitism," pp. 5, 8, 11; Jean-Paul Sartre, *Anti-Semite and Jew* (New York: Schocken Books, 1948), p. 7; David J. Jacobson, *The Affairs of Dame Rumor* (New York: Rinehart & Co., 1948), p. 312; Katz, *From Prejudice*, p. 320; Otto Fenichel, "Psychoanalysis of Antisemitism," *The American Imago*, I (March, 1940), 35; Gossett, *Race*, p. 10; Trachtenberg, *The Devil*, pp. 50–51, 101, 210; Maccoby, *The Sacred*, p. 163; "The Inquisition, the Reformation, and the Jews," CC, 44 (June 9, 1927), 715; Poliakov, *History*, I, 142; Ruether, "Anti-Semitism," p. 88.

48. Quoted in Gossett, *Race*, p. 11.

49. Trachtenberg, *The Devil*, pp. 12, 13, 155.

50. Poliakov, *History*, pp. 124, 126–127; Marcus, *Colonial American Jew*, III, 1117; Louise A. Mayo, *The Ambivalent Image: Nineteenth-Century America's Per-*

ception of the Jew (Rutherford, N.J. : Fairleigh Dickinson University Press, 1988), p. 83; Schiff, *From Stereotype*, p. 12; Sol Liptzin, *The Jew in American Literature* (New York: Bloch Publishing Co., 1966), p. 69.

51. Schiff, *From Stereotype*, p. 5.

52. Quoted in Trachtenberg, *The Devil*, p. 187.

53. Alan Dundes, "Study of Ethnic Slurs: the Jew and the Polack in the United States," *Journal of American Folklore*, 84 (April, 1971), 187.

54. See also Nathan Hurvitz, "Blacks and Jews in American Folklore," *Western Folklore Quarterly*, 33 (October, 1974), 322.

55. Josef Rysan, "Defamation in Folklore," *Southern Folklore Quarterly*, 19 (September, 1955), 143; Dundes, "Ritual Murder," p. 360; Rudolf Glanz, *The Jew in the Old American Folklore* (New York: n. p., 1961), pp. 9–10; Nathan Hurvitz, "Jews and Jewishness in the Street Rhymes of American Children," *Jewish Social Studies*, 16 (1954), 135.

Chapter 1

1. Rufus Learsi, *The Jews in America* (Cleveland: The World Publishing Co., 1954), p. 289; Robert Michael, "America and the Holocaust," *Midstream*, 31 (February, 1985), 14.

2. Charles Herbert Stember, et al., *Jews in the Mind of America* (New York: Basic Books, 1966), p. 305.

3. William Warren Sweet, *Religion in the Development of American Culture, 1765–1840* (Gloucester, Mass.: Peter Smith, 1963, reprinted from New York: Charles Scribner's Sons, 1952), p. 45.

4. Jacob R. Marcus, *The Colonial American Jew, 1492–1776*. (3 volumes; Detroit: Wayne State University Press, 1970), III, 1113, 1119; Anson Phelps Stokes, *Church and State in the United States* (3 volumes; New York: Harper and Brothers, 1950), I, 854.

5. Carl Wittke, review of *Jewish Pioneers in America, 1492–1848* by Anita Libman Lebeson, *Mississippi Valley Historical Review*, 19 (September, 1932), 290–291.

6. Marcus, *Colonial American Jew*, III, 1113.

7. Edwin Wolf, 2nd, and Maxwell Whiteman, *The History of the Jews of Philadelphia from Colonial Times to the Age of Jackson* (Philadelphia: JPS, 1957), pp. 10–11; Stokes, *Church*, I, 854–855; Avram Vossen Goodman, *American Overture* (Philadelphia: JPS, 1947), pp. 74, 86, 88; Jacob Rader Marcus, *Early American Jewry: The Jews of Pennsylvania and the South, 1655–1790* (Philadelphia: JPS, 1953), 384; Max J. Kohler, "Phases of Jewish Life in New York Before 1800," PAJHS, 2(1894), 91; J. H. Innes, *New Amsterdam and Its People* (New York: Charles Scribner's Sons, 1902), p. 5; Richardson Wright, *Hawkers and Walkers in Early America* (New York: J. P. Lippincott Co., 1927), p. 92.

8. Quoted in J. F. Jameson, *Original Narratives of New Netherlands* (New York: Charles Scribner's Sons, 1909), pp. 392–393.

9. Patricia U. Bonomi, *Under the Cape of Heaven: Religion, Society, and Politics in Colonial America* (New York: Oxford University Press, 1986), pp, 23, 24; Charles M. Andrews, *The Colonial Period of American History* (New Haven: Yale University Press, 1936), II, 311.

10. Goodman, *American*, pp. 5, 59; J. H. Hollander, 'The Civil Status of the

Jews in Maryland, 1634–1776," PAJHS, 41; Walter Brownlow Posey, *Religious Strike on the Southern Frontier* (Baton Rouge: Louisiana State University Press, 1965), pp. xv, xvi; Leo Hershkowitz, "Some Aspects of the New York Jewish Merchant Community, 1654–1820," AJHQ, 64 (September, 1976), 16; Emberson Edward Proper, *Colonial Immigration Laws* (New York: AMS Press, Inc., 1967), p. 13; Sanford H. Cobb, *The Rise of Religious Liberty in America* (originally published in 1902; New York: Burt Franklin, 1970), p. 444.

11. Anita Libman Lebeson, *Jewish Pioneers In America, 1492–1848* (New York: Behrman's Jewish Book House, 1938), p. 74; Wesley Frank Craven, *The Colonies in Transition, 1660–1713* (New York: Harper and Row, 1968), p. 102; Sweet, *American Culture*, p. 10; Gilbert Klaperman, "The News in New York City During the Revolution," in Joseph T. P. Sullivan, ed., *The Sheriff's Jury and the Bicentennial, January 30, 1976* (New York: Second Panel Sheriff's Jury, July, 1976), pp. 162–163, 164; Hershkowitz, "Some Aspects," p. 15; Albion Morris Dyer, "Site of the First Synagogue of the Congregation Shearith Israel of New York," PAJHS, 8 (1900), 39; Sidney Ahlstrom, *A Religious History of the American People* (New Haven: Yale University Press, 1972), p. 572.

12. Marcus, *Early American Jewry*, pp. 383, 436; Morris Jastrow, Jr., "Notes on the Jews of Philadelphia from Published Annals," PAJHS, 1 (1893), 49; Wolf and Whiteman, *History of the Jews of Philadelphia*, p. 15; Goodman, *American*, p. 123; Sanford H. Cobb, *The Rise of Religious Liberty in America* (New York: Cooper Square Publishers, Inc., 1902; reprinted by Cooper Square Publishers, 1968), p. 444; David T. Morgan, "Judaism in Eighteenth Century Georgia," *The Georgia Historical Quarterly*, 58 (Spring, 1974), 44; Charles C. Jones, Jr., "The Settlement of the Jews in Georgia," PAJHS, 1 (1893), 7; Louis B. Wright, *The Colonial Civilization of North America, 1607–1763* (London: Eyre and Spottsiwoode, 1949), p. 239.

13. Stokes, *Church and State*, I, 858; Goodman, *American*, pp. 153, 154; Marcus, *Early American Jewry*, p. 232.

14. Marcus, *Early American Jewry*, pp. 389–390; David C. Adelman, "Strangers: Civil Rights of Jews in the Colony of Rhode Island," *Rhode Island History*, 13 (July, 1954), 70; Ahlstrom, *Religious History*, p. 572; Lebeson, *Jewish Pioneers*, p. 86.j

15. Hershkowitz, "Some Aspects," p. 11; Sweet, *American Culture*, p. 11; Lebeson, *Jewish Pioneers*, pp. 78, 182, 187, 191; Goodman, *American*, p. 148; Martin E. Marty, *Pilgrims in Their Own Land: 500 Years of Religion in America* (Boston: Little, Brown and Co., 1984), pp. 285–286; Maxwell Whiteman, *Copper for America: The Hendricks Family and a National Industry, 1755–1939* (New Brunswick: Rutgers University Press, 1971), p. 4.

16. Marcus, *Colonial American Jewry*, III, 1119.

17. Ibid., III, 1560, n. #9.

18. Wold and Whiteman, *History of the Jews of Philadelphia*, p. 13.

19. "A Visitation to the Jews," in William Penn, *A Collection of the Works of William Penn* (2 volumes; London, n.p., 1726), II, 848, 852.

20. Ibid., II, 852.

21. Marcus, *Colonial American Jewry*, III, 1120.

22. Marcus, *Early American Jewry*, p. 525; James A. Gelin, *Starting Over: The Formation of the Jewish Community of Springfield, Massachusetts, 1840–1905* (Lonham, Md.: University Press of America, 1984), p. 22; Stephen G. Mostov, "A

Sociological Portrait of German Jewish Immigrants in Boston: 1845–1861," *AJS Review*, 3 (1978), 128.

23. William A. Clebsch, *American Religious Thought: A History* (Chicago: University of Chicago Press, 1973), p. 13.

24. Increase Mather, *The Mystery of Israel's Salvation* (London: n. p., 1669), pp. 174, 175–176.

25. Ibid., pp. 174, 178; Marcus, *Colonial American Jewry*, III, 1119.

26. Philip L. White, *The Beekmans of New York: In Politics and Commerce, 1647–1877* (New York: New-York Historical Society, 1956), p. 224; Morton Borden, *Jews, Turks, and Infidels* (Chapel Hill: University of North Carolina Press, 1984); Herbert W. Kline, " The Jew That Shakespeare Drew?" *American Jewish Archives*, 23 (April, 1971), 63; Marcus, *Colonial American Jews*, III, 1118; Stephen Bloore, "The Jew in American Dramatic Literature (1794–1930)," PAJHS, 40 (June, 1951), 353; Louis Harap, "Jews in American Drama, 1900–1918," *American Jewish Archives*, 36 (November, 1984), 149; Ellen Schiff, *From Stereotype to Metaphor: The Jew in Contemporary Drama* (Albany: State University of New York Press, 1982), pp. 2, 5.

27. Lebeson, *Jewish Pioneers*, pp. 85, 115–116; Kloperman, "The Jews in New York City," pp. 162, 171; Jacob R. Marcus, ed., *American Jewry: Documents, Eighteenth Century*. Cincinnati: Hebrew Union College Press, 1959), p. 11; Allon Schoener, *The American Jewish Album: 1654 to the Present* (New York: Rizzoli, 1983), pp. 16–17; Bertram Wallace Korn, *The Early Jews of New Orleans* (Waltham: AJHS, 1969), p. 229; Wolf and Whiteman, *The History of the Jews of Philadelphia*, p. 149; Leo Hershkowitz, "Some Aspects of the New York Jewish Merchant In Colonial Trade," in Aubrey L. Newman, ed., *Migration and Settlement* (London: Jewish Historical Society of England, 1971), p. 102.

28. Whiteman, *Copper for America*, p. 18; Michael Kammen, *Colonial New York* (New York: Charles Scribner's Sons, 1975), p. 232; Korn, *Early Jews of New Orleans*, p. 212.

29. Marcus, *Early American Jewry*, p. 392; David C. Adelman, "Strangers: Civil Rights of Jews in the Colony of Rhode Island," *Rhode Island History*, 13 (July, 1954), 66; Ahlstrom, *A Religious History*, p. 572; Vamberto Morais, *A Short History of Anti-Semitism* (New York: W. W. Norton & Co., 1976), p. 161; Goodman, *American*, p. 10; Stanley Ayling, *The Elder Pitt, Earl of Chatham* (London: Williams Collins Sons & Co., 1976), p. 127; Basil Williams, *The Life of William Pitt, Earl of Chatham* (New York: Longmans, Green & Co., 1914), pp. 174–175; Hannah Adams, *The History of the Jews* (London: A. Macintosh, 1818), p. 392.

30. Thomas W. Perry, *Public Opinion, Propaganda, and Politics in Eighteenth-Century England: A Study of the Jew Bill of 1753* (Cambridge: Harvard University Press, 1962), p. 5.

31. E. B. Benson, *Queen Victoria* (London: Longmans, Green & Co., 1935), p. 175.

32. Adams, *History of the Jews*, pp. 390–391.

33. Whiteman, *Copper For America*, p. 18; John Stuart Conning, "The Jewish Situation in America," *International Review of Missions* (London), 16 (January, 1927), 66; Marcus, *Early American Jewry*, p. 390.

34. Conning, "The Jewish Situation," p. 66; Marcus, *Early American Jewry*, p. 436; Sydney V. James, *Colonial Rhode Island* (New York: Charles Scribner's Sons, 1975), p. 209; Adams, *History of the Jews*, pp. 464–467; Ahlstrom, *A Religious History*, p. 573; Goodman, *American*, p. 3; Doris Groshen Daniels, "Colo-

nial Jewry: Religion, Domestic and Social Relations," AJHQ, 66 (March, 1977), 380; Ira Rosenwaike, "The Jews of Baltimore to 1810," AJHQ, 64 (June, 1975), 292.

35. Herhkowitz, "Some Aspects," in Newman, ed, *Migration and Settlement*, pp. 103–104; Kammen, *Colonial New York*, p. 209.

36. Stanely F. Chyet, "The Political Rights of the Jews in the United States: 1776–1840," *American Jewish Archives*, 10 (April, 1958), 14.

37. Marcus, *Early American Jewry*, p. 523.

38. Quoted in Wolf and Whiteman, *History of the Jews of Philadelphia*, p. 106; see also "Notes and Documents: The Autobiography of Peter Stephen Du Ponceau," *Pennsylvania Magazine of History and Biography*, 63 (1939), 328.

39. *Diary of John Quincy Adams* (Cambridge: The Belknap Press of Harvard University, 1981), I, 59.

40. Quoted in Stokes, *Church and State*, I, 76.

41. Ibid., p. 857.

42. Quoted in Borden, *Jews, Turks*, pp. 9–10.

Chapter 2

1. Robert Ernst, *Immigrant Life in New York City, 1825–1863* (New York: King's Crown Press, 1949), p. 46; Louis Ruchames, "The Abolitionists and The Jews: Some Further Thoughts," in Bertram Wallace Korn, ed., *A Bicentennial Festschrift for Jacob Rader Marcus* (Waltham, Mass.: American Jewish Historical Society, 1976), p. 514; Jonathan D. Sarna, "The 'Mythical Jew' and the 'Jew Next Door' in Nineteenth-Century America," in David A. Gerber, ed., *Anti-Semitism in American History* (Urbana: University of Illinois Press, 1986), p. 57ff.; Martin E. Marty, *Righteous Empire: The Protestant Experience in America* (New York: The Dial Press, 1970), p. 124; Martin E. Marty, *Pilgrims in Their Own Land: 500 Years of Religion in America* (Boston: Little, Brown & Co., 1984), p. 285.

2. Morton Borden, *Jews, Turks, and Infidels* (Chapel Hill: University of North Carolina Press, 1984), p. 22.

3. Ira Rosenwaike, "An Estimate of the Jewish Population of the United States In 1790," PAJHS, 50 (September, 1960), 34; John J. Appel, "Popular Graphics as Documents for Teaching and Studying Jewish History," in Korn, ed., *A Bicentennial Festschrift*, p. 539; H. G. Reissner, "The German-American Jews," in Leo Baeck Institute *Year Book*, 10 (1965), 69; Henry L. Feingold, *Zion in America* (New York: Hippocrene Books, Inc., 1974), pp. 68–69; Naomi W. Cohen, *Encounter with Emancipation: The German Jews in the United States, 1830–1914* (Philadelphia: JPS, 1984), p. 9; Gerald Sorin, *A Time for Building: The Third Migration, 1880–1920* (Baltimore: Johns Hopkins University Press, 1992), p. 7; Eli Faber, *A Time for Planting: The First Migration* (Baltimore: Johns Hopkins Press, 1992), p. 111; Israel Goldstein, *A Century of Judaism in New Yok* (New York: Congregation B'nai Jeshurun, 1930), pp. 45, 47; Ira Rosenwaike, "The Jews of Baltimore: 1820 to 1830," AJHQ, 67 (March, 1978), 246; Ann Deborah Michael, "The Origins of the Jewish Community of Cincinnati, 1817–1860," *Cincinnati Historical Society Bulletin*, 30 (Fall–Winter, 1972), 156; Ira Rosenwaike, "The First Jewish Settlers In Louisville," *Filson Club Historical Quarterly*, 53 (1979), 37; Charles Reznikoff, *The Jews of Charleston* (Philadelphia: JPS, 1950), p. 68.

4. Borden, *Jews, Turks, and Infidels*, pp. ix, 18, 19; Anson Phelps Stokes, *Church and State in the United States* (3 volumes; New York: Harper and Brothers, 1950), I, 599; Joseph L. Blau, ed., *Cornerstones of Religious Freedom in America* (Boston: The Beacon Press, 1950), p. 89; Maxine S. Seller, "Isaac Leeser: A Jewish Christian Dialogue in Antebellum Philadelphia," *Pennsylvania History*, 35 (July, 1968), 237; Joseph L. Blau and Salo W. Baron, eds., *The Jews of the United States, 1790–1840, a Documentary History* (3 volumes; New York: Columbia University Press, 1963), I, 17; Jackson T. Main, *The Antifederalists* (Chapel Hill: University of North Carolina Press, 1961), p. 159; Leonard Dinnerstein and Mary Dale Palsson, eds., *Jews in the South* (Baton Rouge: Louisiana State University Press, 1973), p. 43ff.; "The Disabilities in North Carolina," AI, March 22, 1961, p. 300; Naomi W. Cohen, "Pioneers of American Jewish Defense," *AJA*, 29 (November, 1977), 125.

5. Borden, *Jews, Turks and, Infidels*, pp. 23, 28, 141; Jacob Rader Marcus, *Early American Jewry: The Jews of Pennsylvania and the South, 1655–1790* (Philadelphia: JPS, 1953), p. 537; Stokes, *Church and State*, I, 439, 858–859; Blau and Baron, *The Jews of the United States*, I, 3; Richard B. Morris, *The Forging of the Union, 1781–1789* (New York: Harper & Row, 1987), p. 359 fn. 28; Seller, "Isaac Leeser," p. 237.

6. Malcom H. Stern, "The 1820s: American Jewry Comes of Age," in Korn, *A Bicentennial Festschrift*, p. 543; Blau, *Cornerstones of Religious Freedom*, pp. 89–90; Borden, *Jews, Turks, and Infidels*, pp. 37, 38; Isaac M. Fein, *The Making of an American Jewish Community: The History of Baltimore Jewry from 1773 to 1920* (Philadelphia: JPS, 1971), p. 35; Stokes, *Church and State*, I, 874; Edward Eitches, "Maryland's 'Jew Bill,'" AJHQ, 60 (March, 1971), 277–278; Jonathan D. Sarna, "American Anti-Semitism," in David Berger, ed., *History and Hate: The Dimensions of Anti-Semitism* (Philadelphia: JPS, 1986), p. 119; Isaac M. Fein, "Niles' Weekly Register on the Jews," PAJHS, 50 (September, 1960), 4.

7. Fein, "Niles' Weekly Register," p. 4; Robert J. Parker, "A Yankee in North Carolina: Observations of Thomas Oliver Larkin, 1821–1826," *North Carolina Historical Review*, 14 (October, 1937), 33.

8. Jacob R. Marcus, ed., *American Jewry: Documents, Eighteenth Century* (Cincinnati: Hebrew Union College Press, 1959), p. 52.

9. Edmund Wilson, "Notes on Gentile Pro-Semitism, New England's Good Jews," *Commentary*, 22 (October, 1956), 331–332; Egal Feldman, *Dual Destinies: The Jewish Encounter with Protestant America* (Urbana: University of Illinois Press, 1990), 52, 55.

10. Edwin Wolf 2nd and Maxwell Whiteman, *The History of the Jews of Philadelphia from Colonial Times to the Age of Jackson* (Philadelphia: JPS, 1957), Borden, *Jews, Turks, and Infidels*, pp. 24–25; Bertram W. Korn, *American Jewry and the Civil War* (Philadelphia: JPS, 1951), p. 156; Morris U. Schappes, review of *Myer Myers, Goldsmith, 1723–1795* by Jeanette W. Rosenbaum, *New York History*, 35 (April, 1954), 185; Blau, *Cornerstones of Freedom*, p. 106.

11. Stern, "The 1820s," p. 546; Jonathan D. Sarna, *Jacksonian Jew: The Two Worlds of Mordecai Noah* (New York: Holmes and Meier, 1981), pp. 53–54; Donald M. Scott, review of Sarna, *Jacksonian Jew*, in *New York History*, 63 (October, 1982), 478, 479.

12. Quoted in Ruchames, "The Abolitionists and the Jews," pp. 509, 510.

13. Donovan Fitzpatrick and Saul Saphire, *Navy Maverick: Uriah Phillips Levy* (Garden City: Doubleday & Co., 1963), pp. 62–63, 73, 95, 115, 210, 222;

Jacob Rader Marcus, ed., *Memoirs of American Jews, 1775–1865* (Philadelphia: JPS, 1955), p. 84ff.; Abram Kanof, "Uriah Phillips Levy: The Story of a Pugnacious Commodore," PAJHS, 39 (September, 1949), 2, 25, 27; Jerry Rand, "The Ordeal of Uriah Levy," *American Mercury,* 56 (June, 1943), 728, 730; James E. Valle, *Rocks and Shoals: Order and Discipline in the Old Navy, 1800–1861* (Annapolis: Naval Institute Press, 1980), p. 53.

14. John Higham, "Social Discrimination Against Jews in America, 1830–1930," PAJHS, 47 (September, 1957), 4; Seller, "Isaac Leeser," p. 237; see also Naomi W. Cohen, "Antisemitism in the Gilded Age: the Jewish View," JSS, 41 (Summer/Fall, 1979), 190.

15. Quoted in Fein, *"Niles' Weekly Register,"* p. 4.

16. Allan Tarshish, "Jew and Christian in a New Society: Some Aspects of Jewish-Christian Relationships in the United States, 1848–1881," in Korn, ed., *A Bicentennial Festschrift,* p. 574; George L. Berlin, *Defending The Faith: Nineteenth-Century American Jewish Writings on Christianity and Jesus* (Albany: State University of New York Press, 1989), p. 4; R. M. Healy, "From Conversion to Dialogue: Protestant American Missions to the Jews in the 19th and 20th centuries," *Journal of Ecumenical Studies,* 18 (Summer, 1981), 377; Feldman, *Dual Destinies,* p. 63; Borden, *Jews, Turks, and Infidels,* pp. 22, 75; Jay P. Dolan, "In Whose God Do We Trust?" *The New York Times Book Review,* May 10, 1992, pp. 7–8; Martin E. Marty, *The Righteous Empire: The Protestant Experience in America* (New York: The Dial Press, 1970), p. 124; Robert T. Handy, "The Protestant Quest for a Christian America, 1830–1930," *Church History,* 22 (March, 1953), 10, 11.

17. "The Modern Jews," *North American Review,* 60 (1845), 330, 368; Wolf and Whiteman, *The History of the Jews of Philadelphia,* pp. 241–242, 243; Berlin, *Defending the Faith,* pp. 7, 12; Jonathan D. Sarna, "The American Jewish Response to Nineteenth-Century Christian Missions," *Journal of American History,* 68 (June, 1981), 36, 38, 39; Philip L. White, *The Beekmans of New York: In Politics and Commerce, 1647–1877* (New York: New York Historical Society, 1956), pp. 626–627; Hasia Diner, *A Time for Gathering: The Second Migration* (Baltimore: Johns Hopkins University Press, 1992), p. 178; Solomon Henry Jackson, *The Jew; Being a Defence of Judaism Against All Adversaries* (New York, n. p., 1823), p. vii; Robert T. Handy, *A Christian America* (2nd edition; New York: Oxford University Press, 1984), p. 48.

18. Ruth Miller Elson, *Guardians of Tradition: American Schoolbooks of the Nineteenth Century* (Lincoln: University of Nebraska Press, 1964), pp. 82, 85; Glenn T. Miller, *Religious Liberty in America* (Philadelphia: The Westminster Press, 1976), pp. 90–91; Ruchames, "The Abolitionists and The Jews," p. 511; Christian C. Johnstone, *Stories from the History of the Jews adopted for Young Persons* (New York: C. S. Francis and Co., 1853), pp. 206, 215; Louise A. Mayo, *The Ambivalent Image: Nineteenth Century America's Perception of the Jew* (Rutherford, N.J.: Fairleigh Dickinson University Press, 1988), pp. 21–22; see also Alfred Moritz Myers, *The Young Jew* (Philadelphia: American Sunday-School Union, 1848), p. 7.

19. Mayo, *Ambivalent Image,* p. 22.

20. Quoted in Sigmund Livingstone, *Must Men Hate?* (New York: Harper and Brothers, 1944), p. 4.

21. Elizabeth Peabody, *Sabbath Lessons* (Salem: Joshua Cushing, 1813), p. 49.

22. Ibid., p. 60.

23. Myers, *The Young Jew*, p. 7.

24. Ibid., pp. 7, 8. In 1923, philospher Horace Kallen wrote, "Attitudes that Sunday-schools the world over impart automatically to children at five may be deep buried and forgotten at five and fifty, but they are not extirpated, nor translated. They make a sub-soil of preconceptions upon which other interests are nourished and from which they gather strength." Horace Kallen, "The Roots of Anti-Semitism," *The Nation*, 116 (February 28, 1923), 241.

25. Johnstone, *Stories From the History of the Jews*, p. 232.

26. Sarah S. Baker, *The Jewish Twins* (New York: Robert Carter & Bros., 1860), p. 32; Jacob Rader Marcus, *United States Jewry, 1776–1985: The Germanic Period* (Detroit: Wayne State University Press, 1991), II, 288; Hannah Adams, *The History of the Jews from the Destruction of Jerusalem to the Present Time* (London: n.p., 1818), p. 43.

27. Quoted in Samuel E. Karff, "Anti-Semitism In The United States, 1844–1860," p. 32 in "Anti-Semitism II," box 2072, AJA; see also Stokes, *Church and State*, p. 195; Max J. Kohler, "Phases in the History of Religious Liberty in America with Particular Reference to the Jews—II," PAJHS, 13 (1905), 19ff.

28. Floyd S. Fierman, "The Jews and the Problem of Church and State in America Prior to 1881," *The Educational Forum*, 15 (1951), 335–336; Diane Ravich, *The Great School Wars: New York City, 1805–1973* (New York: Basic Books, 1974), chapters 4–7.

29. Max Vorspan and Lloyd P. Gartner, *History of the Jews of Los Angeles* (Philadelphia: JPS, 1970), p. 53; Marcus, *United States Jewry*, II, 288; Elson, *Guardians of Tradition*, pp. 82, 85; Miller, *Religious Liberty in America*, pp. 90–91; John Higham, *Strangers in the Land* (New York: Atheneum, 1974; originally published, New Brunswick: Rutgers University Press, 1955), p. 27; Mayo, *The Ambivalent Image*, p. 84; David A. Gerber, "Cutting Out Shylock: Elite Anti-Semitism and the Quest for Moral Order in the Mid-Nineteenth Century American Market Place," *Journal of American History*, 69 (December, 1982), 629–630; Leslie A. Fiedler, "What Can We Do About Fagin?: The Jew Villain in Western Tradition," *Commentary*, 7 (May, 1949), 413; Diner, *A Time For Gathering*, p. 198.

30. Kate E. R. Pickard, "The Kidnapped and the Ransomed," *American Jewish Archives*, 9 (April, 1957), 16.

31. Gerber, "Cutting Out Shylock," pp. 629–630; Selma C. Berrol, "In Their Image: German-Jews and the Americanization of the *Ost Juden* in New York City," *New York History*, 63 (October, 1982), 427; Rudolf Glanz, *Jew and Irish* (New York: n. p., 1966), p. 128; Nathan Hurvitz, "Blacks and Jews in American Folklore," *Western Folklore Quarterly*, 33 (October, 1974), 324; Parker, "A Yankee in North Carolina," p. 338; "Boston," AI, September 2, 1881, p. 74; Herbert W. Kline, "The Jew That Shakespeare Drew?" *American Jewish Archives*, 23 (April, 1971), 64; Naomi W. Cohen, "Antisemitic Imagery: The Nineteenth-Century Background," JSS, 47 (Summer/Fall, 1985), 309; *Niles' Weekly Register* (Baltimore) 19 (October 20, 1820), 114.

32. Cohen, "Antisemitic Imagery," p. 309; Stephen G. Mostov, "Dun and Bradstreet Reports as a Source of Jewish Economic History: Cincinnati, 1840–1875," AJH, 72 (March, 1983), 339, 349, 350; Peter R. Decker, "Jewish Merchants in San Francisco: Social Mobility on the Urban Frontier," PAJHS, 68 (June, 1979), 398, 399; Gerber, "Cutting Out Shylock," pp. 631, 632; Marc Lee Raphael, ed., *Jews and Judaism in the United States: A Documentary History* (New York:

Behrman House, Inc., 1983), p. 34ff.; Diner, *A Time for Gathering*, pp. 77–78, 188–189; Elliott Ashkenazi, *The Business of Jews in Louisiana, 1840–1875* (Tuscaloosa: University of Alabama Press, 1988), pp. 75, 146, 152–53; Bertram Wallace Korn, *The Early Jews of New Orleans* (Waltham: AJHS, 1969), p. 74.

33. John Higham, "Anti-Semitism in the Gilded Age," *Mississippi Valley Historical Review*, 43 (1958), 568; John Reeves, *The Rothschilds: The Financial Rulers of Nations* (Chicago: A. C. McClurg & Co., 1887), p. 1.

34. *Niles' Weekly Register*, 37 (November 28, 1829), 214.

35. New York *Herald* as quoted in *Niles' Weekly Register*, 49 (September 19, 1835), 41.

36. See, for example, Till van Rahden, "Beyond Ambivalence: Variations of Catholic Antisemitism In Turn of the Century Baltimore" (unpublished M.A. thesis, Johns Hopkins University, 1992), p. 31ff.

37. L. Maria Child, *Letters from New York* (New York: Charles S. Francis and Co., 1843), p. 33.

38. "Editor's Easy Chair," *Harper's Magazine*, 17 (1858), 268.

39. Fein, "Niles' Weekly Register," p. 4; Vorspan and Gartner, *History of the Jews of Los Angeles*, pp. 15–16; Marcus, *United States Jewry*, II, 287, 288; quote from Harold David Brackman, "The Ebb and Flow of Conflict: A History of Black-Jewish Relations Through 1900" (unpublished Ph. D., Department of History, University of California, Los Angeles, 1977), pp. 301–302.

40. John Higham, *Send These to Me: Immigrants in Urban America* (revised edition; Baltimore: The Johns Hopkins Press, 1984), p. 101; Cohen, *Encounter with Emancipation*, pp. 26, 32–33; Irving Howe, *World of Our Fathers: The Journey of the East European Jews to America and the Life They Found and Made* (New York: Harcourt Brace Jovanovich, 1976), p. 630. In 1931, 13-year-old John F. Kennedy wrote to his father and described how he negotiated a deal with a classmate, "Have jewed sled down to $3.00." Quoted in Nigel Hamilton, *JFK: Reckless Youth* (New York: Random House, 1992), p. 86.

41. Marcus, *United States Jewry*, II, 286; "Editor's Easy Chair," 17 (July, 1858), 267; John J. Appel, "Popular Graphics as Documents for Teaching and Studying Jewish History," in Korn, ed., *A Bicentennial Festschrift*, p. 31; Diner, *A Time for Gathering*, p. 171.

42. Myron Berman, *Richmond's Jewry* (Charlottesville: University Press of Virginia, 1979), p. 109ff.; Marcus, *United States Jewry*, II, 287; Henry Schwartz, "The Uneasy Alliance: Jewish-Anglo Relations in San Diego," *Journal of San Diego History*, 20 (Summer, 1974), 59; Maxwell Whiteman, *Copper for America: The Hendricks Family and a National Industry, 1755–1939* (New Brunswick: Rutgers University Press, 1971), pp. 168, 218; Robert T. Handy, *A Christian America: Protestant Hopes and Historical Realities* (New York: Oxford University Press, 1971), p. 59; Reznikoff, *The Jews of Charleston*, p. 124; E. Digby Baltzell, Allen Glicksman, and Jacquelyn Litt, "The Jewish Communities of Philadelphia and Boston: A Tale of Two Cities," in Murray Friedman, ed., *Jewish Life in Philadelphia, 1830–1940* (Philadelphia: ISHI Publications, 1983), p. 292; E. Digby Baltzell, *Philadelphia Gentlemen: The Making of a National Upper Class* (Glencoe, Ill.: The Free Press, 1958), p. 279.

43. Arthur Goren, *The American Jews* (Cambridge: The Belknap Press of Harvard University Press, 1982), p. 24; Israel Goldstein, *A Century of Judaism in New York* (New York: Congregation B'nai Jeshrun, 1930), p. 98; Rollin G. Osterweis, *Three Centuries of New Haven, 1638–1938* (New Haven: Yale Univer-

sity Press, 1953), p. 217; Selig Adler, "Zebulon B. Vance and the 'Scattered Nation,'" *Journal of Southern History*, 7 (August, 1941), 358, 361; Moses Rischin, review of *The Early Jews of New Orleans* by Bertram Korn, *Journal of American History*, 57 (June, 1970), 152; Blau and Baron, *The Jews of the United States*, III, 803; see also Korn, *The Early Jews of New Orleans*, p. 214.

44. Cohen, *Encounter with Emancipation*, pp. 7ff., 13, 14; Diner, *A Time for Gathering*, pp. 37ff., 56; Stephen G. Mostov, "A 'Jerusalem' on the Ohio: The Social and Economic History of Cincinnati's Jewish Community, 1840–1875" (unpublished Ph.D., Department of Near Eastern and Judaic Studies, Brandeis University, 1981) p. 45; Leon A. Jick, *The Americanization of the Synagogue, 1820–1870* (Hanover, NH: Brandeis University Press, 1976), pp. 16, 50; Ashkenazi, *The Business of Jews in Louisiana, 1840–1875*, pp. 7, 8; Barry E. Supple, "A Business Elite: German Jewish Financiers in Nineteenth-Century New York," *Business History Review*, 31 (Summer, 1957), 149; Reissner, "The German-American Jews," p. 69; Blau and Baron, *The Jews of the United States*, III, 803; Goren, *The American Jews*, p. 1; Stephen G. Mostov, "A Sociological Portrait of German Jewish Immigrants in Boston: 1845–1861," p. 125; AI, August 18, 1854, p. 44; Sylvan Morris Dubow, "Identifying the Jewish Serviceman in the Civil War: A Re-appraisal of Simon Wolf's *The American Jews as Patriot, Soldier and Citizen*," AJHQ, 59 (March, 1970), 359; Marc Lee Raphael, "Intra-Jewish Conflict in the United States, 1869–1915" (unpublished Ph.D., Department of History, University of California at Los Angeles), pp. 12–16.

45. "Memorandum on Lincoln and the Jews," n.d., folder, "Jews, Lincoln and [sic]," box 16, James G. Randall mss., LC; Robert Ernst, "Economic Nativism in New York City During The 1840's," *New York History*, 29 (Spring, 1948), 173; John Bodnar, review of *The Peoples of Philadelphia: A History of Ethnic Groups and Lower-Class Life, 1790–1940*, edited by Allen F. Davis and Mark H. Haller, *Pennsylvania History*, 41 (October, 1974), 470.

46. Gustav Gottheill, "The Position of the Jews in America," *North American Review*, 126 (March–April, 1878), 306.

47. Sidney Ahlstrom, *A Religious History of the American People* (New Haven: Yale University Press, 1972), pp. 574–575; Mostov, "A Sociological Portrait of German Jewish Immigrants in Boston," pp. 125, 151; Reissner, "The German-American Jews," p. 73; Jick, *The Americanization of the Synagogue*, pp. 33, 39; Ashkenazi, *The Business of Jews in Louisiana*, pp. 106, 107, 116,125, 132, 133; Stephen G. Mostov, "The Early Years of the Jewish Community," in *Germans in Boston* (Boston: Goethe Society of New England, 1981), pp. 79–81; Jacob M. Sable, "Some American Jewish Organizational Efforts To Combat Anti-Semitism" (unpublished Ph.D., Yeshiva University, 1964), p. 229.

48. Jick, *The Americanization of the Synagogue*, pp. 59, 67; Wolf and Whiteman, *The History of the Jews of Philadelphia*, pp. 372–373; Berlin, *Defending the Faith*, p. 25; Michael A. Meyer, *Response to Modernity* (New York: Oxford University Press, 1988), p. 242; Winthrop S. Hudson, *Religion in America* (2nd edition; New York: Charles Scribner's Sons, 1973), p. 331; Ahlstrom, *A Religious History of the American People*, p. 579.

49. Goren, *The American Jews*, pp. 27–28; Cohen, "Pioneers of American Jewish Defense," p. 116; Joseph Jacobs, "The Damascus Affair of 1840 and the Jews of America," PAJHS, 10 (1902), 119, 122; Goldstein, *A Century of Judaism in New York*, pp. 65–66; Jon Adland, "American Jewish Reaction to the Damascus Affair, 1840" (unpublished essay, Hebrew Union College, 1978), p. 6, Small

Collection, AJA; Sarna, *Jacksonian Jew*, p. 123ff.; Cohen, *Encounter with Emancipation*, pp. 213–215; Alan Dundes, "The Ritual Murder or Blood Libel Legend: A Study of the Anti-Semitic Victimization through Projective Inversion," in Alan Dundes, ed., *The Blood Libel Legend: A Casebook in Anti-Semitic Folklore* (Madison: University of Wisconsin Press, 1991), p. 343.

50. Feldman, *Dual Destinies*, pp. 77, 78; Jick, *The Americanization of the Synagogue*, p. 67; Seller, "Isaac Leeser," pp. 231, 234, 239, 240; Berlin, *Defending the Faith*, pp. 25–26; Sarna, "The American Jewish Response to Nineteenth-Century Christian Missions," p. 49; Cohen, "Pioneers of American Jewish Defense," p. 130.

51. Quoted in Lloyd P. Gartner, "Temples of Liberty Unpolluted: American Jews and Public Schools, 1840–1875," in Korn, ed., *A Bicentennial Festschrift*, p. 172.

52. Seller, "Isaac Leeser," pp. 231, 232, 234, 240.

53. Meyer, *Response to Modernity*, p. 242.

54. Quoted in Mostov, "A 'Jerusalem' on the Ohio," p. 172.

55. Jick, *The Americanization of the Synagogue*, p. 135.

56. Robert S. Wistrich, *Antisemitism: The Longest Hatred* (London: Metheun London, 1991), p. 115.

57. Hudson, *Religion in America*, p. 331; Meyer, *Response to Modernity*, pp. 116–117, 120, 122, 133, 138, 144, 150; Isaac M. Wise, *Reminiscences*, translated and edited by Daniel Philipson (2nd edition; New York: Central Synagogue of New York, 1945; originally published, 1901), p. 273; Cohen, *Encounter with Emancipation*, pp. 262–263; Louis Ginsberg, *Chapters on the Jews of Virginia, 1658–1900* (Petersburg, Va.: n. p., 1969), p. 40; see also Cohen, "Pioneers of American Jewish Defense," *passim*.

58. "Memorandum on Lincoln and the Jews," folder, "Jews, Lincoln and," Randall mss., LC.

59. Isaac M. Wise, *Reminiscences* (Cincinnati: Leo Wise and Co., 1901), p. 272.

60. Robert A. Rockaway, "Anti-Semitism in an American City: Detroit, 1850–1914," AJHQ, 64 (September, 1974), 42.

61. Sarna, *Jacksonian Jew*, p. 199, fn. 3; Higham, "Social Discrimination," PAJHS, p. 4; Brackman, "The Ebb and Flow," p. 452.

62. Rockaway, "Anti-Semitism in an American City," p. 43; "Strange and Startling," *The Occident*, 14 (1856), 146–147; Lloyd P. Gartner, *History of the Jews of Cleveland* (Cleveland: Western Reserve Historical Society, 1978), p. 25; Morton Rosenstock, *Louis Marshall, Defender of Jewish Rights* (Detroit: Wayne State University Press, 1965), p. 37; Marcus, *United States Jewry*, II, 282.

63. Harvey W. Scott, "The Pioneer Character of Oregon Progress," *Oregon Historical Quarterly*, 18 (December, 1917), 252; Diner, *A Time for Gathering*, pp. 146, 151–152; Decker, "Jewish Merchants in San Francisco," p. 397; "Intolerance in California," *The Occident*, 13 (June, 1855), 123, 124; Borden, *Jews, Turks, and Infidels*, p. 51; "Anti-Jewish Sentiment In California—1855," *American Jewish Archives*, 12 (April, 1960), 15; Robert E. Levinson, *The Jews in the California Gold Rush* (New York: KTAV, 1979), pp. 71, 72; Vorspan and Gartner, *History of the Jews of Los Angeles*, p. 15; *Asmonean* quoted in Cohen, "Pioneers of American Jewish Defense," 138.

64. "City Intelligence," *Daily Alta California*, December 11, 1851; Anne Loftis, *California: Where the Twain Did Meet* (New York: Macmillan Publishing Co.,

1973), p. 151; Levinson, *The Jews in the California Gold Rush*, pp. 13, 14, 68, 79, 144; Schwartz, "The Uneasy Alliance," p. 55; Norman Mason, "The Know Nothing Party in California" (unpublished M. A. thesis, Department of History, San Diego State College, n.d.), p. 43.

65. Loftis, *California*, p. 116.

66. *The Works of Hubert Howe Bancroft*. Volume 35: *California Inter Pocula* (San Francisco: The History Company, 1888), p. 373; see also Hilton Rowan Helper, *The Land of Gold* (Baltimore: n.p., 1855), p. 54.

67. Mostov, "A 'Jerusalem' on the Ohio," p. 236; Isidor Bush, "The Jews of St. Louis," *Missouri Historical Society Bulletin*, 8 (October, 1951), p. 68; originally published in *The Jewish Tribune*, 1883.

68. Harold M. Hyman, *Oleander Odyssey: The Kempners of Galveston, Texas, 1854–1990s* (College Statetion: Texas A&M University Press, 1990), p. 10.

69. Mayo, *The Ambivalent Image*, p. 38.

70. Cohen, *Encounter with Emancipation*, p. 101ff.; Cohen, "Pioneers of American Jewish Defense," p. 131; Diner, *A Time for Gathering*, pp. 150, 151.

71. Bertram Wallace Korn, *The American Reaction to the Mortara Case: 1858–1859* (Cincinnati: The American Jewish Archives, 1957), p. 1; Borden, *Jews, Turks, and Infidels*, pp. 55, 87; Ernst, *Immigrant Life in New York City*, p. 138; Josef L. Altholz, "A Note on the English Catholic Reaction to the Mortara Case," JSS, 23 (1961), 111, 116; Diner, *A Time for Gathering*, p. 154; Thomas P. O'Neill, Jr., et al to Edward H. Levi, March 4, 1975, folder, "Subject Files: Jewry, Correspondence, 1975–1976," Bella Abzug mss., Columbia University, New York City; Cohen, *Encounter with Emancipation*, p. 217.

72. Ernst, *Immigrant Life in New York City*, p. 138.

73. Hyman B. Grinstein, *The Rise of the Jewish Community of New York, 1654–1860* (Philadelphia: JPS, 1945), pp. 432–433; Ann G. Wolfe, ed., *A Reader in Jewish Community Relations* (New York: KTAV, 1975), p. 9; Maxwell Whiteman, "The Legacy of Isaac Leeser," in Murray Friedman, ed., *Jewish Life in Philadelphia, 1830–1940* (Philadelphia: ISHI Publications, 1983), pp. 43–44; Goldstein, *A Century of Judaism in New York*, pp. 100–101; Cohen, *Encounter with Emancipation*, pp. 46, 218; Allan Tarshish, "The Board of Delegates of American Israelites (1859–1878), in Karp, ed., *The Jewish Experience in America*, III, 138.

74. Korn, *American Jewry and the Civil War*, pp. 168, 187; AI, August 2, 1861, p. 36, August 16, 1861, p. 52, February 28, 1862, p. 28; see also Bertram Wallace Korn, *Eventful Years and Experiences* (Cincinnati: The American Jewish Archives, 1954), pp. 142, 145; Rockaway, "Anti-Semitism in an American City," p. 44; Diner, *A Time for Gathering*, pp. 158–159, 186–187; Naomi W. Cohen, "Antisemitism in the Gilded Age: The Jewish View," in Naomi W. Cohen, ed., *Essential Papers on Jewish-Christian Relations: Imagery and Reality* (New York: New York University Press, 1990), p. 131; Bertram W. Korn, "Lincoln and the Jew," *Illinois State Historical Society Journal*, 48 (Summer, 1955), 187; Steven V. Ash, "Civil War Exodus: The Jews and Grant's General Orders No. 11," *Historian*, 44 (August, 1982), 517.

75. Quoted in Cohen, "Antisemitism in the Gilded Age," p. 131.

76. Goldstein, *A Century of Judaism in New York*, p. 101; Korn, "Lincoln and the Jews," p. 183; "The Jews of the Union," *American Jewish Archives*, 13 (November, 1961), 169–170, 176–177; Ash, "Civil War Exodus," p. 517; Korn, *American Jewry and the Civil War*, pp. 159–160; "Religious Prejudices," *The Occident and American Jewish Advocate*, 22 (November, 1964), 368.

77. Quoted in Yuri Suhl, *Ernestine L. Rose and the Battle for Human Rights* (New York: Reynal and Co., 1959), p. 220.

78. Anita Libman Lebeson, *Pilgrim People: A History of Jews in America from 1492 to 1974* (revised edition; New York: Minerva Press, 1975), p. 258; Yuri Suhl, *Eloquent Crusader: Ernestine Rose* (New York: Julian Messner, 1970), p. 154; Korn, *The Early Jews of New Orleans*, p. 223.

79. Quoted in Robert A. Rockaway, *The Jews of Detroit: From the Beginning, 1762–1914* (Detroit: Wayne State University Press, 1986), p. 26.

80. Korn, *American Jewry and the Civil War*, pp. 160, 161; Rockaway, "Anti-Semitism in an American City," p. 44; Irving Katz, *August Belmont* (New York: Columbia University Press, 1968), pp. 144, 146, 165.

81. Ash, "Civil War Exodus," pp. 508, 509, 511; *The Papers of Ulysses S. Grant*, edited by John Y. Simon (12 volumes; Carbondale: Southern Illinois University Press, 1970–1984), VII, 50; AI, August 2, 1861, p. 36, January 2, 1863, p. 202; "Renewed Illiberality," *The Occident*, 23 (October, 1865), 316; Korn, "Lincoln and the Jews," p. 185; Ashkenazi, *The Business of Jews in Louisiana*, pp. 76–77; Shields McIlwaine, *Memphis Down in Dixie* (New York: E. P. Dutton and Company, Inc., 1948), pp. 120, 124; Cohen, "Antisemitism in the Gilded Age," p. 129.

82. "Renewed Illiberality," p. 316.

83. Ash, "Civil War Exodus," p. 511.

84. Lebeson, *Pilgrim People*, p. 290 ff.

85. Quoted in Korn, "Lincoln and the Jews," p. 186.

86. Korn, *American Jewry and the Civil War*, pp. 163, 176, 177, 178, 179, 181, 184; Feuer, "America's First Jewish Professor," p. 186; "Renewed Illiberality," p. 316; Ella Lonn, *Foreigners in the Confederacy* (Chapel Hill: University of North Carolina Press, 1940), pp. 336, 392; Allan Peskin, "The Origins of Southern Anti-Semitism," *The Chicago Jewish Forum*, 14 (Winter, 1955–56), 86; Lewis M. Killian, *White Southerners* (revised edition; Amherst: University of Massachusetts Press, 1985), p. 74; Berman, *Richmond's Jewry*, pp. 182, 183–184; Diner, *A Time for Gathering*, p. 158.

87. Korn, *American Jewry and the Civil War*, p. 177; Berman, *Richmond's Jewry*, pp. 181–184; Lonn, *Foreigners in the Confederacy*, p. 336; Max J. Kohler, "Judah P. Benjamin: Statesman and Jewish," PAJHS, 12 (1904), 85.

88. AI, December 5, 1862, p. 172.

89. Quoted in Robert Douthat Meader, *Judah P. Benjamin: Confederate Statesman* (New York: Oxford University Press, 1943), p. 280.

90. Korn, *American Jewry and the Civil War*, p. 177.

91. Sarna, "The 'Mythical Jew,'" in Gerber, ed., *Anti-Semitism in American History*, p. 57 ff.

Chapter 3

1. Naomi W. Cohen, "Antisemitism in the Gilded Age: The Jewish View," in Naomi W. Cohen, ed., *Essential Papers on Jewish-Christian Relations: Images and Reality* (New York: New York University Press, 1990), p. 142; Louise A. Mayo, *The Ambivalent Image: Nineteenth-Century America's Perception of the Jew* (Rutherford, N.J.: Fairleigh Dickinson University Press, 1988), pp. 37–38; Matthew Hale Smith, *Sunshine and Shadow in New York* (Hartford: J. B. Burr and

Co., 1869), p. 458; W. H. Rosenblatt, "The Jews: What They Are Coming To," *Galaxy*, 13 (January, 1872,) 54; Anna Laurens Dawes, *The Modern Jew* (Boston: D. Lathrop & Co., 1886; originally published 1884), pp. 10–11, 41–42; Hasia R. Diner, *In the Almost Promised Land: American Jews and Blacks, 1915–1935* (Westport, Conn.: Greenwood Press, 1977), 14; Leonard A. Greenberg, "Some American Anti-Semitic Publications of the Late 19th Century," PAJHS, 37 (1947), 423, 424; Joakim Isaacs, "Candidate Grant and the Jews," AJA, 17 (April, 1965), 7.

2. Irving Weingarten, "The Image of the Jew in the American Periodical Press, 1881–1891" (unpublished Ph.D., School of Education, Health, Nursing, and Arts Professions, New York University, 1979), p. 254; John J. Appel, "Popular Graphics as Documents for Teaching and Studying Jewish History," in Bertram Wallace Korn, ed., *A Bicentennial Festschrift for Jabob Rader Marcus* (Waltham, Mass.: American Jewish Historical Society, 1976), p. 33; see also Michael Dobkowski, *The Tarnished Dream* (Westport, Conn.: Greenwood Press, 1979). Dobkowski's book gives the fullest treatment of antisemitism during the Gilded Age and the Progressive era.

3. Smith, *Sunshine and Shadow*, pp. 452–453.

4. Ibid., p. 452.

5. Rosenblatt, "The Jews," p. 47.

6. Quoted in Elmer Gertz, "Charles Dana and the Jews," *The Chicago Jewish Forum*, 8 (Spring, 1950), 201.

7. Maxwell Whiteman, *Copper for America: The Hendricks Family and a National Industry, 1755–1939* (New Brunswick: Rutgers University Press, 1971), pp. 204–205.

8. Whiteman, *Copper*, p. 205; Mayo, *Ambivalent*, p. 92; Naomi W. Cohen, *Encounter with Emancipation: The German Jews in the United States, 1830–1914* (Philadelphia: JPS, 1984), p. 25; Jacob M. Sable, "Some American Jewish Organizational Efforts to Combat Anti-Semitism" (unpublished Ph. D., Yeshiva University, 1964), p. 5.

9. Fedora Small Frank, *Beginnings on Market Street* (Nashville: n. p., 1976), pp. 9, 59; Fedora Small Frank, "Nashville Jewry During the Civil War," *Tennessee Historical Quarterly*, 39 (Fall, 1980), 321; Whiteman, *Copper*, pp. 205, 207; AI, May 10, 1867, p. 4; Mayo, *Ambivalent*, p. 92ff.; Allan Tarshish, "Jews and Christians in a New Society: Some Aspects of Jewish-Christian Relationships in the United States, 1848–1881," in Korn, ed., *Bicentennial*, p. 572; Lloyd P. Gartner, *History of the Jews of Cleveland* (Cleveland: Western Reserve Historical Society, 1978), pp. 25–26; Cohen, *Encounter*, p. 26.

10. "The Jews and the Underwriters," *Banking and Insurance Chronicle*, 2 (May 2, 1867), 138.

11. Jonathan D. Sarna, "The Pork on the Fork: A Nineteenth Century Anti-Jewish Ditty," JSS, 44 (Spring, 1982), 169.

12. Quoted in Lloyd P. Gartner, "Temples of Liberty Unpolluted: American Jews and Public Schools, 1840–1875," in Korn, ed., *Bicentennial*, pp. 175–176.

13. Frances Nathan Wolff, *Four Generations* (New York: n. p., 1939), p. 28; Abraham Cronbach, "Autobiography," AJA, 11 (April, 1959), 9; Mark Scott, "The Little Blue Books in the War on Bigotry and Bunk," *Kansas History*, 1 (Autumn, 1978), 157; Nina Morais, "Jewish Ostracism in America," *North American Review*, 133 (September, 1881), 268–269; Gartner, "Temples," p. 176; Jonathan D. Sarna and Nancy H. Klein, *The Jews of Cincinnati* (Cincinnati: Center

for Study of the American Jewish Experience on the Campus of the Hebrew University College—Jewish Institute of Religion, 1989), p. 9; Leonard Dinnerstein, *The Leo Frank Case* (New York: Columbia University Press, 1968), p. 68; see also Hasia Diner, *A Time for Gathering: The Second Migration, 1820–1880* (Baltimore: Johns Hopkins University Press, 1992), pp. 198, 228.

14. Naomi W. Cohen, *A Dual Heritage: The Public Career of Oscar S. Strauss* (Philadelphia: The Jewish Publication Society of America, 1969), p. 10; E. Digby Baltzell, *The Protestant Establishment: Aristocracy and Caste in America* (London: Secker and Warburg, 1964), p. 22.

15. Cohen, "Antisemitism in the Gilded Age," pp. 132, 133; Tarshish, "Jew and Christian," p. 580; Cohen, *Encounter*, p. 256; Naomi W. Cohen, "Antisemitism in the Gilded Age: The Jewish View," JSS, 41 (Summer/Fall, 1979), 192ff; Jon C. Tedford, "Toward a Christian Nation: Religion, Law and Justice Strong," *Journal of Presbyterian History*, 54 (1976), 428; Diner, *A Time for Gathering*, pp. 173–174.

16. Dennis Lynn Pettibone, "Caesar's Sabbath: The Sunday-Law Controversy in the United States, 1879–1892" (unpublished Ph.D., Department of History, University of California, Riverside, 1979), pp. v, 2, 19.

17. Hjalmar H. Boyeson, "Dangers of Unrestricted Immigration," *The Forum*, III (1887), 532.

18. Christopher Hibbert, *Edward VII: A Portrait* (London: Allen Lane, 1976), p. 173; Mayo, *Ambivalent*, p. 182.

19. Anita Libman Lebeson, *Pilgrim People: A History of the Jews in America From 1492 to 1974* (New York: Minerva Press, 1975), p. 383; Cohen, "Antisemitism," JSS, (Summer/Fall, 1979), p. 188; Cohen, *Encounter*, p. 250; Mayo, *Ambivalent*, p. 94 ff.; see also Carey McWilliams, *A Mask for Privilege: Anti-Semitism in America* (New York: Little, Brown, 1948).

20. "Revising a Prejudice," *New York Herald*, July 22, 1879, p. 5.

21. Sable, "Some American," p. 33.

22. Clipping, "Down With the Jews! Meeting of the Society for Suppressing the Jewish Race," n. d., in Nearprint File, Special Topics, "Anti-Semitism (1841–1920)," AJA, Cincinnati.

23. Baltzell, *Protestant Establishment*, pp. 127–128, 138; Peter R. Decker, "Jewish Merchants in San Francisco: Social Mobility on the Urban Frontier," *American Jewish History*, 68 (June, 1979), 405; Sally Priesand "A Study of New York Jewry as Reflected in *The Jewish Messenger*, 1866–1870," term paper, 1971, folder, "New York, New York," box 418, AJA; Marc Lee Raphael, *Jews and Judaism in a Midwestern Community: Columbus, Ohio, 1840–1975* (Columbus: Ohio Historical Society, 1979), p. 234; Barry E. Supple, "A Business Elite: German Jews and Financiers in Nineteenth-Century New York," *Business History Review*, 31 (Summer, 1957), 166; review of Stuart E. Rosenberg, *The Jewish Community in Rochester* (New York: Columbia University Press, 1954) in *New York History*, 35 (July 1954), 320; Max Vorspan and Lloyd P. Gartner, *History of the Jews of Los Angeles* (Philadelphia: JPS, 1970), p. 103; Lloyd P. Gartner, *History of the Jews of Cleveland* (Cleveland: Western Reserve Historical Society, 1978), p. 85; Gustavus H. Wald to Thomas A. Mack, September 8, 1884, Gustavus H. Wall Correspondence File, AJA; Joseph M. Proskauer, *A Segment of My Times* (New York: Farrar, Straus and Co., 1950), p. 12; Hannah R. London, *Shades of My Forefathers* (Springfield, Mass.: The Pond-Ehberg Co., 1941), p. 54; John M. Davis, *The Guggenheims: An American Epic* (New York: William Morrow, 1978),

p. 63; Uchill, *Pioneers*, p. 188; Walda Katz Fishman and Richard L. Zweigenhaft, "Jews and the New Orleans Economic and Social Elites," JSS, 44 (Summer/Fall, 1982), 292; Diner, *A Time for Gathering*, p. 142; Allon Gal, *Brandeis of Boston* (Cambridge: Harvard University Press, 1980), p. 31.

24. Baltzell, *Protestant Establishment*, p. 113.

25. AH, March 30, 1883, p. 74, quoted in Yehezkel Wyszkowski, "The 'American Hebrew' views the Jewish community in the United States, 1879–1884, 1894–1898 and 1903–1906" (unpublished Ph.D., Yeshiva University, 1979), pp. 488–489.

26. Morais, "Jewish Ostracism," pp. 269, 270, 274.

27. Martin E. Marty, *Pilgrims in Their Own Land: 500 Years of Religion in America* (Boston: Little, Brown and Co., 1984), p. 392; Mayo, *Ambivalent*. p. 17; Cohen, "Antisemitism in the Gilded Age," *JSS*, p. 196.

28. Quoted in "The Examiner On Anti-Semitism," AH, 40 (August 23, 1889), 34.

29. Mayo, *Ambivalent*, pp. 180–181; Robert T. Handy, *A Christian America* (2nd edition; New York: Oxford University Press, 1984), p. 142; Rufus B. Spain, *At Ease in Zion: Social History of Southern Baptists, 1865–1900* (Nashville: Vanderbilt University Press, 1967), p. 138.

30. Quoted in Frederick Simon, "Anti-Semitism in America, 1654–1930" (unpublished mss., October, 1969), p. 81, in Oscar Cohen mss., folder 2, box 4, AJA.

31. Letter from S. Plantz, "Anti-Semitism," *The Evening News* (Detroit), April 7, 1893, p. 4.

32. Both quotes from Robert A. Rockaway, *The Jews of Detroit: From the Beginning, 1762–1914* (Detroit: Wayne State University Press, 1986), p. 58.

33. *The Sun* (New York City), June 9, 1896, p. 6.

34. Weingarten, "The Image," pp. 194, 251–252.

35. Pettibone, "Caesar's Sabbath," p. 298.

36. Leonard Dinnerstein and David M. Reimers, *Ethnic Americans* (2nd edition; New York: Harper and Row, 1983), p. 214.

37. Leo P. Ribuffo, "Henry Ford and the International Jew," *American Jewish History*, 69 (June, 1980), 474.

38. Barbara Miller Solomon, *Ancestors and Immigrants* (New York: John Wiley & Sons, Inc., 1965), p. 100ff.; John Higham, *Strangers in the Land: Patterns of American Nativism, 1860–1925* (2nd edition; New Brunswick: Rutgers University Press, 1988; originally published, 1955), p. 102.

39. Solomon, *Ancestors*, pp. 104, 106; E. Digby Baltzell, *The Protestant Establishment: Aristocracy and Caste in America* (London: Secker and Warburg, 1964), p. 108.

40. Solomon, *Ancestors*, pp. 110–111.

41. Quoted in ibid., p. 115.

42. *Congressional Record*, 54th Congress, 1st Session, March 16, 1896, p. 2817.

43. Egal Feldman, *The Dreyfus Affair and the American Conscience* (Detroit: Wayne State University Press, 1981), p. 138.

44. Mayo, *Ambivalent*, p. 129; Mary Dean, "The Doings and Goings-On of Hired Girls," *Lippincott's Monthly Magazine*, 20 (1877), 590; Dennis P. Ryan, *Beyond the Ballot Box: A Social History of the Boston Irish, 1845–1917* (Rutherford, N. J.: Fairleigh Dickinson Press, 1983), p. 69.

45. Hostile writings about Jews were certainly "in the air" at that time. In 1888 a Greek American residing in Boston, Telemachus Thomas Timayenis, characterized by John Higham as "an unprincipled schemer," for some inexplicable reason hated Jews. In his bedeviled mind he imagined that through financial manipulation and political revolution Jews were plotting to overthrow the aryan order. He therefore wrote three books roundly denouncing Jews as born traffickers and liars, "full of cunning and intrigue." "A mercenary by instinct," is what Timayenis called the Jew who allegedly had "no creative facility." Timayenis' condemnations were basically condensed from Edouard Drumont's *La France juive,* which had been published earlier in the decade. He paraphrased, summarized, and adapted the French tome for American audiences by including material relevant to New World readers. Other assumptions made by Timayenis in his books *The Original Mr. Jacobs, The American Jew,* and *Judas Iscariot,* included the supposed facts that Jews were "ill-smelling . . . loud, vulgar and intrusive," that they were "the dirtiest people on the face of the earth," and that they were scoundrels and parasites who lusted after money. Timayenis' books did not make a major impression on the American public but they were forerunners of all too common twentieth century analyses. Moreover, although never before in U.S. history had there been such narratives roundly denouncing Jews, the books contained enough of what people already believed to be publicly acceptable. See John Higham, "Anti-Semitism in the Gilded Age: A Reinterpretation," *Mississippi Valley Historical Review,* 43 (March, 1957), 576; Leonard A. Greenberg and Harold J. Jonas, "An American Anti-Semite in the Nineteenth Century," in Joseph L. Blau, ed., *Essays on American Jewish Life and Thought* (New York: Columbia University Press, 1959), pp. 266–267, 274; Cohen, *Encounter,* p. 229.

46. Isaac M. Fein, *The Making of an American Jewish Community: The History of Baltimore Jewry from 1773 to 1920* (Philadelphia: JPS, 1971), p. 203; A. De-Glequier, "The Antisemitic Movement in Europe," *Catholic World,* 48 (November, 1889), 744.

47. Ibid., pp. 745, 749.

48. Mayo, *Ambivalent,* p. 133.

49. A. DeGlequier, "The Antisemitic Movement in Europe," p. 749; Mayo, *The Ambivalent Image,* p. 133; Ronald Albert Urquhart, "The American Reaction to the Dreyfus Affair: A Study of Anti-Semitism in the 1890's" (unpublished Ph.D., Department of History, Columbia University, 1972), pp. 145–146; Donna Merwick, *Boston's Priests, 1848–1910* (Cambridge: Harvard University Press, 1973), pp. 184–185.

50. Till van Rahden, "Beyond Ambivalence: Variations of Catholic Anti-semitism in Turn of the Century Baltimore" (unpublished M.A. thesis, Johns Hopkins University, 1992), pp. 26, 31ff., 39, 51, 52, 53; Isaac M. Fein, *The Making of an American Jewish Community: the History of Baltimore Jewry from 1773 to 1920* (Philadelphia: JPS, 1971), p. 203.

51. "The Story of an Irish Cook," *The Independent,* 58 (March 30, 1905), 715–716.

52. "The Roots of Political Power," in Robert A. Woods, ed., *The City Wilderness: A Settlement Study* (Boston: Houghton Mifflin & Co., 1898), pp. 134–135.

53. Quoted in Ryan, *Beyond the Ballot,* p. 137.

54. Ibid., *The Christian Advocate* (New York City), 66 (September 10, 1891), 601.

55. Lillian Parker Wallace, *Leo XIII and the Rise of Socialism* (Durham, N.C.: Duke University Press, 1966), p. 370.

56. Baltzell, *Protestant Establishment*, p. 283; Israel Goldstein, *A Century of Judaism In New York* (New York: Congregation B'nai Jeshrun, 1930), pp. 177–178; AI, May 27, 1897, p. 4; Edward J. Bristow, *Prostitution and Prejudice: The Jewish Fight Against White Slavery, 1870–1939* (New York: Schocken Books, 1983), p. 79.

57. E. F. Benson, *King Edward VII: An Appreciation* (London: Longmans, Green & Co., 1933), p. 161.

58. Quoted in Thomas P. O'Neill, Jr., et al. to Edward H. Levi, March 4, 1975, folder, "Subject Files: Jewry, Correspondence, 1975–1976," Bella Abzug Ms., Columbia University, New York City.

59. Anson Phelps Stokes, *Church and State in the United States* (3 volumes; New York: Harper and Brothers, 1950), I, 879ff.; Joseph P. O'Grady, "Politics and Diplomacy: The Appointment of Anthony M. Keiley to Rome in 1885," *Virginia Magazine of History and Biography*, 76 (April, 1968), 191.

60. Urquhart, "American Reaction," pp. xvi, 65, 229.

61. "The Jews and the Workingmen," *The Reformer and Jewish Times* (New York City), June 21, 1878, p. 1.

62. Irwin Unger, *The Greenback Era: A Social and Political History of American Finance, 1865–1876* (Princeton: Princeton University Press, 1964), pp. 210, 211–212.

63. Ibid, p. 210.

64. Walter T. K. Nugent, *The Tolerant Populists* (Chicago: University of Chicago Press, 1963), pp. 83, 111; Ron Chernow, *The House of Morgan: An American Banking Dynasty and the Rise of Modern Finance* (New York: Atlantic Monthly Press, 1990), p. 76; Richard J. Margolis, "Big Lie in Iowa," *The New Leader*, 67 (August 6, 1984), 8; Naomi W. Cohen, "Antisemitism in the Gilded Age: The Jewish View," JSS, 41 (Summer/Fall, 1979), 198; Oscar Handlin, "Reconsidering the Populists," *Agricultural History*, 39 (1965), 69–70.

65. "Are the Populists Anti-Semitic?" *Jewish Voice* (August 14, 1896), 4.

66. Irwin Unger, "Critique of Norman Pollack's 'Fear of Man,'" *Agricultural History*, 39 (1965), 77.

67. Cohen, "Anti-Semitism in the Gilded Age," p. 198; Victor C. Ferkiss, "Populist Influences on American Fascism," *Western Political Quarterly*, 10 (1957), 354; Fine, *The City*, p. 10; Frederick Simon, "Anti-Semitism in America, 1654–1930" (unpublished manuscript, October, 1969), in Oscar Cohen mss., folder 2, box 4, AJA.

68. Oscar Fleishaker, "The Illinois-Iowa Jewish Community on the Banks of the Mississippi River" (unpublished Ph.D., Yeshiva University, 1957), p. 263; Edna Ferber, *A Peculiar Treasure* (New York: Doubleday, Doran and Co., 1939), pp. 9, 41; John Higham, *Send These to Me: Immigrants in Urban America* (revised edition; Baltimore: The Johns Hopkins Press, 1984), p. 114, fn. 40.

69. Gartner, *Cleveland*, pp. 85, 87.

70. Ralph Janis, "Flirtation and Flight: Alternatives to Ethnic Confrontation in White Anglo-American Protestant Detroit," *Journal of Ethnic Studies*, 6 (Summer, 1978), 12; Louis D. Brandeis to Alfred Brandeis, October 10, 1914, as

quoted in Robert A. Rockaway, "Louis Brandeis on Detroit," *Michigan Jewish History,* 17 (July, 1977), 18.

71. *The Evening News (Detroit),* April 3, 5, 7, 20, 1893; Rockaway, "Louis Brandeis on Detroit," p. 19; Robert A. Rockaway, "Anti-Semitism in an American City: Detroit, 1850–1914," AJHQ, 64 (September, 1974), 47.

72. *The Evening News,* April 3, 1893, p. 4.

73. "Anti-Semitism In New York," *The Evening News,* April 20, 1893, p. 4.

74. K. M. Tresfield, "The Jew from a Gentile Standpoint," *Overland Monthly,* 25 (April, 1895), 410.

75. Peter R. Decker, "Jewish Merchants in San Francisco: Social Mobility on the Urban Frontier," in Moses Rischin, ed., *The Jews of the West: The Metropolitan Years* (Waltham: American Jewish Historical Society, 1979), 21ff.

76. Gustav Adolf Danziger, "The Jew in San Francisco: The Last Half Century," *Overland Monthly,* 25 (April, 1895), 381, 388, 410.

77. Diana Hirschler, "Union in Philadelphia," *Arena,* 9 (May 11, 1894), 552; Harlan B. Phillips, "A War on Philadelphia's Slums: Walter Vroonan and the Conference of Moral Workers, 1893," *Pennsylvania Magazine of History and Biography,* 76 (January, 1952), 58.

78. John M. Davis, *The Guggenheims* (New York: William Morrow, 1978), p. 63; see also Gay Talese, *The Kingdom and the Power* (New York: World Publishing Co., 1969), p. 94.

79. See, for example, Kermit Vanderbilt, "Howells Among the Brahmins: Why 'The Bottom Dropped Out' During the Rise of Silas Lapham," *New England Quarterly,* 32 (1962), 303.

80. Jerold S. Auerbach, *Rabbis and Lawyers* (Bloomington: Indiana University Press, 1990), pp. 133–134; Allon Gal, *Brandeis of Boston* (Cambridge: Harvard University Press, 1980), p. 31; E. Digby Baltzell, Allen Glickman, and Jacquelyn Litt, "The Jewish Communities of Philadelphia and Boston: A Tale of Two Cities," in Murray Friedman, ed., *Jewish Life in Philadelphia, 1830–1940* (Philadelphia: ISHI Publications, 1983), p. 312.

81. Rufus Cyrene McDonald, "The Jews of the North End of Boston," *Unitarian Review,* 35 (May, 1891), 368.

82. Ibid., pp. 364, 368.

83. Scott, "The Little Blue Books," p. 157; 157; John F. Stack, Jr., *International Conflict in an American City: Boston's Irish, Italians, and Jews, 1935–1944* (Westport, CT: Greenwood Press, 1979), p. 38; Richard M. Abrams, *Conservatives in a Progressive Era* (Cambridge: Harvard University Press, 1964), p. 50; Ryan, *Beyond the Ballot Box,* p. 137; Rockaway, "Anti-Semitism in an American City," p. 47; Alter F. Landesman, *Brownsville: The Birth, Development and Passing of a Jewish Community in New York* (New York: Bloch Publishing Co., 1969), pp. 59–60; Abraham Cronbach, "Autobiography," *American Jewish Archives,* 11 (April, 1959), 9; AH, 65 (May 19, 1899), 71, "Anti-Semitism Masked as Sociology," AH, 95 (September 18, 1914), 611–612; John Higham, "Anti-Semitism in the Gilded Age: A Reinterpretation," *Mississippi Valley Historical Review,* 43 (March, 1957), 576–577; Higham, *Send These,* p. 113; Edward R. Kantowicz, *Polish-American Politics in Chicago, 1880–1940* (Chicago: University of Chicago Press, 1975), p. 118; Samuel P. Abelow, *History of Brooklyn Jewry* (Brooklyn: Scheba Publishing Co., 1937), p. 12.

84. Landesman, *Brownsville,* pp. 58–59.

85. Sarna and Klein, *The Jews of Cincinnati,* p. 78; Gartner, *Cleveland,* p. 70;

Rockaway, *Jews of Detroit*, p. 90; AH, 65 (May 19, 1899), 71 (July 14, 1899, August 25, 1899), pp. 307, 482.

86. Cohen, *Encounter*, pp. 239, 261; Wyszkowski, "The 'American Hebrew,'" p. 33; Helga Eugenie Kaplan, "Century of Adjustment: A History of the Akron Jewish Community" (unpublished Ph.D., Department of History, Kent State University, 1978), p. 24; Rockaway, "Anti-Semitism in an American City," p. 54; Robert Rockaway, "Ethnic Conflict in an Urban Environment: the German and Russian Jew in Detroit, 1881–1914," AJHQ, 60 (December, 1970), 148; Steven Hertzberg, "The Jewish Community of Atlanta from the End of the Civil War Until the Eve of the Frank Case," AJHQ, 62 (March, 1973), 273; Myron Berman, "Rabbi Edward Nathan Calisch and the Debate over Zionism in Richmond, Virginia," AJHQ, 62 (March, 1973), 295; Selma C. Berrol, "In Their Image: German-Jews and the Americanization of the *Ost Juden* in New York City," *New York History*, 63 (October, 1982), 427; Arnold A. Wieder, *The Early Jewish Community of Boston* (Waltham, Mass.: Brandeis University, 1962), p. 65; Mark K. Bauman and Arnold Shankman, "The Rabbi as Ethnic Broker: The Case of David Marx," *Journal of American Ethnic History*, 2 (Spring, 1983), 52–53; Marc Lee Raphael, "Intra-Jewish Conflict in the United States, 1869–1915" (unpublished Ph.D., Department of History, University of California at Los Angeles, 1972), pp. 8, 89.

87. Gartner, *Cleveland*, p. 85; Ida Libert Uchill, *Pioneers, Peddlers, and Tsadikim* (Denver: Sage Books, 1957), p. 189; John F. Sutherland, "Rabbi Joseph Krauskopf of Philadelphia: The Urban Reformer Returns to the Land," AJHQ, 67 (June, 1978), 356; Arthur Mann, "Solomon Schindler: Boston Radical," *New England Quarterly*, 23 (December, 1950), 459.

88. "Editorial Notes," *Jewish Voice* (St. Louis), 24 (March 18, 1898).

89. Baltzell, *Protestant Establishment*, pp. 114–115.

90. Jonathan D. Sarna, "Anti-Semitism and American History," *Commentary*, 71 (March, 1981), 42; Cohen, *Encounter*, pp. 282–283; Jacob Rader Marcus, "Major Trends in American Jewish Historical Research," *AJA*, 16 (April, 1964), 9; Jacob R. Marcus, "The Quintessential American Jew," AJHQ, 58 (September, 1968), 15.

91. Sylvan Morris Dubow, "Identifying the Jewish Serviceman in the Civil War: A Reappraisal of Simon Wolf's *The American Jew as Patriot, Soldier and Citizen*," AJHQ, 59 (March, 1970), 357.

92. Simon, "Anti-Semitism in America," p. 83; Sable, "Some American Jewish Organizational Efforts," p. 123.

93. AI, July 14, 1898.

94. Urquhart, "American Reaction to the Dreyfus Affair," p. 194; Wyszkowski, "The American Hebrew,"p. 491; Gartner, *Cleveland*, p. 85.

95. Mark Twain, "Concerning the Jews," *Harper's Magazine*, 99(189), 527, 529, 531, 532, 534, 535.

Chapter 4

1. Irving Weingarten, "The Image of the Jew in the American Periodical Press, 1881–1921" (unpublished Ph.D., School of Education, Health, Nursing, and Arts Professions, New York University, 1979), p. 200.

2. Wilhelm Marr, a German journalist, coined the term "antisemitism" in

1879, and also placed antisemitism firmly on the foundation of racism. Cara Gerdel Ryan, "Reviving the Politics of Anti-Semitism," *The Nation*, 233 (December 5, 1981), 633; M. Ginsberg, "Anti-Semitism," *The Sociological Review*, (January–April, 1943), 1.

3. Naomi W. Cohen, *Encounter with Emancipation: The German Jews in the United States, 1830–1914* (Philadelphia: JPS, 1984), p. 302.

4. Edward Alsworth Ross, *The Old World in the New* (New York: The Century Co., 1914), p. 154.

5. Cohen, *Encounter*, p. 302.

6. Quoted in Ida Libert Uchill, *Pioneers, Peddlers, and Tsadikim* (Denver: Sage Books, 1957), p. 158; see also Richard Klayman, *The First Jew: Prejudice and Politics in an American Community, 1900–1932* (Malden, Mass.: Old Suffolk Square Press, 1985), p. x; A. L. Todd, *Justice on Trial* (New York: McGraw-Hill Book Co., 1964), p. 137; Jacob M. Sable, "Some American Jewish Organizational Efforts to Combat Anti-Semitism" (unpublished Ph.D., Yeshiva University, 1964), p. 292.

7. Irving Weingarten, "The Image of the Jew in the American Periodical Press, 1881–1921" (unpublished Ph.D., School of Education, Health, Nursing, and Arts Professions, New York University, 1979), p. 248.

8. Sydney Reid, "Because You're a Jew," *The Independent*, 65 (November 26, 1900), 1215.

9. Herman Scheffauer, "The Conquest of America," *The Living Age*, 280 (March 21, 1914), 714; E. J. Kuh, "The Social Disability of the Jew," *Atlantic Monthly*, 101 (April, 1908), 436; "Is a Dreyfus Case Possible in America?" *The Independent*, 61 (July 19, 1906), 167; "Race Prejudice Against Jews," *The Independent*, 65 (December 17, 1908), 1451, 1456; Reid, "Because You're a Jew," p. 1212; David Herman Joseph, "Some More of It, and Why," *The Temple*, I (December 10, 1909), 3; Bernard Drachman, "Anti-Jewish Prejudice in America," *The Forum*, 52 (July, 1914), 31.

10. Cohen, *Encounter*, p. 302.

11. "Secretary Hay's Note and the Jewish Question," *Harper's Weekly*, 48 (October 11, 1902), 1447; "Is a Dreyfus Case"; "Race Prejudice Against Jews,"; Francis J. Oppenheimer, "Jewish Criminality," *The Independent*, 65 (September 17, 1908), 640–642; Charles S. Bernheimer, "Prejudice Against Jews in the United States," *The Independent*, 65 (November 12, 1908), 1105–1108; Reid, "Because You're a Jew"; "The Jews in the United States," *World's Work*, 11 (January, 1906), 7030–7031; Maurice Fishberg, "White Slave Traffic and Jews," *The American Monthly Jewish Review*, IV (December, 1909), 4, 23; "Jews in the White Slave Traffic," *The Temple*, II (February 25, 1910), 176; Joseph Krauskopf, "Is the Jew Getting a Square Deal?" *The Ladies Home Journal*, 27 (November 1, 1910), 12; "Will The Jews Ever Lose Their Racial Identity?" *Current Opinion*, 50 (March, 1911), 292–294; Nathum [sic] Wolf, "Are the Jews an Inferior Race?" *The North American Review*, 195 (April, 1912), 492–495; Edward Alswoth Ross, "The Hebrews of Eastern Europe in America," *The Century Magazine*, 88 (September, 1914), 785–792; Scheffauer, "The Conquest of America."

12. Burton J. Hendrick, "The Great Jewish Invasion," *McClure's Magazine*, 28 (January, 1907), 307–321; Burton J. Hendrick, "The Jewish Invasion of America," *McClure's Magazine*, 40 (March, 1913), 125–165.

13. Hendrick, "The Great," p. 311.

14. Ibid., p. 314.

15. Deborah Dwork, "Health Conditions of Immigrant Jews on the Lower East Side of New York," *Medical History*, 25 (1981), 1; see also Gerald Sorin, *A Time for Building; The Third Migration, 1880–1920* (Baltimore: The Johns Hopkins University Press, 1992), pp. 80–81.

16. Hendrick, "The Great," pp. 314, 316, and 317.

17. Ibid., p. 316.

18. Ibid., p. 318.

19. Ibid, pp. 319–320.

20. Hendrick, "The Jewish Invasion," *passim*.

21. Ross, "The Hebrews of Eastern Europe," p. 785.

22. Ibid., p. 786.

23. Ibid., p. 787.

24. Ibid., pp. 789, 790.

25. Ibid., pp. 786, 787, and 788.

26. Ibid., pp. 788, 791.

27. Ross, *The Old World*, pp. 11, 39, 41, 42, 49, 54, 64, 295.

28. Ibid., pp. 76, 92, 113, 117, 127–128, 130, 138, 140, 285–286, 287, 291, 293.

29. Charles C. Alexander, "Prophet of American Racism: Madison Grant and the Nordic Myth," *Phylon*, 23 (Spring, 1962), 73ff.

30. John Higham, *Strangers in the Land* (New York: Atheneum, 1974; originally published by Rutgers University Press, 1955), p. 155; Thomas F. Gossett, *Race: The History of an Idea* (New York: Schocken Books, 1965), pp. 353ff.

31. Higham, *Strangers*, p. 155.

32. Higham, *Strangers*, pp. 156–157; Gossett, *Race*, p. 355ff.

33. Madison Grant, *The Passing of the Great Race* (New York: Charles Scribner's Sons, 1916), pp. 43–44, 128, 149, 198.

34. Gossett, *Race*, p. 355.

35. Grant, *Passing*, pp. 14, 16, 80, 81; see also Higham, *Strangers*, pp. 155–156; Carey McWilliams, *A Mask for Privilege: Anti-Semitism in America* (New York: Little, Brown and Company, 1948), p. 57.

36. Anson Phelps Stokes, *Church and State in the United States* (3 volumes; New York: Harper & Brothers, 1950), III, 449.

37. E. Digby Baltzell, Allen Glicksman, and Jacquelyn Litt, "The Jewish Communities of Philadelphia and Boston: A Tale of Two Cities," in Murray Friedman, ed., *Jewish Life in Philadelphia, 1830–1940* (Philadelphia: ISHI Publications, 1983), p. 312; Allon Gal, *Brandeis of Boston* (Cambridge: Harvard University Press, 1980), pp. 42–43, 195–196.

38. Norman Hapgood, "The Jews and American Democracy," *The Menorah Journal*, 2 (October, 1916), 205.

39. Todd, *Justice on Trial*, p. 245.

40. Quoted in James Weldon Johnson, "Views and Reviews," *New York Age*, February 3, 1916, p. 4.

41. Dorothy Ross, "The Irish Catholic Immigrant: A Study in Social Mobility" (unpublished M.A. thesis, Department of History, Columbia University, 1959), *passim*.

42. "The Experience's of a Jew's Wife," *The American Magazine*, 78 (December, 1914), 49.

43. Ibid., pp. 51, 85, 86.

44. James F. Richardson, *The New York Police: Colonial Times to 1901* (New York: Oxford University Press, 1970), p. 167.

45. Klayman, *The First American*, pp. 102, 103, 106.

46. Leonard Dinnerstein, "The Funeral of Rabbi Jacob Joseph," in David A. Gerber, ed., *Anti-Semitism in American History* (Urbana: University of Illinois Press, 1986), pp. 275–298, *passim*.

47. Michael Dobkowski, *The Tarnished Dream* (Westport, Conn.: Greenwood Press, 1979), p. 69.

48. Francesco Cordasco and Thomas Monroe Pitkin, *The White Slave Trade and the Immigrants* (Detroit: Blaine Ethridge, 1981), p. 60; Ruth Rosen, *The Lost Sisterhood: Prostitution in America* (Baltimore: The Johns Hopkins University Press, 1982), p. 139ff.

49. G. K. Turner, "The Daughters of the Poor," *McClure's Magazine*, 34 (1909), 45–61; G. K. Turner, "The City of Chicago," *McClure's Magazine*, 28 (April, 1907), 581–582.

50. "Jews in the White Slave Traffic," p. 176.

51. Edward J. Bristow, *Prostitution and Prejudice: The Jewish Fight Against White Slavery, 1870–1939* (New York: Schocken Books, 1983), p. 283.

52. Harold M. Hyman, *Oleander Odyssey: The Kempners of Galveston, Texas, 1854–1990s* (College Station: Texas A & M University Press, 1990), p. 244.

53. Bristow, *Prostitution and Prejudice*, p. 45; see also Hyman, *Oleander Odyssey*, p. 246.

54. Bristow, *Prostitution and Prejudice*, p. 161.

55. Lucy S. Dawidowicz, "Louis Marshall's Yiddish Newspaper, *The Jewish World*" (unpublished M. A. thesis, Department of History, Columbia University, 1961), pp. 39–40; AH, June 29, 1906, pp. 93–94; Cohen, *Encounter*, p. 342.

56. Quoted in Sable, "Some American Jewish," pp. 234–235.

57. AI, September 25, 1913, p. 1.

58. Thomas David Mantel, "The Anti-Defamation League of B'nai B'rith" (unpublished Honors thesis, Department of Government, Harvard University, 1950), pp. 26, 27; Sable, "Some American Jewish," pp. 242, 292.

59. Minutes, IV (September 2, 1918), 703–704.

60. Ibid., III, (December 9, 1917), 589ff.; IV (March 10, 1918), 659.

61. Minutes, IV (January 13, 1918), 66; Morton Rosenstock, *Louis Marshall, Defender of Jewish Rights* (Detroit: Wayne State University Press, 1965), p. 107; "Roosevelt Censures Annapolis Editor," AH, 111 (June 16, 1922), 162, 164; Adolph Sabath to Josephus Daniels, October 14, 1919, folder, "Jewish Relief, 1913–1915," box 524 (reel 40), Josephus Daniels mss., LC.

62. Norman Polmar and Thomas B. Allen, *Rickover* (New York: Simon and Schuster, 1982), p. 47.

63. Louis Marshall to Josephus Daniels, December 27, 1917, file, "December," box 1787, Louis Marshall mss., AJA.

64. Quoted in Sable, "Some American Jewish," p. 256.

65. Minutes, IV (November 19, 1918), 731–732, January 12, 1919, pp. 62–63.

Chapter 5

1. Quoted in Carey McWilliams, *A Mask for Privilege: Anti-Semitism in America* (New York: Little, Brown and Company, 1948), p. 38; see also Leo P.

Ribuffo, *The Old Christian Right* (Philadelphia: Temple University Press, 1983), p. 12.

2. David O. Levine, *The American College and the Culture of Aspiration, 1915–1940* (Ithaca: Cornell University Press, 1985), pp. 147–148.

3. John Higham, *Strangers in the Land* (New Brunswick: Rutgers University Press, 1955; this edition: New York: Atheneum, 1974), pp. 270, 277–278; Carol S. Gruber, *Mars and Minerva: World War I and the Uses of Higher Learning in America* (Baton Rouge: Louisiana State University Press, 1975), p. 241.

4. Joel Williamsoin, *A Rage for Order: Black/White Relations in the American South Since Emancipation* (New York: Oxford University Press, 1986), p. 246; Ribuffo, *Old Christian*, pp. 9–10; Higham, *Strangers*, pp. 265–266, 271; 277–278; E. Digby Baltzell, *Philadelphia Gentlemen: The Making of a National Upper Class* (Glencoe, Ill.: The Free Press, 1958), p. 288; Leo P. Ribuffo, "Henry Ford and *The International Jew*," AJH, 69 (June, 1980), 441.

5. "Are Bolsheviki Mainly Jewish?" *Literary Digest*, 59 (December 14, 1918), 32; "American Jews in the Bolshevik Oligarchy," *Literary Digest*, 60 (March 1, 1919), 32; Henry Feingold, *A Time for Searching: Entering the Mainstream, 1920–1945* (Baltimore: The Johns Hopkins Press, 1992), p. 7.

6. Michael Gerald Rapp, "An Historical Overview of Anti-Semitism in Minnesota, 1920–1960—With Particular Emphasis on Minneapolis and St. Paul" (unpublished Ph.D., Department of History, University of Minnesota, 1977), p. 26.

7. *Minneapolis Journal*, May 26, 1919, quoted in Morton Rosenstock, *Louis Marshall, Defender of Jewish Rights* (Detroit: Wayne State University Press, 1965), p. 216.

8. Jacob M. Sable, "Some American Jewish Organizational Efforts To Combat Anti-Semitism" (unpublished Ph.D., Yeshiva University, 1964), p. 203.

9. Edward R. Kantowicz, *Polish-American Politics in Chicago, 1880–1940* (Chicago: University of Chicago Press, 1975), p. 124.

10. Quoted in ibid., p. 124.

11. Rosemary Radford Ruether, "Anti-Semitism and Christian Theology," in Eva Fleischner, ed., *Aushwitz: Beginning of a New Era? Reflections on the Holocaust* (New York: KTAV Publishing House, Inc., 1977), p. 90; Rosenstock, *Louis Marshall*, p. 119ff.; Baron S. A. Korff, "The Great Jewish Conspiracy," *The Outlook*, 127 (February 2, 1921), 181; Burton J. Hendrick, "The Jews in America," *World's Work*, 45 (1922–23), 266; David Brian Davis, *The Fear of Conspiracy* (Ithaca: Cornell University Press, 1971), p. 26; Ribuffo, "Henry Ford," p. 441.

12. "Mr. Ford Retracts," *Review of Reviews*, 76 (August, 1927), 197; Rosenstock, *Louis Marshall*, p. 145.

13. David Levering Lewis, "Henry Ford's Anti-Semitism and Its Repercussions," *Michigan Jewish History*, 24 (1984), 3; Leonard Dinnerstein, "When Henry Ford Apologized to the Jews," *Moment*, 15 (February, 1990), 24; Feingold, *A Time for Searching*, pp. 8–9.

14. Ibid., pp. 26–27; quote on page 27.

15. "The International Jew", a reprint of the series appearing in *The Dearborn Independent* from May 22 to October 2, 1920 (no city: n. p., n. d.), pp. 10–11; "Mr. Ford Retracts," p. 197.

16. Albert Lee, *Henry Ford and the Jews* (New York: Stein and Day, 1980), pp. 41, 161; Kantowicz, *Polish-American*, p. 119.

17. Dinnerstein, "When Henry Ford," pp. 27, 54; Robert Lacey, *Ford: The*

Men and the Machine (Boston: Little, Brown, 1986), p. 218; Samuel Walker McCall, *The Patriotism of the American Jew* (New York: Plymouth Press, Inc., 1924), p. 36; Rosenstock, *Louis Marshall*, p. 197.

18. Quote in Lee, *Henry Ford*, p. 47; see also Lacey, *Ford*, p. 218.

19. T, July 31, 1938, p. 31, December 1, 1938, p. 12; Lacey, *Ford*, p. 386; Lee, *Henry Ford*, p. 113.

20. Rosenstock, *Louis Marshall*, p. 125; Bernard G. Richards Oral History Memoir, COHC, p. 183; "Anti-Jewish Propaganda," *The Outlook*, 127 (January 26, 1921), 125; Herbert Adams Gibbon, "The Jewish Problem," *The Century*, 102 (September, 1921), 786–787.

21. McCall, *The Patriotism of the American Jew*, p. 28.

22. "Reaction and the Jew," *The Nation*, 111 (November 3, 1920), 493.

23. Quoted in T, December 27, 1920, p. 13.

24. "Anti-Jewish Propaganda," p. 125; Rosenstock, *Louis Marshall*, p. 154ff.; "Wilson and Harding Defend Jews," *The Independent*, 105 (January 29, 1921), 118.

25. Henry Aaron Yeomans, *Abbot Lawrence Lowell* (Cambridge: Harvard University Press, 1948), p. 209; Feingold, *A Time for Searching*, p. 17; Oliver B. Pollak, "Antisemitism, the Harvard Plan, and the Roots of Reverse Discrimination," *JSS*, 45 (Spring, 1983), 114.

26. Yeomans, *Abbot Laurence Lowell*, p. 209; William T. Ham, "Harvard Student Opinion on the Jewish Question," *The Nation*, 115 (September, 6, 1922), 225; Marcia Graham Synnot, *The Half-Opened Door: Discrimination and Admissions at Harvard, Yale, and Princeton, 1900–1970* (Westport, Conn.: Greenwood Press, 1979), p. 15ff.; Morgan's letter quoted in Ron Chernow, *The House of Morgan: An American Banking Dynasty and the Rise of Modern Finance* (New York: Atlantic Monthly Press, 1990), pp. 214–215.

27. Ray Allen Billington, *Frederick Jackson Turner* (New York: Oxford University Press, 1973), p. 437; Levine, *The American College*, p. 150.

28. "The American University," *The Nation and the Atheneum*, 33 (April 28, 1923), 117.

29. Quoted in Ham, "Harvard Student Opinion," p. 225.

30. Dan A. Oren, *Joining the Club: A History of Jews and Yale* (New Haven: Yale University Press, 1985), p. 47.

31. Lewis S. Gannett, "Is America Anti-Semitic?" *The Nation*, 116 (March 21, 1923), 331.

32. Francis Biddle, *A Casual Past* (Garden City, N.Y.: Doubleday & Co., Inc., 1961), p. 219; see also Heywood Broun and George Britt, *Christians Only* (New York: The Vanguard Press, 1931), p. 60.

33. Quoted in Harold S. Wechsler, *The Qualified Student* (New York: John Wiley and Sons, 1977), p. 135.

34. Lynn D. Gordon, "Annie Nathan Meyer and Barnard College: Mission and Identity in Women's Higher Education, 1889–1950," *History of Education Quarterly*, 26 (Winter, 1986), 508; Wechsler, *The Qualified Student*, p. 148; Ralph Philip Boas, "Who Shall Go to College? *The Atlantic*, 130 (October, 1922), 446; Levine, *The American College*, pp. 148–149; Synott, *The Half-Opened Door*, p. 41.

35. Synott, *Half-Opened Door*, p. 15ff.

36. Jerold Auerbach, "From Rags to Robes: The Legal Profession, Social Mobility and the American Jewish Experience," *AJHQ*, 66 (October, 1976), 252; Wechsler, *The Qualified Student*, p. 163; "May Jews Go to College?" *The Nation*,

114 (June 14, 1922), 708; Broun and Britt, *Christians Only*, pp. 89–90, 118–119; Ruth Marcus Platt, *The Jewish Experience at Rutgers* (East Brunswick, N.J.: Jewish Historical Society of Central Jersey, 1987), p. 21; Oren, *Joining the Club*, pp. 51, 54–55; Garry T. Greenebaum, "The Jewish Experience in the American College and University" (unpublished Ph.D., 1978, Small Collections, AJA), p. 110; *The Nation and the Atheneum*, 32 (January 27, 1923), 635; Minutes, V (1932), pp. 1608–1609; McWilliams, *Mask for Privilege*, p. 135; Higham, *Strangers*, p. 278; Levine, *The American College*, pp. 146, 154ff; Michael Greenberg and Seymour Zenchelsky, "Private Bias and Public Responsibility: Anti-Semitism at Rutgers in the 1920s and 1930s," *History of Education Quarterly*, 33 (Fall, 1993), 311.

37. Broun and Britt, *Christians Only*, pp. 74, 88, 92; Harvey Strum, "Louis Marshall and Anti-Semitism at Syracuse University," *American Jewish Archives*, 35 (April, 1983), 8, 10; Paula S. Fass, *The Damned and the Beautiful: American Youth in the 1920's* (New York: Oxford University Press, 1977), p. 152; L. B. Rose, "Secret Life of Sarah Lawrence," *Commentary*, 75 (May, 1983), 54; Pollak, "Antisemitism," p. 1129.

38. Quoted in Levine, *The American College*, p. 156.

39. Synott, *Half-Opened Door*, pp. 19–20; "The Jew in America: A Jewish University?" *The Nation*, 116 (May 16, 1923), 573; Ronald Steel, "Walter Lippmann's Harvard," in Mitza Rosovsky, *The Jewish Experience at Harvard and Radcliffe* (Cambridge: Harvard University Press, 1986), p. 73; Laura Z. Hobson, *Laura Z: A Life* (New York: Arbor House, 1983), p. 357; Boas, "Who Shall Go," p. 446; see also Laura Kalman, *Abe Fortas* (New Haven: Yale University Press, 1990), p. 14.

40. Quoted in Oren, *Joining the Club*, p. 69.

41. Quoted in Broun and Britt, *Christians Only*, p. 62; see also Victor A. Kramer, "What Lowell Said," AH, 112 (January 26, 1923), 394.

42. Minutes, V (November 10, 1928) 1410.

43. Norman Polmar and Thomas B. Allen, *Rickover* (New York: Simon and Schuster, 1982), pp. 51, 52–53; Robert Wallace, "A Deluge of Honors for an Exasperating Admiral," *Life*, 45 (September 8, 1958), 109.

44. Quoted in Elizabeth Drew, "Profile of Robert Strauss," *The New Yorker*, May 17, 1979, p. 117.

45. Quoted in Oren, *Joining the Club*, p. 120.

46. Lewis S. Feuer, "The Stages of Social History of Jewish Professors in American Colleges and Universities," AJH, 71 (June, 1982), 455; Susanne Klingenstein, *Jews in the American Academy, 1910–1940* (New Haven: Yale University Press, 1991), p. xi; Irving Howe, *World of Our Fathers: The Journey of the East European Jews to America and the Life They Found and Made* (New York: Harcourt Brace Jovanovich, 1976), p. 412; Diana Trilling, "Lionel Trilling, a Jew at Columbia," *Commentary*, 67 (March, 1979), 44, 46.

47. Daniel J. Kevles, *The Physicists* (New York: Alfred A. Knopf, 1978), pp. 213, 215.

48. Feuer, "Stages," p. 455.

49. Quoted in Peter Novick, *That Noble Dream* (Cambridge: Cambridge University Press, 1988), pp. 172–173.

50. Quoted in Nathan Reingold, "Refugee Mathematicians in the United States of America, 1933–1941: Reception and Reaction," in *Annals of Science*, 38 (May, 1981), 320, 321.

51. A. L. Severson, "Nationality and Religious Preferences as Reflected in

Newspaper Advertisements," *American Journal of Sociology*, 44 (1939), 541, 542; Laura E. Weber, "'Gentiles Preferred': Minneapolis Jews and Employment, 1920–1950," *Minnesota History*, 52 (Spring, 1991), 174; Stanley Feldstein, *The Land That I Show You* (Garden City: Anchor Press/Doubleday, 1978), p. 250; Feingold, *A Time for Searching*, p. 2.

52. Judd L. Teller, *Strangers and Natives* (New York: Delacorte Press, 1968), p. 97; Lee J. Levinger, "Jews in Liberal Professions in Ohio," JSS, 2 (October, 1940), 429; Broun and Britt, *Christians Only*, p. 234 ff; Feuer, "Stages," p. 458; Rosenstock, *Louis Marshall*, p. 260; Jacob Rader Marcus, "Zionism and the American Jew," *American Scholar*, 2 (1933), 287; Helga Eugenie Kaplan, "Century of Adjustment: A History of the Akron Jewish Community" (unpublished Ph.D., Department of History, Kent State University, 1978), pp. 435–436; Daniel Pope and William Toll, "We Tried Harder: Jews in American Advertising," AJH, 72 (September, 1982), 41; Max Vorspan and Lloyd P. Gartner, *History of the Jews of Los Angeles* (Philadelphia: JPS, 1970), 205–206; "The Jew and the Club," *Atlantic Monthly*, 134 (October, 1924), 455; Rapp, "An Historical Overview," p. 32; Maurice J. Karpf, *Jewish Community Organization in the United States* (New York: Bloch Publishing Co., 1938), pp. 20–21; Lloyd P. Gartner, "Assimilation and American Jews," in Bela Vago, ed., *Jewish Assimilation in Modern Times* (Boulder, Colo.: Westview Press, 1981), 179; Marc Lee Raphael, *Jews and Judaism in a Midwestern Community: Columbus, Ohio, 1840–1975* (Columbus: Ohio Historical Society, 1979), p. 235.

53. Quoted in Spencer Klaw, "The Wall Street Lawyers," *Fortune*, 57 (February, 1958), 192.

54. Elaine H. Maas, "The Jews of Houston: An Ethnographic Study" (unpublished Ph.D., Sociology Department, Rice University, 1973), p. 196; Judith E. Endelman, *The Jewish Community of Indianapolis: 1849–Present* (Bloomington: Indiana University Press, 1984), pp. 117, 172; Teller, *Strangers and Natives*, p. 97.

55. Kirk Douglas, *The Ragman's Son* (London: Pan Books, 1988), pp. 34–35.

56. Keith Sward, *The Legend of Henry Ford* (New York: Rinehart & Co., 1948), p. 137; Sable, "Some American Jewish," p. 50; Rapp, "An Historical Overview," p. 40; Broun and Britt, *Christians Only*, p. 273; Deborah Dash Moore, *At Home in America: Second Generation New York Jews* (New York: Columbia University Press, 1981), p. 21; Henry L. Feingold, *Zion in America* (New York: Twayne Publishers, Inc., 1974), p. 266; Johan J. Smertenko, "Hitlerism Comes to America," *Harper's Magazine*, 167 (November, 1933), 661.

57. Quoted in Richard Kluger, *The Paper: The Life and Death of the New York Herald Tribune* (New York: Alfred A. Knopf, 1986), p. 386.

58. Kalman, *Abe Fortas*, p. 13; Greenebaum, "The Jewish Experience," p. 114.

59. Moore, *At Home*, pp. 45, 51.

60. Samuel B. Hand, *Counsel and Advise: A Political Biography of Samuel I. Rosenman* (New York: Garland, 1979), p. 233.

61. Folder, "Thomas G. Corcoran," box 7, Benjamin V. Cohen mss., LC; Maxwell Whiteman, *Copper for America: The Hendricks Family and a National Industry, 1755–1939* (New Brunswick: Rutgers University Press, 1971), p. 232; E. Digby Baltzell, *The Protestant Establishment: Aristocracy and Caste in America* (London: Secher and Warburg, 1964), pp. 86, 371; Polmer and Allen, *Rickover*,

p. 193; McWilliams, *A Mask for Privilege*, p. 122; Osborn Elliott, *Men at the Top* (New York: Harper and Brothers, 1959), p. 158.

62. Whiteman, *Copper for America*, p. 218; Baltzell, *Protestant Establishment*, pp. 138, 206; E. Digby Baltzell, Allen Glicksman, and Jacquelyn Litt, "The Jewish Communities of Philadelphia and Boston: A Tale of Two Cities," in Murray Friedman, ed., *Jewish Life in Philadelphia, 1830–1940* (Philadelphia: ISHI Publications, 1983), p. 292; Neal Gabler, *An Empire of Their Own: How the Jews Invented Hollywood* (New York: Crown Publishers, Inc., 1988), p. 273.

63. Baltzell, *Protestant Establishment*, p. 120; Frederick Simon, "Anti-Semitism in America, 1654–1930" (unpublished mss., October, 1969), p. 81, folder, 2, box 4, Oscar Cohen mss., AJA; Bertram Wallace Korn, *The Early Jews of New Orleans* (Waltham, Mass.: AJHS, 1969), p. 228; Leonard Dinnerstein, "From Desert Oasis to Desert Caucus: The Jews of Tucson," in Moses Rischin and John Livingston, eds., *Jews of the American West* (Detroit: Wayne State University Press, 1991), p. 140; Proskauer, *A Segment of My Times*, p. 12; Allen duPont Breck, *The Centennial History of the Jews of Colorado, 1859–1959* (Denver: The Hirschfield Press, 1960), p. 70; William Barton McCash and June Hall McCash, *The Jekyll Island Club: Southern Haven for America's Millionaires* (Athens: University of Georgia Press, 1989), p. 10.

64. Raphael, *Jews and Judaism*, p. 236; Endelman, *The Jewish Community*, p. 172; Lloyd P. Gartner, *History of the Jews of Cleveland* (Cleveland: Western Reserve Historical Society, 1978), p. 300; Ida Libert Uchill, *Pioneers, Peddlers, and Tsadikim* (Denver: Sage Books, 1957), pp. 162–163; Vorspan and Gartner, *History of the Jews*, p. 144; John Cooney, *The Annenbergs* (New York: Simon and Schuster, 1982), p. 185; Robert A. Rockaway, "Anti-Semitism in an American City: Detroit, 1850–1914," *AJHQ*, 64 (September, 1974), 50–51; Eli N. Evans, *The Provincials: A Personal History of Jews in the South* (New York: Atheneum, 1973), p. 288.

65. Dwight W. Hoover, "To be a Jew in Middletown: A Muncie Oral History Project," *Indiana Magazine of History*, 81 (1985), 151; Amy Hill Siewers, "Judaism in the Heartland: The Jewish Community of Marietta, Ohio (1895–1940)," *The Great Lakes Review*, 5 (Winter, 1979), 31; Rapp, "An Historical Overview," p. 29.

66. Gartner, "Assimilation, p. 178; Cleveland Amory, *The Last Resorts* (New York: Harper & Bros., 1952), p. 48; Baltzell, *Protestant Establishment*, pp. 33, 325; X, "The Jew and the Club," *The Atlantic*, 134 (October, 1924), 451.

67. "Resort Advertising in The New York Herald Tribune 1926 through 1940," and "Discrimination [Advertising], 1946, 1951" in folders 7 and 7a, box 7, Brooklyn Jewish Community Council (1941–1959) mss., Mss. Collection #164, AJA; Severson, "Nationality and Religious Preferences," 545; "Article on Anti-Semitism *ca* 1925," box 37, Charles E. Russell mss., LC; Broun and Britt, *Christians Only*, p. 251; Feldstein, *The Land*, pp. 240–241; Anon., "A Half-Jew Speaks," *The American Mercury*, 51 (October, 1940), 179; Minutes, IV (January 25, 1920), 863; Jeffrey Gurock, "The 1913 New York State Civil Rights Act," *AJS Review*, 1 (1976), 111; Thomas David Mantel, "The Anti-Defamation League of B'nai B'rith" (Honors Thesis, Department of Government, Harvard University, 1950), pp. 22–23, Small Collection, AJA; Douglas, *The Ragman's Son*, p. 50; "Summer Hotels That Bar Jews," *AH*, 110 (April 14, 1922), 613; Maud Nathan, *Once Upon a Time and Today* (New York: G. P. Putnam's Sons, 1933), p. 91; Blaine

Peterson Lamb, "Jewish Pioneers in Arizona, 1850–1920" (unpublished Ph.D., Department of History, Arizona State University, 1982), pp. 268–269.

68. Moore, *At Home*, pp. 36, 38; Nathan, *Once Upon a Time*, p. 91; Carol A. O'Connor, *A Sort of Utopia: Scarsdale, 1891–1981* (Albany: SUNY Press, 1983), 98; W. Gunther Plaut, *The Jews in Minnesota* (New York: AJHS, 1959), p. 170n; Minutes, V (October 5, 1924), 1143, V (1931), 1539–1540; Vorspan and Gartner, *History of the Jews*, p. 205; Robert Kotlowitz, "Baltimore Boy," *Harper's Magazine*, 231 (December, 1965), 62; Francis Russell, "The Coming of the Jews," *Antioch Review*, 15 (Spring, 1955), 23; Breck, *Centennial History*, p. 198.

69. Rosenstock, *Louis Marshall*, p. 240.

70. Quoted in Kluger, *The Paper*, p. 386.

71. Susan Dworkin, *Miss America, 1945: Bess Meyerson's Own Story* (New York: Newmarket Press, 1987), p. 12.

72. Moore, *At Home*, p. 38 and 18–30 *passim*.

73. Kenneth L. Roberts, *Why Europe Leaves Home* (Indianapolis: The Bobbs-Merrill Co., 1922), p. 48; Kenneth T. Jackson, *The Ku Klux Klan in the City, 1915–1930* (New York: Oxford University Press, 1967), p. 243.

74. Ribuffo, *Old Christian*, p. 9; Gino Speranza, "The Immigration Peril," *World's Work*, 47 (November, 1923), 57–65, (December, 1923), 147–160, (January, 1924), 256–270, (February, 1924), 399–409, (March, 1924), 479–490, (April, 1924), 643–648, 48 (May, 1924), 62–68; Rosenstock, *Louis Marshall*, p. 221.

75. Roberts, *Why Europe*, p. 49; Rosenstock, *Louis Marshal*, pp. 220–221.

76. Roberts, *Why Europe*, pp. 15, 46, 48, 49, 50, 78, 117, 118.

77. Hendrick, "The Jews In America," pp. 275, 276, 277, 281, 283, 284, 285, 373, 375, 376, 591, 593, 594.

78. "Another Jewish Menace," *The Freeman*, 7 (July 18, 1923), 436.

79. Emory S. Bogardus, *Immigration and Race Attitudes* (Boston: D. C. Heath & Co., 1923), pp. 19–20.

80. Jackson, *Ku Klux Klan*, p. xv; Gannett, "Is America Anti-Semitic?" p. 330; Ralph Janis, "Flirtation and Flight: Alternatives to Ethnic Confrontation in White Anglo-American Protestant Detroit," *Journal of Ethnic Studies*, 6 (Summer, 1978), 10.

81. Quoted in Arnold S. Rice, *The Ku Klux Klan in American Politics* (originally published, 1962; New York: Haskill House Publishers, Ltd., 1972), p. 132 n. 2.

82. Minutes, IV (March 11, 1923), 1088; Leonard J. Moore, *Citizen Klansmen: The Ku Klux Klan in Indiana, 1921–1928* (Chapel Hill: University of North Carolina Press, 1991), p. 10.

83. Charles C. Alexander, *The Ku Klux Klan in the Southwest* (University of Kentucky Press, 1965), pp. 25, 26.

84. Dr. H. W. Evans, "The Ku Klux Klan and the Jew," *B'nai B'rith News*, 15 (March, 1923), 1–2; Carolyn Lipson-Walker, " 'Shalom Y'All': The Folklore and Culture of Southern Jews (unpublished Ph.D., Department of Folklore in the Program in American Studies, Indiana University, 1986), p. 222.

85. David Chalmers, *Hooded Americans: The First Century of the Ku Klux Klan, 1865–1965* (New York: Doubleday & Co., Inc., 1965), p. 197.

86. Quoted in Emanuel Steiner, "Clippings . . . ," Misc. File, AJA.

87. Henry J. Tobias, *The Jews in Oklahoma* (Norman: University of Oklahoma Press, 1980), p. 60; Minutes, IV (May 27, 1923), 1099.

88. *Oregon Voter*, May 22, 1922, quoted in Malcolm Clark, Jr., "The Bigot

Disclosed: 90 Years of Nativism," *Oregon Historical Quarterly,* 75 (June, 1974), 165.

89. Kathleen M. Blee, *Women of the Klan: Racism and Gender in the 1920s* (Berkeley: University of California Press, 1991), pp. 40, 75.

90. Michael W. Rubinoff, "Rabbi in a Progressive Era: C. E. H. Kanvar of Denver," *The Colorado Magazine,* 54 (Summer, 1977), 233–234; Uchill, *Pioneers,* p. 160; Breck, *Centennial History,* p. 196; Robert Alan Goldberg, *Hooded Empire: The Ku Klux Klan in Colorado* (Urbana: University of Illinois Press, 1981), p. 24. Both of Colorado's United States Senators (Republicans) owed their election to Klan support and one of them, Rice W. Means, allegedly belonged to the KKK. "Klan Victories and Defeats," *Literary Digest,* 83 (November 22, 1924), 16.

91. Goldberg, *Hooded Empire,* pp. 37, 39; Charles Larsen, *The Good Fight* (Chicago: Quadrangle, 1972), p. 192; Jackson, *The Ku Klux Klan in the City,* p. 218; Rubinoff, "Rabbi in a Progressive Era," p. 234.

92. Shih-Shan Henry Tsai, *The Chinese Experience in America* (Bloomington: Indiana University Press, 1986), p. 69; Rubinoff, "Rabbi in a Progressive Era," p. 233; "Ordeal of Colorado's Germans," *Colorado Magazine of History and Biography,* 51 (Fall, 1974), 277–293 *passim.*

93. Chalmers, *Hooded Americans,* p. 245; Larsen, *The Good Fight,* p. 192; Martin E. Marty, *The Righteous Empire: The Protestant Experience in America* (New York: The Dial Press, 1970), p. 230; Broun and Britt, *Christians Only,* p. 44.

94. Minutes, IV (September 28, 1922), 1044, (November 11, 1922), 1061, VII (1936–1943), 558; Philip S. Bernstein, "Unchristian Christianity and the Jew," *Harper's Magazine,* 162 (May, 1931), 661; John T. Pawlikowski, *Catechetics and Prejudice: How Catholic Teaching Materials View Jews, Protestants and Racial Minorities* (New York: Paulist Press, 1973), pp. 19, 81, 83, 87; Kaplan, "Century of Adjustment," p. 467; James Brown, "Christian Teaching and Anti-Semitism," *Commentary,* 24 (December, 1957), 496, 497; Bernhard E. Olson, *Faith and Prejudice; Intergroup Problems in Protestant Curricula* (New Haven: Yale University Press, 1963), p. 1; Claire Huchet Bishop, "Learning Bigotry," *Commonweal,* 80 (May 22, 1964), 264; Bruno Lasker, *Race Attitudes in Children* (originally published, 1929; New York: Greenwood Press, 1968), pp. 179, 181–182, 252, 279; Marty, *The Righteous Empire,* p. 253.

95. Philip Bernstein, "We Are Not Taught To Hate," *Commonweal,* 14 (June 17, 1931), 187–188; Bernstein, "Unchristian Christianity," p. 660.

96. Everett R. Clinchy, "The Borderland of Prejudice," CC, 47 (May 14, 1930), 623, 624; see also Nina Morais, "Jewish Ostracism in America," *North American Review,* 133 (September, 1881), 271.

97. Albert Levitan, "Leave the Jewish Problem Alone!" CC, 51 (April 25, 1934), 555.

98. "Jews and the Crucifixion," CC, 45 (February 2, 1928), 136; see also Nicholas A. Berdyaev, "The Crime of Anti-Semitism," *Commonweal,* 29 (April 21, 1939), 707.

99. Ewa Morawska, *Insecure Prosperity: Small Town Jews in Industrial America, 1880–1940* (London and New York: Basil Blackwell), chapter 5, forthcoming.

100. AJYB, 30 (1928–1929), 276.

101. Harry Schneiderman, "The King's County Hospital Case," in Bruno Lasker, ed., *Jewish Experiences in America* (New York: The Inquiry, 1930), 72ff; Sable, "Some American Jewish," p. 57; Rosenstock, *Louis Marshall,* p. 259; Feldstein, *The Land,* p. 249.

102. "Jews and Christians in the Middle Ages," *Saturday Review* (London), 56 (July 14, 1883), 41; Florence H. Ridley, "A Tale Told Too Often," *Western Folklore*, 26 (1967), 153.

103. Harold David Brackman, "The Ebb and Flow of Conflict: A History of Black-Jewish Relations Through 1900" (unpublished Ph.D., University of California, Los Angeles, 1977), p. 252; New York *Tribune*, April 21, 1913, p. 14; Abraham G. Duker, "Twentieth-Century Blood Libels in the United States," in Leo Landman, ed., *Rabbi Joseph H. Lookstein Memorial Volume* (New York: KTAV Publishing House, Inc., 1980), pp. 89, 90–91, 92, 99, 100, 102; Leslie A. Fiedler, "What Can We Do About Fagin?" *Commentary*, 7 (May, 1949), 417.

104. "The Inquisition, the Reformation, and the Jews," CC, 44 (June 9, 1927), 715; Minutes, V (November 10, 1928), 1408 ff.; Rosenstock, *Louis Marshall*, pp. 264–266; *Jewish Chronicle* (London), September 19, 1980, p. 25; Saul S. Friedman, *The Incident At Massena* (New York: Stein and Day, 1978).

105. Minutes, IV (December 12, 1920), 929.

106. Rosenstock, *Louis Marshall*, pp. 159–160; Jerome C. Rosenthal, "The Public Career of Louis Marshall" (unpublished Ph.D., Department of History, University of Cincinnati, 1983), p. 677.

107. Mantel, "Anti-Defamation League," p. 22.

108. Ribuffo, "Henry Ford," p. 146ff; Rosenstock, *Louis Marshall*, pp. 182–197 *passim*.

109. Minutes, IV (March 12, 1922), 1029.

110. Sable, "Some American Jewish," p. 228; see also, Naomi W. Cohen, *A Dual Heritage: The Public Career of Oscar S. Straus* (Philadelphia: JPS, 1969), p. 290.

111. Rosenstock, *Louis Marshall*, p. 229; Levine, *The American College*, pp. 154–155; Endelman, *The Jewish Community*, p. 122; Gordon, "Annie Nathan Meyer," p. 517.

112. Walter Lippmann, "Public Opinion and the American Jew," AH, 110 (April 14, 1922), 575.

113. Rosenstock, *Louis Marshall*, p. 277.

Chapter 6

1. Arthur Liebman, *Jews and the Left* (New York: John Wiley, 1979), p. 425; Nearptint File—Special Topics, Anti-Semitism (1921–1959), folder, "Anti-Semitism, 1921–1929," box 1 (1921–1938), AJA.

2. Robert T. Handy, *A Christian America: Protestant Hopes and Historical Realities* (New York: Oxford University Press, 1971), p. 66.

3. George Wolfskill and John A. Hudson, *All But the People: Franklin D. Roosevelt and His Critics, 1933–1939* (New York: The Macmillan Co., 1969), pp. 66, 72; Egal Feldman, *Dual Destinies: The Jewish Encounter with Protestant America* (Urbana: University of Illinois Press, 1990), p. 244; George M. Marsden, *Fundamentalism and American Culture: The Shaping of Twentieth-Century Evangelism, 1870–1925* (New York: Oxford University Press, 1980), pp. 207, 210.

4. John Sheridan Zelie, "Why Do The Gentiles Rage?" CC, 48 (October 7, 1931), 1239, 1241.

5. Nathan Glazer, "Social Characteristics of American Jews, 1654–1954," AJYB, 56 (1955), 20 ff.; Deborah Dash Moore, *At Home in America: Second Gen-*

eration New York Jews (New York: Columbia University Press, 1981), passim; Lawrence H. Fuchs, "American Jews and the Presidential Vote," *The American Political Science Quarterly*, 49 (July, 1955), 385.

6. Henry L. Feingold, *A Time for Searching: Entering the Mainstream* (Baltimore: The Johns Hopkins University Press, 1992), pp. 148–149.

7. Leo P. Ribuffo, "Henry Ford and *The International Jew*," AJH, 69 (June, 1980), 440; M. Ginsberg, "Anti-Semitism," *The Sociological Review*, 35 (January–April, 1943), 7; X, "The Jew and the Club," *The Atlantic*, 134 (October, 1924), 451; Ralph Philip Boas, "Jew-Baiting in America," *The Atlantic*, 127 (May, 1921), 659; "Anti-Semitism Is Here," *The Nation*, 147 (August 25, 1938), 167; Minutes, V (January 13, 1929–February 14, 1932), passim, VI (March 13, 1932–December 23, 1933), 66; Isabel Cohen, "The Reign of Anti-Semitism," *New Statesman and Nation*, 4 (July 23, 1932), 97; Subject File: "Anti-Semitism," reel 77, Felix Frankfurter mss., LC; Conrad Hoffmann, "Modern Jewry and the Christian Church," *International Review of Missions*, 12 (April, 1934), 189; Aaron Goldman, "The Resurgence of Antisemitism in Britain during World War II," JSS, 46 (Winter, 1984), 37–38.

8. Minutes, V (January 13, 1929–February 14, 1932), 1497.

9. Minutes, V (January 13, 1929–February 14, 1932), 1454, "Memorandum on Conference with Charles A. Davila, Romanian Minister to the United States, December 18, 1929," and "Abstract of Statement of Dr. Bernhard Kahn at Meeting of the Executive Committee, December 14, 1930"; Philip S. Bernstein, "Unchristian Christianity and the Jew," *Harper's Magazine*, 162 (May, 1931), 660, 665; William Zuckerman, "The Jews—A Nation Trapped," *The Nation*, 131 (August 20, 1930), 200–201; "Miscellaneous, 1911–1939," folder, 6, box 6, of A. J. Sabath mss. Collection #43, AJA; Cohen, "Reign of Anti-Semitism," p. 97.

10. Cohen, "Reign of Anti-Semitism," p. 96.

11. Max Vorspan and Lloyd P. Gartner, *History of the Jews of Los Angeles* (Philadelphia: JPS, 1970), p. 206.

12. William Manchester, *The Glory and the Dream* (Boston: Little, Brown, 1974), p. 8; Thomas Karfunkel and Thomas W. Ryley, *The Jewish Seat: Anti Semitism and the Appointment of Jews to the Supreme Court* (Hicksville: Exposition Press, 1978), p. 61; Johan J. Smertenko, "Hitlerism Comes to America," *Harper's Magazine*, 167 (November, 1933), p. 660; Donald S. Strong, *Organized Anti-Semitism in America: The Rise of Group Prejudice During the Decade, 1930–1940* (Washington: American Council on Public Affairs, 1941), p. 174; John F. Stack, Jr., *International Conflict in an American City: Boston's Irish, Italians, and Jews, 1935–1944* (Westport, Conn.: Greenwood Press, 1979), p. 92; AJYB, 37 (September 28, 1935–September 16, 1936), p. 156; Paul S. Holbo, "Wheat or What? Populism and American Fascism," *Western Political Quarterly*, 14 (September, 1961), 735; Leo P. Ribuffo, *The Old Christian Right* (Philadelphia: Temple University Press, 1983), p. 18; J. A. Rogers, "Negroes Suffer More in U. S. Than Jews in Germany," *The Philadelphia Tribune*, September 21, 1933, p. 3.

13. Richard F. Nelson, "Nothing Will Save Us but a Pogrom!" CC, 50 (June 28, 1933), 850.

14. Maud Nathan, *Once Upon a Time and Today* (New York: G. P. Putnam's Sons, 1933), p. 275.

15. Minutes, VI (1932–1938), 178; Ron Chernow, *The House of Morgan: An American Banking Dynasty and the Rise of Modern Finance* (New York: Atlantic

Monthly Press, 1990), p. 394; Laura Z. Hobson, *Laura Z.: A Life* (New York: Arbor House, 1983), p. 115.

16. Leonard Dinnerstein, "Jews and the New Deal," AJH, 72 (June, 1983), 463ff.

17. Jerold S. Auerbach, *Unequal Justice: Lawyers and Social Change in Modern America* (New York: Oxford University Press, 1976), pp. 187–188.

18. Dinnerstein, "Jews and the New Deal," p. 463.

19. W. M. Kiplinger, "The Facts About Jews in Washington," *Reader's Digest*, 41 (September, 1942), 3; *The New Dealers*, by Unofficial Observer (New York: The Literary Guild, 1934), p. 322.

20. [signature illegible] to FDR, October 21, 1934, OF 76C, FDR mss.

21. AJYB, 37 (1935–1936), 153–154; Wolfskill and Hudson, *All But the People*, p. 86; newspaper clipping, May 20, 1934, box 160, Raymond Clapper mss., LC; Zosa Szajkowski, "The Attitude of American Jews to Refugees from Germany in the 1930's," AJHQ, 61 (December, 1971), 106; *The New Dealers*, p. 332; Myron Scholnick, "The New Deal and Anti-Semitism in America (unpublished Ph.D., Department of History, University of Maryland, 1971), pp. 33, 76–77; "Washington Notes," *The New Republic*, 77 (January 10, 1934), 250.

22. Quoted in Scholnick, "The New Deal," p. 77; see also Richard Yaffe, "The Roosevelt Magic," *The National Jewish Monthly*, 87 (October, 1972), 31; Bruce Allen Murphy, *The Brandeis/Frankfurter Connection* (New York: Oxford University Press, 1982), pp. 133, 288; Marquis W. Childs, "They Still Hate Roosevelt," *The New Republic*, 96 (September 14, 1938), 148; T. R. B., "Washington Notes," *The New Republic*, 77 (January 10, 1934), 250; "Anti-Semitism Is Here," p. 167; Wolfskill and Hudson, *All But the People*, 65ff; E. Digby Baltzell, *The Protestant Establishment* (New York: Random House, 1964), p. 248.

23. W. M. Kiplinger, "The Facts About Jews in Washington," *Reader's Digest*, 41 (September, 1942), 2ff; see also, Frank Freidel, *Franklin D. Roosevelt: Launching the New Deal* (Boston: Little, Brown & Co., 1973), p. 393.

24. Wolfskill and Hudson, *All But the People*, p. 66; see also Franklin Thompson, *America's Ju-Deal* (Woodhaven, N.Y.: Community Press, 1935).

25. Scholnick, "The New Deal," p. 79; Michael Gerald Rapp, "An Historical Overview of Anti-Semitism In Minnesota, 1920–1960—With Particular Emphasis on Minneapolis and St. Paul" (unpublished Ph.D., Department of History, University of Minnesota, 1977), p. 61; Wolfskill and Hudson, *All But the People*, pp. 66–67, 70; Paul W. Ward, "Washington Weekly," *The Nation*, 143 (November 7, 1936), 540; copy of Winrod's article, "Roosevelt's Jewish Ancestry," *The Review*, October 15, 1936, in folder, "Edmundsen Service [NYC]," PPF #1632, FDR mss.

26. Quoted in Wolfskill and Hudson, *All But the People*, p. 87.

27. Feldman, *Dual Destinies*, p. 201.

28. Franklin H. Littell, "American Protestantism and Antisemitism," in Naomi W. Cohen, ed., *Essential Papers on Jewish-Christian Relations in the United States* (New York: NYU Press, 1990), p. 184.

29. "Jews and Jesus," CC, 50 (May 3, 1933), 582.

30. "The Jewish Problem," CC, 53 (April 29, 1936), 625; "Jews, Christians and Democracy," CC, 53 (May 13, 1936), 697; "Jews and Democracy," CC, 53 (June 9, 1937), 734–735; "Tolerance Is Not Enough!" CC, 53 (July 1, 1936), 928.

31. "Shall Christians Let the Jews Alone?" *Missionary Review of the World*, 54 (July, 1931), 517.

32. "The Jewish Question in America," *Missionary Review of the World*, 58 (September, 1935), 388.

33. Marsden, *Fundamentalism and American Culture*, p. 207.

34. William R. Glass, "Fundamentalism's Prophetic Vision of the Jews: The 1930s," JSS, 47 (Winter, 1985), 67, 69; Jimmy Harper, "Alabama Baptists and the Rise of Hitler and Fascism, 1930–1938," *Journal of Reform Judaism*, 32 (Spring, 1985), 7, 8, 9 (quote on page 8).

35. The Alabama Baptist, December 15, 1938, p. 4.

36. Ribuffo, *Old Christian*, pp. 56–57; Strong, *Organized Anti-Semitism*, pp. 40, 46.

37. Wolfskill and Hudson, *All But the People*, pp. 68, 72.

38. Quoted in ibid., p. 72.

39. Smertenko, "Hitlerism," p. 663.

40. Stanley High, "Star Spangled Fascists," *The Saturday Evening Post*, 211 (May 27, 1939), 7.

41. Ibid., pp. 5, 6, 70; Strong, *Organized Anti-Semitism*, pp. 16, 146; Alvin Johnson, "The Rising Tide of Anti-Semitism," *Survey Graphic*, 28 (February, 1939), 115; Norton Belth, "Problems of Anti-Semitism in the United States," *Contemporary Jewish Record*, 2 (May-June, 1939), 6 ff.

42. Quoted in David J. Jacobson, *The Affairs of Dame Rumor* (New York: Rinehart & Co., 1948), pp. 316–317.

43. Ribuffo, *Old Christian*, pp. 17, 73, 104, 117; David H. Bennett, *Demagogues in the Depression* (New Brunswick: Rutgers University Press, 1969), p. 279; Stack, *International Conflict*, p. 53; William C. Kernan, "Coughlin, the Jews, and Communism," *The Nation*, 147 (December 17, 1938), 655.

44. Quoted in Walter Goodman, *The Committee* (New York: Farrar, Straus and Giroux, 1968), p. 94; see also David S. Wyman, *Paper Walls: America and the Refugee Crisis, 1938–1941* (University of Massachusetts Press, 1968), pp. 15–17.

45. Wolfskill and Hudson, *All But the People*, pp. 66, 72; Glass, "Fundamentalism's Prophetic Vision," p. 69; Marsden, *Fundamentalism and American Culture*, p. 207.

46. Robert W. Ross, *So It Was True: The American Protestant Press and the Nazi Persecution of the Jews* (Minneapolis: University of Minnesota Press, 1980), pp. 40, 45, 57, and passim; Chuck Badger, "The Response of Christian Fundamentalists to the Holocaust, 1933–1945" (unpublished seminar paper, Department of History, University of Arizona, December, 1987), p. 12 ff., copy in possession of the author; Glass, "Fundamentalism's Prophetic Vision," pp. 67, 69; Gerhard Falk, "The Reaction of the German-American Press to Nazi Persecutions, 1933–1941," *Journal of Reform Judaism*, 32 (Spring, 1985), 22. See also Deborah E. Lipstadt, *Beyond Belief: The American Press and the Coming of the Holocaust, 1933–1935* (New York: The Free Press, 1986).

47. Edward C. McCarthy, "The Christian Front Movement in New York City, 1938–1940" (unpublished M. A. thesis, Department of History, Columbia University, 1965), pp. 171, 177; Feingold, *A Time for Searching*, p. 198ff.; William C. Kernan, "Coughlin, the Jews, and Communism," *The Nation*, 147 (December 17, 1938), 655; Stack, *International Conflict*, p. 101; Ronald H. Bayor, *Neighbors in Conflict: The Irish, Germans, Jews, and Italians of New York City, 1929–1941* (2nd edition; University of Illinois Press, 1988), pp. 90, 148–149.

48. *A Time for Searching*, pp. 193 and 221.

49. John P. Diggins, *Mussolini and Fascism: The View from America* (Princeton: Princeton University Press, 1972), pp. 336–337.

50. Budd Schulberg, *Moving Pictures: Memories of a Hollywood Prince* (New York: Stein and Day, 1981), p. 5; Kirk Douglas, *The Ragman's Son* (London: Pan Books, 1988), p. 21; Ann Birstein, *The Rabbi on Forty-Seventh Street: The Story of Her Father* (New York: Dial Press, 1982), p. 88; Diggins, *Mussolini and Fascism*, pp. 185, 192; William Foote Whyte, "Race Conflicts in the North End of Boston," *New England Quarterly*, 12 (December, 1939), 642; Bayor, *Neighbors in Conflict*, p. 94ff; Ron Avery, "From A to 'Zink': Philadelphia Jews in Sports," in Murray Friedman, ed., *Philadelphia Jewish Life, 1940–1985* (Ardmore, Penn.: Seth Press, 1986), p. 295; Charles I. Cooper, "The Jews of Minneapolis and their Christian Neighbors," *JSS*, 8 (1946), 36; *New Republic*, 95 (May 25, 1938), 66–67; Carolyn F. Ware, *Greenwich Village, 1920–1930* (Boston: Houghton Mifflin Co., 1935), pp. 139, 140.

51. McCarthy, "The Christian Front," pp. 161, 170; Arthur Liebman, "The Ties That Bind: The Jewish Support for the Left in the United States," *AJHQ*, 66 (December, 1976), 305; Vorspan and Gartner, *History of the Jews of Los Angeles*, p. 202; Neal Gabler, *An Empire of their Own: How The Jews Invented Hollywood* (New York: Crown Publishers, Inc., 1988), pp. 330–331.

52. McCarthy, "The Christian Front," p. 161.

53. Quoted in J. David Valaik, "In the Days Before Ecumenism," *Journal of Church and State*, 13 (Autumn, 1971), 469.

54. George N. Shuster, "The Conflict Among Catholics," *American Scholar*, 10 (Winter, 1940–1941), 11.

55. Valaik, "In The Days Before," p. 468.

56. Quoted in Oswald Garrison Villard, "Issues and Men," *The Nation*, 148 (April 22, 1939), 470.

57. Catholic Telegraph-Register clippings, June 4, 1936, September 17, 1936, March 25, 1937, June 5, 1937, June 26, 1937 in folder 12, "*Catholic Telegraph-Register*, 1936–1966," box 42 JCRC mss., collection #202, AJA.

58. Abigail McCarthy, "An Ugly Resurgence," *Commonweal*, 109 (December 17, 1982), 679; see also John T. Pawlikowski, *Catechetics and Prejudice: How Catholic Teaching Materials View Jews, Protestants, and Racial Minorities* (New York: Paulist Press, 1973), pp. 8, 13; James O'Gara, "Christian Anti-Semitism," *Commonweal*, 80 (May 22, 1964), 252; James O'Gara, "Catholics and Jews," *Commonweal*, 80 (May 29, 1964), 286.

59. Sheldon Marcus, *Father Coughlin: The Tumultuous Life of the Priest of the Little Flower* (Boston: Little, Brown & Co., 1973), p. 11.

60. Handy, *Christian America*, p. 78; Stack, *International Conflict*, p. 53; George N. Shuster, "The Conflict Among Catholics," *Contemporary Jewish Record*, 4 (February, 1941), 48; John Cooney, *The American Pope: The Life and Times of Francis Cardinal Spellman* (New York: Dell Books, 1986), pp. 100–101.

61. Shuster, "Conflict," *Contemporary Jewish Record*, p. 48; Ronald Modras, "Father Coughlin and Anti-Semitism: Fifty Years Later," *Journal of Church and State*, 31 (Spring, 1989), 234; Alan Brinkley, *Voices of Protest: Huey Long, Father Coughlin and the Great Depression* (New York: Alfred A. Knopf, 1982), pp. 266, 269ff.

62. George Seldes, "Father Coughlin: Anti-Semite," *The New Republic*, 96 (November 2, 1938), 353.

63. Hyman Berman, "Political Antisemitism in Minnesota During the Great Depression," JSS, 38 (Summer/Fall, 1976), 261, 263.

64. Quoted in Wyman, *Paper Walls*, p. 73.

65. Charles J. Tull, *Father Coughlin and the New Deal* (Syracuse, N.Y.: Syracuse University Press, 1965), p. 197; "Persecution—Jewish and Christian," in Charles E. Coughlin, "Am I an Anti-Semite?" *Anti-Semitism In America, 1878–1939* (New York: Arno Press, 1977), 36–37. This essay is the text of the radio address that he made on November 20, 1938.

66. Kernan, "Coughlin, the Jews, and Communism," p. 658; Coughlin, "Am I an Anti-Semite?" pp. 37–38, 41, 42.

67. Coughlin, "Am I an Anti-Semite?" pp. 44, 46.

68. "Slap," *Life*, 32 (November 28, 1938), 65; Modras, "Father Coughlin," p. 246.

69. Ibid.

70. T, November 21, 1938, p. 7; Kernan, "Coughlin, the Jews, and Communism," p. 655 ff.; Alan Brinkley, *Voices of Protest: Huey Long, Father Coughlin and the Great Depression* (New York: Vintage, 1983), p. 266.

71. Francis L. Broderick, *Right Reverend New Dealer: John A. Ryan* (New York: The Macmillan Co., 1963), p. 253; John A. Ryan, "Anti-Semitism in the Air," *Commonweal*, 29 (December 30, 1938), 260, 261.

72. "Week by Week," *Commonweal*, 29 (December 9, 1938), 169.

73. Hans Ascar, "Catholics and Anti-Semitism," *Catholic World*, 150 (November, 1939), 175–176.

74. Arnold Benson, "The Catholic Church and the Jews," *The American Jewish Chronicle*, 1 (February 15, 1940), 7.

75. Quoted in Ralph L. Kolodny, "Catholics and Father Coughlin: Misremembering the Past," *Patterns of Prejudice*, 19 (October, 1985), 19.

76. Wyman, *Paper Walls*, p. 17; Bayor, *Neighbors in Conflict*, p. 88; High, "Star-Spangled Fascists," p. 72; Kolodny, "Catholics and Father Coughlin," p. 18; Morris L. Ernst to FDR, May 14, 1942, in subject file, "Ernst, Morris L., 1940–1942," PSF, box 132, FDR Library.

77. Shuster, "The Conflict Among Catholics," *American Scholar*, p. 12.

78. Quoted in "Anti-Semitism Is Here," p. 167.

79. Quoted in Herbert Samuel Rutman, "Defense and Development: A History of Minneapolis Jewry, 1930–1950" (unpublished Ph.D., University of Minnesota, 1970), p. 120.

80. Both quotes in Alfred Winslow Jones, *Life, Liberty, and Property* (Philadelphia: J. B. Lippincott Co., 1941), pp. 216, 274.

81. Burton Alan Boxerman, "Reaction of the St. Louis Jewish Community to Anti-Semitism, 1933–1945" (unpublished Ph. D., St. Louis University, 1967), p. 64; Burton Alan Boxerman, "Rise of Anti-Semitism in St. Louis, 1933–1945," *YIVO Annual of Jewish Social Science*, 14 (1969) 252.

82. Wyman, *Paper Walls*, pp. 71–72.

83. David S. Wyman, *The Abandonment of the Jews* (New York: Pantheon Books, 1984), p. 8; Charles H. Stember, *Jews in the Mind of America* (New York: Basic Books, 1966), p. 138.

84. These are quotes randomly selected from hundreds of letters in four folders marked, "Jewish Refugees and Immigration Laws, November, 1938," "Jewish Refugeees and Immigration Laws, 1938," box 766, folder, "Jewish Ref-

ugees and Immigration Laws, Dec,1938," box 767, and folder, "Nomination of Felix Frankfurter, 1939, Jan. I," box 774, William E. Borah mss., LC.

85. Mrs. Wm. Schweigert (city illegible) to W. E. Borah, December 28, 1938, folder "Jewish Refugees and Immigration Laws, Nov., 1938," box 766, Borah mss.

86. John Roy Carlson, *Under Cover* (New York: E. P. Dutton & Co., 1943), p. 54; Martin E. Marty, *Pilgrims in Their Own Land: 500 Years of Religion in America* (Boston: Little, Brown & Co., 1984), p. 399.

87. Marcus, *Father Coughlin*, p. 156; Stack, *International Conflict*, pp. 129, 142; McCarthy, "The Christian Front," pp. 158, 177, 182, 192; Alson J. Smith, "The Christian Terror," CC, 56 (August 23, 1939), 1017; Gordon W. Allport and Bernard M. Kramer, "Some Roots of Prejudice," *The Journal of Psychology*, 22 (1946), 28; Carlson, *Under Cover*, p. 56; Theodore Irwin, "Inside the 'Christian Front,'" *The Forum*, 103 (March, 1940), 103–104; "Letters to the Editors," *The Nation*, 149 (August 12, 1939), 180; Stanley Feldstein, *The Land That I Show You* (Garden City, NY: Anchor Press/Doubleday, 1978), p. 346; Wallace Stegner, "Who Persecutes Boston?" *The Atlantic Monthly*, 174 (July, 1944), 48; Holbo, "Wheat or What?" p. 735; Tull, *Father Coughlin*, pp. 207–208, 244; Wyman, *Paper Walls*, p. 18. For a succinct discussion of the Christian Front in New York City, see Bayor, *Neighbors in Conflict*, p. 97ff.

88. McCarthy, "Christian Front," p. 76.

89. McCarthy, "Christian Front," p. 90; Irwin, "Inside the 'Christian Front,'" pp. 106, 107; James Wechsler, "The Coughlin Terror," *The Nation*, 149 (September 22, 1939), 96; Smith, "The Christian Terror," p. 1017.

90. Smith, "The Christian Terror," pp. 1017, 1018; McCarthy, "Christian Front," p. 43; Selden Menefee, *Assignment: U.S.A.* (New York: Reynal and Hitchcock, Inc., 1943), p. 11; Irwin, "Inside the 'Christian Front,'" pp. 102, 107; Feldstein, *The Land That I Show You*, p. 346; Marcus, *Father Coughlin*, p. 156.

91. Jay Dolan, *The American Catholic Experience: A History from Colonial Times to the Present* (Garden City, NY: Doubleday & Co., Inc., 1985), p. 404; "Letters to the Editors," p. 180; Stack, *International Conflict*, p. 128; David H. Bennett, *Demagogues in the Depression* (New Brunswick: Rutgers University Press, 1969), p. 280; Smith, "Christian Terror," p. 1017; Wechsler, "The Coughlin Terror," p. 96; Stegner, "Who Persecutes Boston?" p. 50; McCarthy, "Christian Front," p. 36; Irwin, "Inside the 'Christian Front,'" pp. 103, 108; Tull, *Father Coughlin*, pp. 207–208; Marcus, *Father Coughlin*, p. 156; Naomi W. Cohen, *Not Free To Desist: The American Jewish Committee, 1906–1966* (Philadelphia: JPS, 1972), p. 217; Thomas Kessner, *Fiorello H. LaGuardia and the Making of Modern New York* (New York: McGraw-Hill Publishing Co., 1989), p. 523.

92. Stack, *International Conflict*, pp. 58, 98; George Britt, "Poison in the Melting Pot," *The Nation*, 148 (April 1, 1939), 374; Cooney, *American Pope*, p. 101; Jonathan Daneils, *White House Witness, 1942–1945* (Garden City, NY: Doubleday & Co., Inc., 1975), p. 80; McCarthy, "Christian Front," pp. 13, 45, 78, 90, 95, 96, 116, 119–120.

93. Quoted in Kessner, *Fiorello*, p. 523.

94. McCarthy, "Christian Front," p. 100.

95. High, "Star-Spangled Fascists," pp. 7, 72–73; Britt, "Poison in the Melting Pot," p. 374; "The Nazis Are Here," *The Nation*, 148 (March 4, 1939), 253.

96. Britt, "Poison in the Melting Pot," 376.

97. Wechsler, "The Coughlin Terror," p. 96; "Century Marks," CC, 106 (May 3, 1989), 462.

98. Henry Pratt Fairchild, "New Burdens for America," *The Forum*, 101 (June, 1939), 317; "The American Pattern for Anti-Semitism," *Interracial Review*, July, 1939, p. 99; "The Shadow of Anti-Semitism," *The American Magazine*, 128 (November, 1939), 91, 92.

99. Quoted in David Brody, "American Jewry, The Refugees and Immigration Restriction (1932–1942)," in Abraham J. Karp, ed., *The Jewish Experience In America: Selected Studies From The Publications of the American Jewish Historical Society* (5 volumes; New York: KTAV Publishing House, Inc., 1969), V, 336.

100. Kurt Lewin, "Bringing Up the Child," *Menorah Journal*, 28 (Winter, 1940), 29–30, 43–44.

101. "Current Proceedings of Anti-Semitism," pp. 14–15, in folder 4a, box 50, JCRC mss.; "Harvest of Violence," *The Economist*, 189 (October 25, 1958), 324; Arnold M. Eisen, *The Chosen People in America: A Study in Jewish Religious Ideology* (Bloomington: Indiana University Press, 1983), 70; Helga Eugenie Kaplan, "Century of Adjustment: A History of the Akron Jewish Community" (unpublished Ph.D., Department of History, Kent State University, 1978), pp. 437, 441–442; Cohen, *Not Free To Desist*, p. 203; Judith E. Endelman, *The Jewish Community of Indianapolis: 1849–to the Present* (Bloomington: Indiana University Press, 1984), pp. 122, 177; Philip Rosen, Robert Tabak, David Gross, "Philadelphia Jewry, the Holocaust, and the Birth of the Jewish State," in Friedman, ed., *Philadelphia Jewish Life*, p. 31; Brody, "American Jewry," p. 344; see also Solomon Lowenstein, "The American Principle of Tolerance," *Proceedings of the National Conference of Social Work*, 66th Annual Conference, Buffalo, New York, June 18–24, 1939 (New York: Published for the National Conference of Social Work by Columbia University Press, 1939), p. 36.

102. "Anti-Semitism Is Here," p. 168.

103. Lewis Browne, "What Can the Jews Do?" *Virginia Quarterly Review*, 15 (Spring, 1939), 225; Albert Levitan, "Leave the Jewish Problem Alone!" CC, 51 (April 25, 1934), 555.

104. Anonymous, "I Was a Jew," *The Forum*, 103 (March, 1940) 10.

105. Ibid., p. 8; see also "I Married a Jew," *The Atlantic Monthly*, 163 (January, 1939), 38–46; "I Married a Gentile," *The Atlantic Monthly*, 163 (March, 1939), 321–326.

106. Gay Talese, *The Kingdom and the Power* (New York: World Publishing Co., 1969), p. 59; Edward S, Shapiro, *A Time for Healing: American Jewry Since World War II* (Baltimore: The Johns Hopkins University Press, 1992), p. 109; Richard Kluger, *The Paper: The Life and Death of the New York Herald Tribune* (New York: Alfred A. Knopf, 1986), p. 386; David Halberstam, *Summer of '49* (New York: William Morrow and Co., Inc., 1989), p. 151.

107. Morris Freedman, "The Jewish College Student: 1951 Model," *Commentary*, 12 (October, 1951), 311–312; Nitza Rosovsky, *The Jewish Experience at Harvard and Radcliffe* (Cambridge: Harvard University Press, 1986), p. 31; Jean Baer, *The Self-Chosen: "Our Crowd" Is Dead; Long Live Our Crowd* (New York: Arbor House, 1982), p. 254; Natalie Gittelson, "AS: It's Still Around," *Harper's Bazaar*, 105 (February, 1972), 134; Leonard Room, Helen P. Beem, Virginia Harris, "Characteristics of 1,107 Petitioners for Change of Name," *American Sociological Review*, 20 (February, 1955), 34–35; Neil C. Sandberg, *Jewish Life in Los Angeles* (New York: University Press of America, 1986), p. 37; see also Hershel

Shanks, "Irving Shapiro: 'You'll Never Build a Career with a Name like Shapiro,'" *Moment*, 13 (September, 1988), 33ff.

108. James Greenberg, "Our Crowd," *American Film*, 13 (July–August, 1988), 42.

109. Abraham G. Duker, "Emerging Cultural Patterns In American Jewish Life," PAJHS, 39 (June, 1950), 386; Vamberto Morais, *A Short History of Anti-Semitism* (New York: W. W. Norton & Co., 1976), pp. 64 and 262n.

110. "Anti-Semitism Is Here," p. 167; Joseph P. Lash, *Dealers and Dreamers* (New York: Doubleday, 1988), p. 386; Michael E. Parrish, *Felix Frankfurter and His Times: The Reform Years* (New York: The Free Press, 1982), p. 276; Ferdinand M. Isserman, "FDR & Felix Frankfurter," *The National Jewish Monthly*, 80 (November, 1965), 16; *Diaries of Harold Ickes*, pp. 2967–2968, LC; Max Freedman, ed., *Roosevelt and Frankfurter: Their Correspondence—1928–1945* (Boston: Little Brown, 1967), pp. 481–482.

111. Quoted in Scholnick, "The New Deal," p. 160.

112. Karfunkel and Ryley, *The Jewish Seat*, pp. 93, 96.

113. Letters in folder, "Nomination of Felix Frankfurter, 1939, Jan. I," box 774, Borah mss.

114. High, "Star-Spangled Fascists," p. 70; Shuster, "The Conflict Among Catholics," *American Scholar*, p. 6.

115. "Jews in America," *Fortune*, February, 1936, in Leonard Dinnerstein and Frederic Cople Jaher, eds., *The Aliens: A History of Ethnic Minorities in America* (New York: Appleton-Century-Crofts, 1970), p. 230.

116. Milton Steinberg, "First Principles for American Jews," *Contemporary Jewish Journal*, 4 (December, 1941), 587, 588.

117. "Confidential Report on Investigation of Anti-Semitism in the United States in the Spring of 1938," pp. 12, 38, Blaustein; *Public Opinion, 1935–1946*, under the direction of Hadley Cantril (Princeton: Princeton University Press, 1951), p. 385; Wyman, *Paper Walls*, p. 22.

Chapter 7

1. John F. Stack, Jr., *International Conflict in an American City: Boston's Irish, Italians, and Jews, 1935–1944* (Westport, Conn.: Greenwood Press, 1979), p. 118.

2. Wayne S. Cole, *America First: The Battle Against Intervention, 1940–1941* (Madison: University of Wisconsin Press, 1953), pp. 8, 132; Charles J. Tull, *Father Coughlin and the New Deal* (Syracuse University Press, 1965), p. 229; Bert Lanier Stafford, "The Emergence of Anti-Semitism in the America First Committee, 1940–1941" (unpublished M. A. thesis, New School, 1948), p. 12.

3. William E. Leuchtenburg, *Franklin D. Roosevelt and the New Deal* (New York: Harper and Row, 1963), p. 311; Morris Janowitz, "Black Legions on the March," in Daniel Aaron, ed., *America in Crisis* (New York: Alfred A. Knopf, 1952), p. 316; *The Modern View* (St. Louis), May 1, 1941; Stafford, "The Emergence," pp. 10, 13, 14, 27; "*Catholic Telegraph Register*, 1936–1966," folder 12, box 42, JCRC mss.; Victor C. Ferkiss, "Populist Influences on American Fascism," *Western Political Quarterly*, 10 (1957), 368; Cole, *America First*, pp. 137, 138; Wayne S. Cole, *Senator Gerlad P. Nye and American Foreign Relations* (Minneapolis: University of Minnesota Press, 1962), pp. 186–188; AJYB, 44

(September 12, 1942–September 29, 1943), 152–153; Michael Straight, "The Anti-Semitic Conspiracy," *New Republic*, 105 (September 22, 1941), 362.

4. "Civil Liberties: Jew Baiting," *Time*, 38 (September 22, 1941), 17; Ferkiss, "Populist Influence," p. 369.

5. G. G. Miller (San Francisco) to FDR, September 11, 1991, Jack Thompson (Massena, N. Y.) to FDR, September 13, 1941, and Leon Anthony Sielig (St. Louis) to FDR, October 11, 1941, in folder, "Charles A. Lindberg, Aug.–Dec., 1941," OF 92, FDR mss.

6. Quoted in AJYB, 44 (1942–1943), 161.

7. *The Modern View*, September 25, 1941, p. 12, October 2, 1941, p. 7.

8. "Civil Liberties," p. 17; Henry L. Feingold, *A Time for Searching: Entering the Mainstream, 1920–1945* (Baltimore: The Johns Hopkins University Press, 1992), p. 208; Cole, *America First*, pp. 145, 148, 154; Henry L. Feingold, *Zion in America* (New York: Twayne Publishers, Inc., 1974), p. 273.

9. Joseph King (St. Paul) to FDR, n.d., folder, "Charles A. Lindberg, Aug.–Dec., 1941," OF 92, FDR mss.

10. Hildegard Johnson (Stanford, Ct.) to Steve Early [FDR's press secretary], September 22, 1941, folder, "Charles A. Lindbergh, Aug.–Dec., 1941," OF 92, FDR mss.

11. Cole, *Senator Gerald P. Nye*, p. 190; Neil Gabler, *An Empire of Their Own: How The Jews Invented Hollywood* (New York: Crown Publishers, Inc., 1988), p. 347.

12. Wallace Stegner, "Who Persecutes Boston?" *The Atlantic Monthly*, 174 (July, 1944), 51; see also AJYB, 47 (1945–1946), 282.

13. Leonard Dinnerstein, Roger L. Nichols, and David M. Reimers, *Natives and Strangers: Blacks, Indians, and Immigrants in America* (2nd edition; New York: Oxford University Press, 1990), p. 250ff.

14. Stanley High, "Jews, Anti-Semites, and Tyrants," *Harper's Magazine*, 185 (June, 1942), 22ff; Harry W. Flannery, "The Secret Enemy at San Francisco," *Free World*, 10 (July, 1945), 55; Milton Steinberg, "First Principles for American Jews," *Contemporary Jewish Record*, IV (December, 1941), 587; Kirk Douglas, *The Ragman's Son* (London: Pan Books, 1988), p. 78.

15. Charles H. Stember, *Jews in the Mind of America* (New York: Basic Books, 1966), p. 128.

16. *Public Opinion, 1935–1946*, under the editorial direction of Hadley Cantril (Princeton: Princeton University Press, 1951), p. 477.

17. Hazel Gaudet Erskine, "The Polls: Religious Prejudice, Part 2: Anti-Semitism," *Public Opinion Quarterly*, 29 (Winter 1965–1966), 651.

18. Stafford, "The Emergence," pp. 4–5, 7; *New Masses*, 33 (November 21, 1939), 4; Tull, *Father Coughlin*, pp. 229ff., 233.

19. Gerald P. Fogarty, *The Vatican and the American Hierarchy from 1870 to 1965* (Stuttgart: Anton Hiersemann, 1982), p. 278; Sheldon Marcus, *Father Coughlin: The Tumultuous Life of the Priest of the Little Flower* (Boston: Little, Brown and & Co., 1973), p. 216; Tull, *Father Coughlin*, p. 234.

20. *Catholic Telegraph Register*, folder 12, box 42, JCRC mss.

21. Arnold Beichman, "Christian Front Hoodlums Terrorize Boston Jews," *PM* (New York City), October 18, 1943, p. 5; Arnold Beichman, "Saltonstall Orders Anti-Semitic Attacks Investigated," *PM*, October 20, 1943, p. 4; *PM*, November 28, 1943, p. 6; "In re: Commonwealth v. Jacob Hodus, et al., October 11, 1943, October 25, 1943, October 27, 1943," Boston, Box 1788, AJA; Stegner,

"Who Persecutes Boston?" pp. 45, 46; Robert T. Bushnell (Attorney-General of Massachusetts) to Leverett Saltonstall, November 23, 1943, folder, "Anti-Semitism, 1943," in Leverett Saltonstall mss., Massachusetts Historical Society, Boston; "Doing Hitler's Work," *The Crisis*, 51 (February, 1944), 39; "Race Fight May Oust Boston Police Head," New York *Post*, November 11, 1943, p. 4; "Timity Fired as Head of Boston Police," New York *Post*, November 26, 1943, p. 4; T, November 3, 1943, p. 22, November 10, 1943, p. 19; Stack, *International Conflict*, pp. 110, 134–135, 137.

22. *Christian Science Monitor*, October 20, 1943, p. 18.

23. Jonathan Daniels, *White House Witness, 1942–1945* (Garden City, N.Y.: Doubleday & Co., 1975), p. 106; see also Anson Phelps Stokes, *Church and State in the United States* (3 volumes; New York: Harper & Bros., 1950), III, 641.

24. Stack, *International Conflict*, pp. 130, 151.

25. "More Police at Coney Hunt Anti-Semites," *PM*, August 10, 1944, p. 12; "Anti-Semitism, 1939–1944," in folder 1, box 2, of Brooklyn, New York, Brooklyn Jewish Community Council, 1941–1959 mss., collection #164, AJA; Michael Gerald Rapp, "An Historical Overview of Anti-Semitism In Minnesota, 1920–1960—With Particular Emphasis on Minneapolis and St. Paul" (unpublished Ph.D., Department of History, University of Minnesota, 1977), pp. 50–51; Philip Roth, "My Life as a Boy," *The New York Times Book Review*, October 18, 1987, p. 47; Edward C. McCarthy, "The Christian Front Movement in New York City, 1938–1940" (unpublished M. A. thesis, Department of History, Columbia University, 1965), p. 60; Steven M. Lowenstein, *Frankfurt on the Hudson: The German-Jewish Community of Washington Heights, 1933–1983* (Detroit: Wayne State University Press, 1989), p. 222; T, January 11, 1944, p. 24; Samuel G. Friedman, "From Neil Simon: A New Film, a New Play," T, March 24, 1985, II, 31; Ronald H. Bayor, *Neighbors in Conflict: The Irish, Germans, Jews, and Italians of New York City, 1929–1941* (2nd edition; Urbana: University of Illinois Press, 1988), p. 156.

26. Stegner, "Who Persecutes Boston, p. 52; "Anti-Semitism, 1939–1944," folder 1, box 2, Brooklyn JCC mss.; see also John Cooney, *The American Pope: The Life and Times of Francis Cardinal Spellman* (New York: Dell Books, 1986), p. 161; Marcus, *Father Coughlin*, pp. 173–174.

27. Studs Terkel, *"The Good War"* (New York: Pantheon Books, 1984; this edition, Ballantine, 1985), p. 233.

28. Leo P. Ribuffo, *The Old Christian Right* (Philadelphia: Temple University Press, 1983), p. 127; AJYB, 44 (1942–1943), 158, 159.

29. Ribuffo, *Old Christian Right*, pp. 167, 169, 172, 175; Glen Jeansonne, *Gerald L. K. Smith: Minister of Hate* (New Haven: Yale University Press, 1988), pp. 7, 28–29, 44, 74, 92; Burton Alan Boxerman, "Reaction of the St. Louis Jewish Community to Anti-Semitism, 1933–1945" (unpublished Ph.D., St. Louis University, 1967), p. 94; Marshall Field Stevenson, Jr., "Points of Departure, Acts of Resolve: Black-Jewish Relations in Detroit, 1937–1962 (unpublished Ph.D., Department of History, University of Michigan, 1988), pp. 155–56.

30. "Voices of Defeat," *Life*, 12 (April 13, 1942), 88, 90; Rapp, "An Historical Overview," pp. 50–51; AJYB, 46 (September 18, 1944–September 7, 1945), 138ff; AJYB, 47 (1945–1946), 274; "Report of Public Relations Committee, 1942–1943," in "NCRAC, Overt Anti-Semitism, 1941–54," folder 4, box 50, JCRC mss.

31. AJYB, 47 (1945–1946), 274–275; Max Vorspan and Lloyd P. Gartner, *History of the Jews of Los Angeles* (Philadelphia: JPS, 1970), p. 245; Diaries of

Harold Ickes, September 19, 1942, p. 6992, LC; Dominic J. Capeci, Jr., *Race Relations in Wartime Detroit: The Sojourner Truth Housing Controversy of 1942* (Philadelphia: Temple University Press, 1984), p. 64; Susan Dworkin, *Miss America, 1945: Bess Myerson's Own Story* (New York: Newmarket Press, 1987), p. 91; Selden C. Menefee, "What Americans Think," *The Nation*, 156 (May 5, 1943), 765; Roi Ottley, *"New World A-Coming"* (Boston: Houghton Mifflin, 1943; reprinted New York: Arno Press, 1968), p. 128; David Brinkley, *Washington Goes to War* (New York: Ballantine Books, 1989), p. 18; Louis Ruchames, "Danger Signals in the South," *Congress Weekly*, 11 (April 21, 1944), 8, 9.

32. High, "Jews, Anti-Semites, and Tyrants," p. 26.

33. *"Turn To The South,"* edited by Nathan M. Kaganoff and Melvin I. Urofsky (published for the American Jewish Historical Society by the University Press of Virginia, 1979), p. 154; "Not in New England Alone," *The Christian Register* (Unitarian), 124 (May, 1945), 175; Oliver C. Cox, "Race Prejudice and Intolerance—A Distinction," *Social Forces*, 24 (December, 1945), 219.

34. Kenneth Leslie, "Christianity and Anti-Semitism," *New Currents*, 1 (September, 1943), 21.

35. *Christian Science Monitor* clipping, October 27, 1943, in "Miscellaneous, 1942–1945," folder 7, box 6, A. J. Sabath mss., collection #43, AJA; Lois J. Meltzer, "Anti-Semitism in the United States Army During World War II" (unpublished M. A. thesis, Baltimore Hebrew College, 1977), p. 47, copy in box 1066, AJA; "Report on National Conference on Anti-Semitism Under the Auspices of the American Jewish Congress, New York, New York, February 14, 1944," in Anti-Semitism III, AJA.

36. See, for example, John T. Pawlikowski, *Catechetics and Prejudice: How Catholic Teaching Materials View Jews, Protestants and Racial Minorities* (New York: Paulist Press, 1973).

37. Jeansonne, *Gerlad L. K. Smith*, p. 87; AJYB, 44 (1942–1943), 160, 45 (September 30, 1943–September 17, 1944), 183; Ferkiss, "Populist Influences," p. 369, fn. 79; Ickes Diaries, March 7, 1942, p. 6400; *PM*, February 19, 1945, p. 8; Murray Frank, "Washington Notes," *The Chicago Jewish Forum*, 4 (Fall, 1945), 65.

38. T. R. B., "Thorkelsonianism," *The New Republic*, 101 (November 18, 1939), 13; Gordon Sager, "Swastika Over Philadelphia," *Equality*, 1 (July, 1939), 5–6; Ernest Yolkman, *A Legacy of Hate: Anti-Semitism in America* (New York: Franklin Watts, 1982), p. 42; T, June 5, 1941, p. 24; Howard Suber, "Politics and Popular Culture: Hollywood at Bay, 1933–1953," AJH, 68 (June, 1979), 527; *Contemporary Jewish Record*, 6 (March 24 and 29, April 2, 1943); "Report of Public Relations Committee for 1942–1943," June, 1943, "NCRAC—Committee Minutes, 1943–1955," folder 5, box 50, JCRC mss.; "Jewish Community Relations and the NCRAC, 1944–1957," an address by Bernard H. Trager, June 22, 1957, in NCRAC, Nearprint File, Special Topics, AJA.

39. Selden Menefee, *Assignment U.S.* (New York: Reynal and Hitchock, Inc., 1943), p. 14; John Morton Blum, *V Was for Victory: Politics and American Culture During World War II* (New York: Harcourt Brace Jovanovich, 1976), p. 174; Capeci, *Race Relations*, p. 62; K. R. M. Short, "Hollywood Fights Anti-Semitism, 1940–1945," in K. R. M. Short, ed., *Film and Radio Propaganda in World War II* (Knoxville: University of Tennessee Press, 1983), p. 164; "The New Wave of Anti-Semitism," *New Currents*, 1 (June, 1943), 3; Aaron Goldman, "The

Resurgence of Antisemitism in Britain During World War II," JSS, 46 (Winter, 1984), 41; Flannery, "The Secret Enemy," p. 56.

40. Morris Freedman, "The Knickerbocker Case," *Commentary*, 8 (August, 1949), 122, 123.

41. "Report of the Public Relations Committee for 1943," JCRC mss.; Louis Ruchames, review of *Overcoming Antisemitism* by Solomon Andhil Feinberg," *New Currents*, 1 (August, 1943), 23; Harold J. Laski, "A Note On Anti-Semitism," *The New Statesman and Nation*, 25 (February 13, 1943), 107; Short, "Hollywood Fights Anti-Semitism," p. 164.

42. Lloyd P. Gartner, *History of the Jews of Cleveland* (Cleveland: Western Reserve Historical Society, 1978), p. 317; Sidney Bolkosky, *Harmony and Dissonance: Voices of Jewish Identity in Detroit, 1914–1967* (Detroit: Wayne State University Press, 1991), p. 257.

43. Menefee, *Assignment*, p. 102.

44. Maury Paul, "Letter to Editor," *Time*, 41 (April 12, 1943), 10; see also Gordon Allport and Leo Postman, *The Psychology of Rumor* (New York: Henry Holt and Co., 1947), p. 11; Karl M. C. Chworowsky, "Are Jews Like That?" *The Christian Register*, 124 (May, 1945), 191; Flannery, "The Secret Enemy," p. 56.

45. Gartner, *History of the Jews of Cleveland*, p. 317; Louis Ruchames, review of Solomon Andhil Fineberg, *Overcoming Anti-Semitism*, *New Currents*, 1 (August, 1943) 23; AJYB, 46 (1944–1945), p. 141; *Contemporary Jewish Record*, 6 (1943), 396.

46. Hobart B. Pillsbury, Jr., "Raising the Armed Forces," *Armed Forces and Society*, 14 (Fall, 1987), 69–70.

47. Quoted in James Atlas, *Delmore Schwartz: The Life of an American Poet* (New York: Farrar Straus Giroux, 1977), p. 165.

48. George Q. Flynn, *The Mess in Washington: Manpower Mobilization in World War II* (Westport, Conn.: Greenwood Press, 1979), pp. 162–163; AJYB, 46 (1944–1945), 141; *Contemporary Jewish Record*, 6 (1943), 396.

49. Meltzer, "Anti-Semitism," p. 35ff; "Literature and Jews," p. 20, folder 6, box 12, Oscar Cohen mss., AJA; letter from seaman to his mother, February 23, 1945, in "Anti-Semitism, 1945," folder 2, box 2, Brooklyn Jewish Community Council mss.; interview with Edith Schaller, Tucson, April 30, 1990; Friedman, "From Neil Simon," p. 31; Terkell, *"The Good War,"* p. 278.

50. Quoted in George Seldes, *Never Tire of Protesting* (New York: Lyle Stuart, 1968), p. 48.

51. Clipping from *PM*, July 1, 1945, in "Miscellaneous, News Clippings, 1909–1948," folder 1, box 7, A. J. Sabath mss.

52. "Army, Navy Taught Aryanism Until 2 Months Ago," *PM*, June 24, 1945, p. 3.

53. Meltzer, "Anti-Semitism," pp. 34, 74; Roland B. Gittelsohn, "Brothers All?" *The Reconstructionist*, 12 (February 7, 1947), 10; Edward T. Sandrow, "Journal of Jewish Chaplain in the United States Army," recorded in 1944 in "Chaplaincy (Journal) 1944," Ms. Collection #137, AJA.

54. Quoted in Meltzer, "Anti-Semitism," p. 33.

55. Gittelsohn, "Brothers All?" p. 10.

56. Quoted in Sandrow, "Journal of a Jewish Chaplain," p. 21.

57. David Max Eichhorn, "General Patton's Harassment," in Louis Barish, ed., *Rabbis in Uniform* (New York: Jonathan David, 1962), pp. 279, 280; Gittelsohn, "Brothers All?" p. 10; see also *Contemporary Jewish Record*, 6 (1943), 279.

58. Harold U. Ribalow, "The Jewish GI in American Fiction," *The Menorah Journal*, 37 (Spring, 1949), 266–67; Blum, *V Was for Victory*, p. 81ff.

59. Meltzer, "Anti-Semitism," p. 38; Rapp, "An Historical Overview," p. 55; Menefee, "What Americans Think," p. 763; Menefee, *Assignment*, p. 102; Flannery, "The Secret Enemy," p. 57; *PM*, December 22, 1942, June 17, 1945, p. 20; the version distributed by McDonald Trailer Sales in Detroit added a line: "Long Live the Irish!," in folder 6, box 4, of Oscar Cohen mss.

60. Meltzer, "Anti-Semitism," p. 38; Feingold, *A Time for Searching*, p. 258.

61. Meltzer, p. 98.

62. Ibid., p. 99.

63. Ibid., p. 101.

64. Ibid., p. 42.

65. "The Field of Clinical Psychology: Past, Present and Future: A Critical Survey in 1945," *Journal of Clinical Psychology*, 1 (1945), 12–13, 166.

66. T, January 23, 1946, p. 20; Nathan Zukerman, "Democracy vs. Dartmouth," *Congress Weekly*, 12 (September 7, 1945), 10; Dan W. Dodson, "College Quotas and American Democracy," *American Scholar*, 15 (July, 1046), 270.

67. Philip Wylie, "Memorandum on Anti-Semitism," *The American Mercury*, 60 (January, 1945), 68.

68. David L. Cohn, "What Can the Jews Do?" *Saturday Review of Literature*, 28 (January 27, 1945), 9; see also Lewis Browne, "What Can the Jews Do?" *Virginia Quarterly Review*, 15 (Spring, 1939), 225.

69. Wyman, *Paper Walls*, pp. 75, 94–95; Gerald S. Berman, "Reaction to the Resettlement of World War II Refugees in Alaska," JSS, 44 (Summer–Fall, 1982), 274–275.

70. Robert Dallek, *Franklin D. Roosevelt and American Foreign Policy, 1932–1945* (New York: Oxford University Press, 1979), p. 444ff.

71. H. G. Nicholas, ed., *Wartime Despatches, 1941–1945* (London: Weidenfeld and Nicolson, 1981), p. 117; Dallek, *Franklin D. Roosevelt*, p. 446.

72. Sheldon Morris Neuringer, "American Jewry and United States Immigration Policy, 1881–1953" (unpublished Ph.D., Department of History, University of Wisconsin, 1969), pp. 260–261; T, November 10, 1943, p. 19; Henry Morgenthau, Jr., "The Refugee Run-Around," *Collier's*, November 1, 1947, p. 65; Leonard Dinnerstein, *America and the Survivors of the Holocaust* (New York: Columbia University Press, 1982), pp. 4–5.

73. "Memo to Stephen Early," March 8, 1944, subject file, "Refugees," box 158, PSF, FDR mss.

74. Grace Tully, "Memorandum for the President," May 22, 1944, in subject file, "Refugees," box 158, PSF, FDR mss.

75. Sharon R. Lowenstein, *Token Refuge* (Bloomington: Indiana University Press, 1986), p. 111.

76. Neuringer, "American Jewry," pp. 231, 235, 237, 271–272; Dinnerstein, *America and the Survivors*, p. 5; see also, Zosa Szajkowski, "The Attitudes of American Jews to Refugees from Germany in the 1930's," AJHQ, 61 (December, 1971), 101–102, 111–112; Robert Michael, "America and the Holocaust," *Midstream*, 31 (February, 1985), 14–15.

77. Alvin Johnson to Nelson Glueck, March 9, 1960, Correspondence File, AJA.

78. Barbara Tuchman, "When Assimilation Was a Norm," *The Jerusalem Post*, February 6, 1977, p. 9.

79. Flannery, "The Secret Enemy," p. 5.

80. Leonard Dinnerstein, "What Should American Jews Have Done to Rescue Their European Brethren?" in *Simon Wiesenthal Center Annual*, 3 (1986), 278–279; Feingold, *A Time for Searching*, pp. 225, 250 and n. 1 on p. 295.

81. Feingold, *A Time for Searching*, p. 250.

82. David S. Wyman, *The Abandonment of the Jews* (New York: Pantheon Books, 1984), p. 9.

83. Dorothy Thompson to Laura Z. Hobson, August 28, 1944, in Scrapbook, box 21, Laura Z. Hobson, mss., Columbia University.

84. British Public Record Office, Kew, FO 371 44538/AN3069.

85. Cited in Leonard Dinnerstein, "Anti-Semitism Exposed and Attacked, 1945–1950," AJH, 71 (September, 1981), 135; Samuel H. Flowerman and Marie Jahoda, "Polls on Anti-Semitism," *Commentary*, 1 (April, 1946), 83.

86. "Minutes of the Administrative Committee," November 4, 1947, AJC.

87. Jacob M. Sable, "Some American Jewish Organizational Efforts to Combat Anti-Semitism," (unpublished Ph.D., Yeshiva University, 1964), p. 293.

88. Thomas David Mantel, "The Anti-Defamation League of B'nai B'rith" (unpublished honor's thesis, Department of Government, Harvard University, 1950), pp. 26, 27, copy in Small Collection, AJA.

89. Frederick A. Lazin, "The Response of the American Jewish Committee to the Crisis of German Jewry, 1933–1939," AJH, 68 (March, 1979), 290; Morris Frommer, "The American Jewish Congress, a History, 1914–1950" (unpublished Ph.D., Department of History, Ohio State University, 1978), p. 518; Mantel, "The Anti-Defamation League," p. 26; AJYB, 47 (1945–46), 280–281; Vorspan and Gartner, *History of the Jews of Los Angeles*, pp. 206, 222–223; Gabler, *An Empire of their Own*, p. 341; Marc Lee Raphael, *Jews and Judaism In A Midwestern Community: Columbus, Ohio, 1840–1975* (Columbus: Ohio Historical Society, 1979), p. 435, n. 31; Judith E. Endelman, *The Jewish Community of Indianapolis: 1849 to the Present* (Bloomington: Indiana University Press, 1984), p. 178; Boxerman, "Reaction of the St. Louis Jewish Community," pp. 234–235; Burton A. Boxerman, "The St. Louis Jewish Coordinating Council: Its Formative Years," *Missouri Historical Review*, 65 (October, 1970), 53–54; Frank Rich, "Richmond's Enigmatic Ethnics," p. 9, in folder #11, box 20, Oscar Cohen mss., AJA; Paul Lyons, "Philadelphia Jews and Radicalism," in Murray Friedman, ed., *Philadelphia Jewish Life, 1940–1985* (Ardmore, PA: Seth Press, 1986), p. 112; Elaine H. Maas, "The Jews of Houston: An Ethnographic Study" (unpublished Ph.D., Department of Sociology," Rice University, 1973), p. 54; Rapp, "An Historical Overview," pp. 205–207.

90. Haim Genizi, "American Interfaith Cooperation on Behalf of Refugees from Naziism, 1933–1945," AJH, 70 (March, 1981), 350ff., 360; Mantel, "The Anti-Defamation League," p. 26; AJYB, 47 (1945–1946), 280–281; Richard C. Rothschild, "The American Jewish Committee's Fight Against Anti-Semitism, 1938–1950" (unpublished essay), Small Collections, AJA; Dinnerstein, *America and the Survivors*, p. 119.

91. Wylie, "Memorandum on Anti-Semitism," p. 66ff.

92. Neuringer, "American Jewry," p. 231.

93. David W. Petegorsky, "The Anatomy of Hatred," *Contemporary Jewish Record*, V (June, 1942), 325.

94. Feingold, *A Time for Searching*, p. 261; Justine Wise Polier, "The Middle Years," *Congress Bi-Weekly*, 35 (May 6, 1968), 11.

Chapter 8

1. Naomi W. Cohen, *Not Free To Desist: The American Jewish Committee, 1906–1966* (Philadelphia: JPS, 1972), p. 345; Harold S. Wechsler and Paul Ritterband, "Jewish Learning in American Universities: The Literature of the Field," *Modern Judaism*, 3 (October, 1983), 274; Murray Friedman, "Black Anti-Semitism on the Rise," *Commentary*, 68 (October, 1979), 31; NCRAC, "Overt Anti-Semitism, 1941–1954," folder 5, box 50, "Discrimination—Country and Sports Clubs, 1949–1950," folder 15, box 14, "Annual Report for 1950–1951—The Jewish Community Council of Cleveland," p. 7, folder, "A-S, 1950," "AS, March–June, 1951," NCRAC, "Minutes of the Committee on Overt Anti-Semitism," February 9, 1951, box 4, "1947 Survey of Anti-Semitism, Cincinnati Area," folder 1, box 3, JCRC mss.; C. H. Stember, *Jews in the Minds of America* (New York: Basic Books, 1966), p. 302; Minutes of the AJC Administrative Committee, March 1, 1955; Michael Rapp, "An Historical Overview of Anti-Semitism In Minnesota, 1920–1960—With Particular Emphasis on Minneapolis and St. Paul" (unpublished Ph.D., Department of History, University of Minnesota, 1977), pp. 171, 178, 179; Arthur Liebman, "The Ties That Bind: The Jewish Support for the Left in the United States," AJHQ, 66 (December, 1976), 318; Judith E. Endelman, *The Jewish Community of Indianapolis: 1849 to the Present* (Bloomington: Indiana University Press, 1984), p. 211; "Anti-Semitism," *Life*, 23 (December 1, 1947) 44; William Attwood, "The Position of the Jews in America Today," Leonard Dinnerstein, "Anti-Semitism Exposed and Attacked, 1945–1950," AJH, 71 (September, 1981), 146; *Look*, 19 (November 29, 1955), 34; Herbert J. Gans, "The Future of American Jewry," *Commentary*, 21 (June, 1956), 557; Oscar Handlin, "The American Jewish Committee," *Commentary*, 23 (January, 1957), 9; see also Emory S. Bogardus, "Racial Distance Changes in the United States During The Past Thirty Years," *Sociology and Social Research*, 43 (1958), 130; Samuel I. Cohen, "Jewish-American Defense Agencies: A Study," *The Chicago Jewish Forum*, 20 (Fall, 1961), 2.

2. Minutes of the AJC Administrative Committee, March 1, 1955, p. 8; see also Friedman, "Black Anti-Semitism," p. 31.

3. Hazel Gaudet Erskine, "The Polls: Religious Prejudice, Part 2: Anti-Semitism," *Public Opinion Quarterly*, 29 (Winter 1965–1966), 651.

4. Joe McCarthy, "GI Vision of a Better America," *New York Times Magazine*, August 5, 1945, p. 10; E. Digby Baltzell, "Foreword," in Murray Friedman, ed., *Philadelphia Jewish Life, 1940–1985* (Ardmore, Penn.: Seth Press, 1986), p. ix.

5. Attwood, "The Position of the Jews," p. 34; "My Best Friend's a Jew," *The Economist*, 222 (February 25, 1967), 730.

6. This was certainly not true, however, of ideological differences as evidenced by the hysteria of the "Red Scare" that lasted from approximately 1946 or 1947 through the mid-1950s, and even later in some parts of the country.

7. T, November 12, 1949, p. 4.

8. Dore Schary, "Letter from a Movie-Maker," *Commentary*, 4 (October, 1947), 344; Paul Johnson, *A History of the Jews* (New York: Harper and Row, 1987), p. 31; Lester D. Friedman, *The Jewish Image in American Film* (Secaucus, NJ: Citadel Press, 1987), p. 50.

9. Elliott E. Cohen, "Letter to the Movie-Maker," *Commentary*, 4 (August, 1947), 111; John Mason Brown, "If You Prick Us," *Saturday Review of Literature*, 30 (December 6, 1947), 69.

10. Friedman, *The Jewish Image*, pp. 142, 145.

11. T. W. Adorno, Else Frenkel-Brunswick, Daniel J. Levinson, and R. Nevitt Sanford, *The Authoritarian Personality* (abridged edition; New York: W. W. Norton, 1969; originally published in 1950); Henry L. Feingold, *A Time for Searching: Entering the Mainstream, 1920–1945* (Baltimore: The Johns Hopkins University Press, 1992), p. 256; Edward S. Shapiro, *A Time for Healing: American Jewry since World War II* (Baltimore: The Johns Hopkins Press, 1992), pp. 32–33.

12. AJYB, 47 (1945–1946), 285, 48 (1946–1947), 191, 51 (1950), 103; Milton R. Konvitz, *A Century of Civil Rights; With a Study of State Laws Against Discrimination* by Theodore Leskes (New York: Cornell University Press, 1961), pp. 199, 201; Ida Libert Uchill, *Pioneers, Peddlers, and Tsadikim* (Denver: Sage Books, 1957), p. 164.

13. Susan Dworkin, *Miss America, 1945: Bess Myerson's Own Story* (New York: Newmarket Press, 1987), p. 180; "Fair Employment Practices—Various Cities, 1946–1952," folder 8, box 14, "*Cincinnati Enquirer*—Correspondence and Clippings, 1944–1974," folder 1, box 35, JCRC mss.; "Study of Manhattan Commercial Employment Agencies," folder 1a, box 8, Brooklyn Jewish Community Council (1941–1949) mss., Manuscript Collection #164, AJA.

14. Vernon Keenan and Willard A. Kerr, "Unfair Employment Practices as Viewed by Private Employment Counselors," *Journal of Applied Psychology*, 36 (December, 1952), 362; Abraham K. Korman, *The Outsiders: Jews and Corporate America* (Lexington, Mass.: Lexington Books, 1988), pp. 48–49; Arthur Gilbert, "The Contemporary Jew in America," *Thought*, 43 (Summer, 1968), 215; Benjamin R. Epstein, "Anti-Semitism in the United States: 1957 Report," *Journal of Human Relations*, 6 (Autumn, 1957), 55; AJYB, 53 (1952), 94.

15. "The Jewish Law Student and New York Jobs—Discriminatory Effects in Law Firm Hiring Practices," *Yale Law Journal*, 73 (March, 1964), 626; Max Vorspan and Lloyd P. Gartner, *History of the Jews of Los Angeles* (Philadelphia: JPS, 1970), 245; "Fair Employment News," Chicago, November 15, 1954 in folder, "Anti-Semitism, Feb-Dec, 1954," box 5, Anti-Semitism (1921–1959), Nearprint File, Special Topics, AJA; Epstein, "Anti-Semitism," p. 54; Rapp, "An Historical Overview," p. 190.

16. Carey McWilliams, "West Coast Letter," *The Chicago Jewish Forum*, 9 (Winter, 1950–1951), 144; Murray Friedman, "Introduction: From Outsiders to Insiders?" in Friedman, *Philadelphia Jewish Life*, p. 18; Cohen, *Not Free to Desist*, p. 416; *Time*, 85 (June 25, 1965), 34; E. Digby Baltzell, *The Protestant Establishment: Aristocracy and Caste in America* (London: Secker and Warburg, 1964), p. 321; Minutes of the AJC Administrative Committee, February 4, 1964, p. 3; Minutes of the AJC Board of Governors, October 4, 1966, p. 6; Steven Applebaum, "Eliminating Bias Against Jews in the Executive Suite, 1960–1970," Philadelphia Chapter, 1971, AJC; Vance Packard, *The Pyramid Climbers* (New York: McGraw-Hill Book Co., 1962), p. 37; Daniel Pope and William Toll, "We Tried Harder: Jews in American Advertising," AJH, 72 (September, 1982), pp. 34–35, 40.

17. Paul A. Freund, "Oral History Memoir," Tape 6, pp. 185–186, William E. Wiener Oral History Library, AJC.

18. Spencer Klaw, "The Wall Street Lawyers," *Fortune*, 57 (February, 1958), 194.

19. "A Survey of the Employment Experiences of Law School Graduates of Chicago, Columbia, Harvard, and Yale Universities," folder 1b, box 8, Brooklyn Jewish Community Council (1941–1959) mss., AJA.

20. The Jewish Law Student," p. 635; Klaw, "Wall Street Lawyers," p. 192; Jerold S. Auerbach, "From Rags to Robes: The Legal Profession, Social Mobility and the American Jewish Experience," AJHQ, 66 (December, 1966), 273.

21. "Current Proceedings of Anti-Semitism," p. 8, cosponsored by ADL and National Council of Churches of Christ, October, 1962, folder 4a, box 50, JCRC mss.; Gilbert, "The Contemporary Jew," p. 214; "Jews and Gentlemen," *The Spectator*, 203 (July 25, 1959), 90; Korman, *The Outsiders*, p. 34; Baltzell, *Protestant Establishment*, pp. 36, 357, 368, 370; Vance Packard, *The Status Seekers* (New York: David McKay Co., Inc., 1954), p. 188; Osborn Elliott, *Men at the Top* (New York: Harper & Bros., 1959), p. 162; Mark Rich, "Richmond's Enigmatic Ethnics," *The Richmond Mercury*, November 8, 1972 in Richmond, Virginia, folder 11, box 20, Oscar Cohen mss.; Whitney H. Gordon, "Jews and Gentiles in Middletown—1961," *American Jewish Archives*, 18 (April, 1966), 52; Walda Katz Fishman and Richard L. Zeigenhaft, "Jews and the New Orleans Economic and Social Elites," JSS, 44 (Summer–Fall, 1982), 296.

22. Packard, *The Status Seekers*, p. 265.

23. Cohen, *Not Free to Desist*, p. 412; Packard, *The Status Seekers*, p. 91; Benjamin R. Epstein and Arnold Forster, "*Some of My Best Friends . . .*" (New York: Farrar, Straus and Cudahy, 1962), pp. 102, 103; "Restricted," *Jewish Digest*, 11 (December, 1965), 1; Harry Gersh, "Gentlemen's Agreement in Bronxville," *Commentary*, 27 (February, 1959), 109; "Current Proceedings of Anti-Semitism," folder 4a, box 50, JCRC mss.; Minutes of the AJC Administrative Board, January 17, 1964, p. 3.

24. Quoted in Packard, *The Pyramid Climbers*, p. 39.

25. "Michigan: Grosse Pointe's Gross Points," *Time*, 75 (April 25, 1960), 25.

26. "Current Proceedings of Anti-Semitism"; Epstein and Forster, "*Some of My Best Friends*," pp. 107, 113; Steven Brill, *The Teamsters* (New York: Pocket Books, 1979), p. 33; Robert Lacey, *Ford: The Men and the Machine* (Boston: Little, Brown, 1986), p. 533; Baltzell, *Protestant Establishment*, p. 126.

27. Rapp, "An Historical Overview," pp. 189–190.

28. T, April 12, 1950, p. 22; Epstein and Forster, "*Some of My Best Friends*," pp. 45, 50; "Current Proceedings of Anti-Semitism," pp. 7–8.

29. Frank Kingdom, "Discrimination in Medical Colleges," *American Mercury*, 56 (October, 1945), 394; Walter R. Hart, "Anti-Semitism in N. Y. Medical Schools," *American Mercury*, 65 (July, 1947), 56; Edward N. Saveth, "Discrimination in the Colleges Dies Hard," *Commentary*, 9 (February, 1950), 119–120; T January 23, 1946, p. 20, December 24, 1946, p. 1, September 29, 1947, p. 8.

30. Minutes of the AJC Administrative Committee, March 4, 1947, p. 4.

31. Hart, "Anti-Semitism in N. Y. Medical Schools," p. 56; T, January 23, 1946, p. 1, January 28, 1946, p. 10; January 31, 1946, p. 23; March 13, 1946, p. 48.

32. T, January 13, 1948, p. 1, February 17, 1948, p. 1, February 22, 1948, IV, 9.

33. T, March 3, 1947, p. 3, December 27, 1947, p. 13, January 4, 1948, p. 1, January 13, 1948, p. 1, January 17, 1948, p. 1, April 6, 1948, pp. 1, 15; Saveth, "Discrimination in the Colleges," p. 121; Konvitz, *A Century of Civil Rights*, pp. 201–202, 225, 227–229; Harold S. Wechsler, *The Qualified Student* (New York: John Wiley, 1977), p. 194 ff.

34. Frank K. Shuttleworth, "Discrimination In College Opportunities and Admissions," *School and Society*, 74 (December 22, 1951), 398ff.

35. Dan A. Oren, *Joining the Club: A History of Jews and Yale* (New Haven: Yale University Press, 1985), pp. 165, 167.

36. Edward C. McDonagh, "Status Level of American Jews," *Sociology and Social Research*, 32 (July, 1948), 949; "A Preliminary Analysis of Discrimination Against Jewish Applicants for Admission to Medical Schools in New York State," p. 19, folder 8a, box 7, Brooklyn Jewish Community Council mss., AJA; Will Maslow, "The Postwar Years," *Congress Bi-Weekly*, 35 (May 6, 1968), 14.

37. Epstein and Forster, *"Some of My Best Friends,"* p. 169ff.

38. T, November 16, 1949, p. 34, November 27, 1949, p. 1; "College Fraternities Are Lowering Race Bars," CC, 68 (July 11, 1951), 812; Harold Whitman, "The College Fraternity Crisis," *Collier's*, 123 (January 8, 1949), 9, 65, (January 15, 1949), 34–35.

39. John Cogley, "A Program for Tolerance," *Commonweal*, 50 (June 10, 1949), 217; Robert W. Ross, *So It Was True: The American Protestant Press and the Nazi Persecution of the Jews* (Minneapolis, 1980), pp. xiii, xv–xvi.

40. Camilia G. Booth to William G. Stratton, May 26, 1947, box 23, Stratton mss., Illinois State Historical Society, Springfield, Illinois; G. D. Minick to Tom Connally, May 26, 1946, box 185, Conally mss., LC; Henry G. Fuller to Pat McCarran, January 6, 1950, S/MC/9/1, Itemized File, McCarran mss., Nevada State Archives, Carson City, Nevada; George Perrine to William Stratton, April 18, 1947, quoted in David Kenney, *A Political Passage: The Career of Stratton of Illinois* (Carbondale: Southern Illinois Press, 1990), pp. 52–53.

41. Leonard Dinnerstein, "America, Britain, and Palestine: The Anglo-American Committee of Inquiry and the Displaced Persons, 1945–46," *Diplomatic History*, 4 (Summer, 1980), 283–302 passim.

42. Leonard Dinnerstein, *America and the Survivors of the Holoaust: The Evolution of a United States Displaced Persons Policy, 1945–1950* (New York: Columbia University Press, 1982), chapters 5–7.

43. Ibid., chapters 7 and 9.

44. See David M. Reimers, *Still the Golden Door* (2nd edition; New York: Columbia University Press, 1992), pp. 159ff, 175ff, and 205–206.

45. Leonard Dinnerstein, "Anti-Semitism Exposed and Attacked, 1945–1950," AJH, 71 (September, 1981), 148–149.

46. T, January 8, 1948, p. 40, October 27, 1948, p. 31, April 12, 1950, p. 22; AJYB, 48 (1946–1947), 190, 193, 49 (1947–1948), 188, 196, 51 (1950), 116; Carey McWilliams, "Minneapolis: The Curious Twin," *Common Ground*, VII (Autumn, 1946), 61; Rapp, "An Historical Overview," p. 178; Aaron Rosenbaum, "Woo and Woe on the Campaign Trail," *Moment*, 6 (January–February, 1981), 52.

47. "Prejudice Subsides," CC, 67 (April 19, 1950), 483; "Incidents 1950–1959," folder 1, box 6, JCRC mss.; Carey McWilliams, *A Mask for Privilege: Anti-Semitism in America* (Boston: Little, Brown & Co., 1948), p. 161; talk by Robert M. McIver, "The Swastika Incidents: Counteraction Via Work with Youth and Youth Groups, pp. 16–17, in NCRAC, "Report of the Plenary Session," June 23–26, 1960, NCRAC, Nearprint File, Special Topics, AJA; Arnold Forster and Benjamin R. Epstein, *The Trouble-Makers* (Garden City, N.Y.: Doubleday, 1952), p. 284ff.

48. Harold E. Quinley and Charles Y. Glock, *Anti-Semitism in America* (New Brunswick: Transaction Books, 1983), pp. ix, xiii; McIver, "The Swastika Incidents," pp. 16–17; Stanley A. Rudin, "The New Anti-Semites: Our Generation of Psychopaths," n. d. (*ca.* 1962), folder 8, box 3, Cohen mss.; Minutes of

the AJC Administrative Committee, June 6, 1961, p. 2; "Anti-Semitic Vandalism," *CC*, 77 (February 17, 1960), 198.

49. Quoted in "Anti-Semitic Vandalism," p. 200.

50. "Anti-Semitic Activity in the United States: A Report and Appraisal," AJC, 1954, Blaustein Library; Vorspan and Gartner, *History of the Jews of Los Angeles*, p. 241.

51. Walter Goodman, *The Committee* (New York: Farrar, Straus and Giroux, 1968), p. 181; David Levering Lewis, "Henry Ford's Anti-Semitism and Its Repercussions," *Michigan Jewish History*, 24 (1984), 6; Vorspan and Gartner, *History of the Jews of Los Angeles*, p. 240; speech by Harley Kilgore in the United States Senate, June 1, 1948, on "A Cruel and Vicious Forgery," in box 3 (1945–1949) of "Anti-Semitism (1921–1959)," Nearprint File, Special Topics, AJA; Burton Alan Boxerman, "Reaction of the St. Louis Jewish Community to Anti-Semitism, 1933–1945" (unpublished Ph. D., Department of History, St. Louis University, 1967), p. 95; Carey McWilliams, "West Coast Letter," *The Chicago Jewish Forum*, 7 (Summer, 1949), 267–268; Arnold Lee Levine, "Anti-Semitism in the United States, 1945 to Date," October, 1968, pp. 1–2, 9, Box #1533, Anti-Semitism III and IV, AJA; James Graham Cook, *The Segregationists* (New York: Appleton-Century-Crofts, 1962), pp. 153–154; "American Jewish Committee [Anti-Semitic Canards]," folder 5, box 1, "AJC (Quarantine of Anti-Semitic Agitators), 1946–67, 72," folder 8, box 2, Solomon Andhil Fineberg mss., Collection #149, AJA; John Benedict, "Who Controls the Whiskey Trust?" *American Mercury*, 89 (December, 1959), 3; "Termites of the Cross, Part VI: Merchants of Hate," *American Mercury*, 90 (February, 1960), 96ff; "Nazi Party, 1960–64," folder 8, box 4, JCRC mss.

52. "NCRAC, Overt Anti-Semitism, 1941–54," folder 5, box 50, JCRC mss; Mark Goldman, *High Hopes: The Rise and Decline of Buffalo, New York* (Albany: SUNY Press, 1983), p. 247.

53. See second page of this chapter.

54. Arnold M. Eisen, *The Chosen People in America: A Study in Jewish Religious Ideology* (Bloomington: Indiana University Press, 1983), p. 143; Henry Cohen, "Jewish Life and Thought in an Academic Community," *American Jewish Archives*, 14 (May, 1962), 108, 119–120; Gordon, "Jews and Gentiles," pp. 43, 47; Benjamin B. Ringer, *The Edge of Friendliness* (New York: Basic Books, 1967), pp. 139–140.

55. Louise Laser (pseudonym), "The Only Jewish Family in Town," *Commentary*, 28 (December, 1959), 493, 495.

56. Sol Lesser, "Two Oral Histories," 1971, p. 1ff., Small Collections, AJA.

57. "American Jewish Committee [Anti-Semitic Canards]," folder 4, box 1, Fineberg mss.

58. Will Herberg, "Anti-Semitism Today," *Commonweal*, 60 (July 16, 1954), 360; Attwood, "The Position of the Jews," p. 33; Epstein, "Anti-Semitism," p. 53.

59. Herberg, "Anti-Semitism Today," p. 360ff; Will Herberg, "Religious Group Conflict in America," in Robert Lee and Martin E. Marty, eds., *Religion and Social Conflict* (New York: Oxford University Press, 1964), p. 156.

60. Kenneth E. Burnham, John F. Connors, III, and Richard C. Leonard, "Religious Affiliation, Church Attendance, Religious Education and Student Attitudes Toward Race," *Sociological Analysis*, 30 (Winter, 1969), 243; Gary M. Maranell, "An Examination of Some Religious and Political Attitude Correlates

of Bigotry," *Social Forces*, 45 (March, 1967), 361–362; Gordon W. Allport and Bernard M. Kramer, "Some Roots of Prejudice," *The Journal of Psychology*, 22 (1946), 26; E. Terry Prothro and John A. Jensen, "Group Differences in Ethnic Attitudes of Louisiana College Students," *Sociology and Social Research*, 34 (March, 1950), 253; Barbara Sandra Blum and John H. Mann, "The Effect of Religious Membership on Religious Prejudice," *Journal of Social Psychology*, 52 (1960), 99.

61. Charles Y. Glock and Rodney Stark, *Christian Beliefs and Anti-Semitism* (New York: Harper and Row, 1966), pp. 113, 207, 212; "The Unforgiving," *Newsweek*, 67 (May 2, 1966).

62. Quotes are from "Judaism and Christian Education," p. 1410; see also O'Gara, "Christian Anti-Semitism," p. 252; Thomas Cooper, "On the Roots of Bias," *CC*, 81 (June 10, 1964), 770; James Brown, "Christian Teaching and Anti-Semitism," *Commentary*, 24 (December, 1957), 496; Eugene J. Fisher, "Research on Christian Teaching Concerning Jews and Judaism: Past Research and Present Needs," *Journal of Ecumenical Studies*, 21 (Summer, 1984), 427; Dean R. Hoge and Jackson W. Carroll, "Christian Beliefs, Nonreligious Factors, and Anti-Semitism," *Social Forces*, 53 (June, 1975), 581; Helga Eugenie Kaplan, "Century of Adjustment: A History of the Akron Jewish Community" (unpublished Ph.D., Department of History, Kent State University, 1978), p. 467; Judith H. Banki, "The Image of Jews in Christian Teaching," *Journal of Ecumenical Studies*, 21 (Summer, 1984), 439; Marty, *The Righteous Empire*, p. 253; Minutes of the AJC Board of Governors, January 3, 1966, p. 1.

63. Bernhard E. Olson, *Faith and Prejudice* (New Haven: Yale University Press, 1963), pp. 24, 84, 178, 204, 257, 258–259, 272–273; "Judaism and Christian Education: The Strober Report," *CC*, 86 (November 5, 1969), 1410; John T. Pawlikowski, *Catechetics and Prejudice: How Catholic Teaching Materials View Jews, Protestants and Racial Minorities* (New York: Paulist Press, 1973), p. 19; Martin E. Marty, *The Righteous Empire: The Protestant Experience in America* (New York: The Dial Press, 1970), p. 253.

64. Pawlikowski, *Catechetics and Prejudice*, pp. 25, 81, 82, 87; James O'Gara, "Christian Anti-Semitism," *Commonweal*, 80 (May 22, 1964), 252.

65. Jay P. Dolan, *The American Catholic Experience: A History from Colonial Times to the Present* (Garden City, N.Y.: Doubleday and Co., Inc., 1985), p. 424; William Roddy, "How The Jews Changed Catholic Thinking," *Look*, 30 (January 25, 1966), 19; Paul Johnson, *Pope John XXIII* (Boston: Little, Brown & Co., 1974), p. 223; clipping from *Catholic Telegraph-Register*, September 25, 1959, in folder 12, box 42, JCRC mss.

66. Jakob Jocz, "The Church of Rome and the Jews," *International Review of Missions*, 51 (July, 1962), 317; John P. Sheerin, "The Myth of Black Anti-Semitism," *The Catholic World*, 209 (May, 1969), 51.

67. Alan Dundes, "The Ritual Murder or Blood Libel Legend: A Study of the Anti-Semitic Victimization Through Projective Inversion," in Alan Dundes, ed., *The Blood Libel Legend: A Casebook in Anti-Semitic Folklore*, (Madison: University of Wisconsin Press, 1991), p. 341.

68. Quotes cited in Dundes, "The Ritual Murder," p. 342.

69. Dundes, "Ritual Murder," p. 343.

70. Quoted in AI, October 21, 1965, in *"Catholic Telegraph Register, 1936–1966,"* folder 12, box 42, JCRC mss.; see also Sheerin, "The Myth of Black Anti-

semitism," p. 51; Jack Winocour, "The Pope and the Jews," *New Statesman*, 77 (January 17, 1969), 68.

71. Leo Pfeffer, *Church, State, and Freedom* (revised edition; Boston: Beacon Press, 1967), p. 280.

72. Minutes of the AJC Administrative Board, October 3, 1961, p. 3; Minutes of the AJC Board of Governors, January 5, 1965, p. 3, February 1, 1966, pp. 2–3; clipping from *Catholic Telegraph Register*, March 19, 1965, in "Catholic/ Jewish Relations, 1963–1972," folder 17, box 42, JCRC mss.; Charles D. Ward, "Anti-Semitism at College: Changes Since Vatican II," *Journal for the Scientific Study of Religion*, 12 (March, 1973), 88.

73. Minutes of the AJC Board of Governors, March 1, 1966, p. 1.

74. Feinberg, "Religion in the Public Schools," in "AJC (Religion in the Public Schools) 1955–64, 1973," in folder 6, box 3 of Fineberg mss., AJA.

75. Endelman, *The Jewish Community*, p. 216.

76. Naomi W. Cohen, *Jews in Christian America: The Pursuit of Religious Equality* (New York: Oxford University Press, 1992), pp. 165–171, 197–201.

77. Stember, *Jews in the Mind of America*, pp. 208, 270.

78. "Combatting Anti-Semitism," folder 10, box 5, Cohen Mss.

79. *Time*, 85 (June 25, 1965), 34; "NCRAC, 'Plenary Session Report,' June 24–27, 1965," NCRAC, Nearprint File, Special Topics, AJA.

80. Gilbert, "The Contemporary Jew," p. 217; Stember, *Jews in the Mind of America*, p. 106; John Shelton Reed, *The Enduring South* (Chapel Hill: University of North Carolina Press, 1972) p. 64.

81. Stember, *Jews in the Mind of America*, pp. 239–240.

82. Marty, *The Righteous Empire*, p. 253; Bruno Bettelheim and Morris Janowitz, *Social Change and Prejudice* (New York: The Free Press of Glencoe, 1964), pp. 4–10; "My Best Friend's a Jew," p. 730; "Shadow of the Pogrom," *The Economist*, 230 (February 15, 1969), 45–46; A. Abbot Rosen, "Anti-Semitism in America Today," p. 2A in "New Anti-Semitism (American Context)," folder 7, box 13, Cohen mss.; Quinley and Glock, *Anti-Semitism in America*, p. 23; Richard Robbins, "'Native Sons,'" *The Catholic World*, November, 1967, p. 81.

83. "Incident in Suburbia," *Newsweek*, 57 (January 23, 1961), 58; *Time*, 77 (January 20, 1961), 51; Carol A. O'Connor, *A Sort of Utopia: Scarsdale, 1891–1981* (Albany: SUNY Press, 1983), pp. 188–189; "Rector Resigns Under Pressure," CC, 80 (February 6, 1963), 165.

84. NCRAC, "A Report from the Proceedings of the Plenary Session," June 23–26, 1966, #3, Oscar Cohen, "Report on the Five-Year Study of Anti-Semitism," pp. 4–11, NCRAC, Nearprint File, Special Topics, AJA.

85. *Time*, 85 (June 25, 1965), 34.

86. "The Trouble in Wayne," *Newsweek*, 69 (February 27, 1967), 80.

87. Rodney Stark and Stephen Steinberg, "Jews and Christians in Suburbia," *Harper's Magazine*, 235 (August, 1967), 73; T, February 9, 1967, p. 41.

88. Stark and Steinberg, "Jews and Christians," pp. 73, 76; "The Trouble in Wayne," p. 80; T, February 9, 1967, p. 41, February 10, 1967, pp. 37, 40, February 12, 1967, p. 47.

89. T, February 15, 1967, p. 1.

90. T, February 15, 1967, p. 32.

91. T, February 15, 1967, p. 32.

92. "The Gentile Beast," CC, 84 (March 8, 1967), 300; T, February 16, 1967, p. 38.

93. T, February 25, 1967, p. 26.

94. Minutes of the AJC Board of Governors, April 4, 1967, p. 3.

95. Stark and Steinberg, "Jews and Christians," pp. 73–74; "My Best Friend's a Jew," p. 730.

96. T, February 10, 1967, p. 40.

97. "My Best Friend's a Jew," p. 730.

98. T, February 22, 1967, p. 31; "The Trouble in Wayne," p. 80.

99. T, February 16, 1967, p. 52.

100. Quoted in Stark and Steinberg, "Jews and Christians," p. 78.

Chapter 9

1. Albert Vorspan, "The South, Segregation, and the Jew: A First-hand Report," *Jewish Frontier*, 23 (November, 1956), 18.

2. Carolyn Lipson-Walker, " 'Shalom Y'All': The Folklore and Culture of Southern Jews" (unpublished Ph.D., 1986, Department of Folklore in the Program in American Studies, Indiana University, 1986), p. 106.

3. Interview with Eli Evans, p. 14, in "Interviews, B-H," folder 13, box 20, Oscar Cohen mss., AJA; Charles C. Alexander, *The Ku Klux Klan in the Southwest* (University of Kentucky Press, 1965), p. 33; Charlton Mosely, "Latent Klanism in Georgia, 1890–1915," *Georgia Historical Quarterly*, 56 (Fall, 1972), 378–379; Leonard Dinnerstein, *The Leo Frank Case* (New York: Columbia University Press, 1968), p. 69ff; John Shelton Reed, *The Enduring South* (Lexington, Mass.: Lexington Books, 1972), p. 63; Leonard Dinnerstein, *Uneasy at Home* (New York: Columbia University Press, 1987), p. 123; Steven Hertzberg, *Strangers Within The Gate City: The Jews of Atlanta, 1845–1915* (Philadelphia: JPS, 1978), p. 200; Alfred O. Hero, Jr., "Southern Jews," in Leonard Dinnerstein and Mary Dale Palsoon, eds., *Jews in the South* (Baton Rouge: Louisiana State University Press, 1973), pp. 245–246.

4. "University of Virginia," *Watchman of the South*, August 5, 1841, p. 198.

5. Philip Alexander Bruce, *History of the University of Virginia, 1819–1919* (New York: The Macmillan Co., 1921), III, 73–74, 75, quote on page 76. See also Susanne Klingenstein, *Jews in the American Academy, 1900–1940* (New Haven: Yale University Press, 1991), p. 2; Lewis S. Feuer, "America's First Jewish Professor: James Joseph Sylvester at the University of Virginia," *American Jewish Archives*, 36 (November, 1984), 155.

6. Eli N. Evans, *The Provincials: A Personal History of Jews in the South* (New York: Atheneum, 1973), p. 42.

7. Elliott Ashkenazi, *The Business of Jews in Louisiana* (Tuscaloosa: University of Alabama Press, 1988), p. 9.

8. Myron Berman, *Richmond's Jewry* (Charlottesville: University Press of Virginia, 1979), p. 184; W. J. Cash, *The Mind of the South* (New York: Alfred A. Knopf, 1941), pp. 305, 342; Evans, *The Provincials*, p. 40; Karen I. Blu, "Varieties of Ethnic Identity: Anglo-Saxons, Blacks, Indians, and Jews in a Southern County," *Ethnicity*, 4 (1977), 276, 277; "The Unforgiving," *Newsweek*, 67 (May 2, 1966), 66; Abraham D. Lavender, "Jewish Values in the Southern Milieu," in Nathan M. Kaganoff and Melvin I. Urofsky, eds., *"Turn to the South"* (Published for the American Jewish Historical Society by the University Press of Virginia, Charlottesville, 1979), p. 131; Oliver C. Cox, "Race Prejudice and Intol-

erance—A Distinction," *Social Forces*, 24 (December, 1945), 219; Ella Lonn, *Foreigners in the Confederacy* (Chapel Hill: University of North Carolina Press, 1940), pp. 336, 392; Jack Nelson, *Terror in the Night: The Klan's Campaign Against Jews* (New York: Simon and Schuster, 1993), pp. 33–34; R. Baird Shuman, "Clifford Odets: A Playwright and His Jewish Background," *South Atlantic Quarterly*, 71 (Spring, 1972), 226, 227.

9. Shuman, "Clifford Odets," p. 226.

10. Berman, *Richmond's Jewry*, pp. 184, 288; Dinnerstein and Palsson, *Jews in the South*, pp. 304–305; Hero, "Southern Jews," pp. 246–247; Lipson-Walker, " 'Shalom Y'All,' " p. 126; Nelson, *Terror in the Night*, p. 34.

11. Berman, *Richmond's Jewry*, p. 184; Cash, *Mind of the South*, p. 342; Harold M. Hyman, *Oleander Odyssey: The Kempners of Galveston, Texas, 1854–1990s* (College Station: Texas A & M University Press, 1990), pp. 10, 12–13; Louis Schmier, "The First Jews of Valdosta," *Georgia Historical Quarterly*, 62 (Spring, 1978), 37; Louis Schmier, ed., *Reflections of Southern Jewry: The Letters of Charles Wessolowsky, 1878–1879* (Mercer University Press, 1982), pp. 39, 42, 51, 67, 71–72, 83, 89, 101, 129.

12. Schmier, "The First Jews," p. 42.

13. Naomi W. Cohen, *Encounter with Emancipation: The German Jews in the United States, 1830–1914* (Philadelphia: JPS, 1984), p. 77.

14. Gordon McKinney and Richard McMurray, eds., *The Papers of Zebulon Vance* (Frederick, Md.: University Publications of America, Inc., 1987), p. xxxi; Harry L. Golden, "The Jewish People of North Carolina," *North Carolina Historical Review*, 32 (1955), 213, fn. 26.

15. "Zebulon Vance," *Dictionary of American Biography* edited by Dumas Malone (10 volumes; New York: Charles Scribner's Sons, 1936), X, 161; Zebulon Vance, "The Scattered Nation," in Thomas B. Reed, et al, eds., *Modern Eloquence* (17 volumes; Philadelphia: John D. Morris & Co., 1900), VI, 1136–1137; Selig Adler, "Zebulon B. Vance and the 'Scattered Nation,' " *Journal of Southern History*, 7 (August, 1941), 358, 362, 370, 373.

16. *South Georgia Times*, November 11 and November 18, 1882 as quoted in Schmier, "The First Jews," p. 37; Berman, *Richmond's Jewry*, p. 247; Leah R. Atkins, "Populism in Alabama: Reuben F. Kolb and the Appeals to Minority Groups," *Alabama Historical Quarterly*, 32 (Fall and Winter, 1970), 178; Joseph C. Kiger, "Social Thought as Voiced in Rural Middle Tennessee Newspapers, 1878–1898," *Tennessee Historical Quarterly*, 9 (1950), 150; Louisville (Tennessee) *Times*, quoted in Washington *National Economist*, July 11, 1891, cited in Roger L. Hart, *Redeemers, Bourbons & Populists: Tennessee, 1870–1896* (Baton Rouge: LSU Press, 1975), p. 230; Dinnerstein, *The Leo Frank Case*, p. 68.

17. William F. Holmes, "Whitecapping: Anti-Semitism in the Populist Era," AJHQ, 63 (March, 1974), 246, 248, 250, 261.

18. Mark Twain, "Concerning the Jews," *Harper's Magazine*, 99 (1899), 530.

19. Mosley, "Latent Klanism," p. 378.

20. Dinnerstein, *The Leo Frank Case*, p. 97ff.

21. Holmes, "Whitecapping," pp. 246, 248, 250, 261.

22. Joseph M. Proskauer, *A Segment of My Times* (New York: Farrar, Straus and Co., 1950), pp. 12–13; Irving M. Engel, "Transcript of Oral Memoir," 1969–1970, tape I, pp. 10,23, Small Collections, AJA.

23. *Jewish Sentiment* (Atlanta), October 5, 1900, p. 3.

24. Saul Jacob Rubin, *Third to None: Saga of Savannah Jewry, 1733–1983* (Savannah, Georgia: Congregation Mikvah Israel, 1983), p. 216.

25. Quoted in William Link, *The Paradox of Southern Progressivism* (Chapel Hill: University of North Carolina Press, 1992), p. 121.

26. Berman, *Richmond's Jewry*, p. 241; Mark K. Bauman and Arnold Shankman, "The Rabbi as Ethnic Broker: The Case of David Marx," *Journal of American Ethnic History*, 2 (Spring, 1983), 53–54; Malcolm H. Stern, "The Role of the Rabbi in the South," in Kaganoff and Urofsky, "Turn to the South," p. 26.

27. Berman, *Richmond's Jewry*, pp. 248ff., 264, 273.

28. David and Adele Bernstein, "Slow Revolution in Richmond, Va: A New Pattern in the Making," in Dinnerstein and Palsson, *Jews in the South*, pp. 254–255.

29. "Rabbi Marx," *The Constitution* (Atlanta), March 11, 1895, p. 5; Bauman and Shankman, "The Rabbi as Ethnic Broker," pp. 53ff.

30. Ibid., p. 60.

31. Quoted in *Jewish Sentiment*, October 5, 1900, p. 3.

32. Bauman and Shankman, "The Rabbi as Ethnic Broker," pp. 57, 59.

33. Dinnerstein, *The Leo Frank Case*, p. 18.

34. Ibid., p. 45.

35. Ibid, chapter two, passim.

36. Ibid., p. 60.

37. Ibid., p. 74.

38. Ibid., p. 99.

39. Ibid., p. 145ff.

40. Evans, *The Provincials*, p. 141; Murray Friedman, "One Episode in Southern Jewry's Response to Desegregation: An Historical Memoir," *American Jewish Archives*, 33 (November, 1981), 171.

41. Earl Raab, "In Promised Dixieland," *Commentary*, 5 (May, 1948), 463.

42. Annie Nathan Meyer, "The Shoe Pinches Mr. Samuels," *The Crisis*, 42 (January, 1935), 6.

43. Dan T. Carter, *Scottsboro* (Baton Rouge: Louisiana State University Press, 1969), pp. 258–259.

44. Ibid., p. 235; Robert Leibowitz, *The Defender: The Life and Career of Samuel S. Leibowitz, 1893–1933* (Englewood Cliffs, N.J.: Prentice-Hall, 1981), p. 242.

45. Carter, *Scottsboro*, pp. 240–241.

46. Ibid., p. 241.

47. Joel Williamson, *A Rage for Order: Black/White Relations in the American South Since Emancipation* (New York: Oxford University Press, 1986), pp. 151, 246.

48. Evans, *The Provincials*, p. 96; Minutes, January 14, 1922, p. 101; E. C. Lindeman, "Sapiro, the Spectacular," *The New Republic*, 50 (April 13, 1927), 217; Arnold Shankman, *Ambivalent Friends: Afro-Americans View the Immigrant* (Westport, Conn.: Greenwood Press, 1982), p. 129.

49. Minutes, March 13, 1932, p. 15; James Lebeau, "Profile of a Southern Jewish Community: Waycross, Georgia," AJHQ, 58 (June, 1969), 430–431.

50. Roi Ottley "New World A-Coming" (Boston: Houghton Mifflin, 1943; reprinted New York: Arno Press, 1968), p. 128; Louis Ruchames, "Danger Signals in the South," *Congress Weekly*, 11 (April 21, 1944), 8; E. Terry Prothro and John A. Jensen, "Group Differences in Ethnic Attitudes of Louisiana College Students," *Sociology and Social Research*, 34 (March, 1950), 258.

51. Benjamin Kaplan, *The Eternal Stranger: A Study of Jewish Life in the Small Community* (New York: Bookman Associates, 1957), p. 157.

52. Ibid., p. 158.

53. Berman, *Richmond's Jewry*, p. 296; Stern, "The Role of the Rabbi," p. 28; Mark Rich, "Richmond's Enigmatic Ethnics," *The Richmond Mercury*, November 8, 1972, II, 8, in "Richmond, Virginia: The Richmond Mercury (Va) 11/8/72," folder #11, box 20, Oscar Cohen mss., AJA; Myron Berman, "Rabbi Edward Nathan Calisch and Debate Over Zionism in Richmond, Virginia," AJHQ, 62 (March, 1973), 305; Lebeau, "Profile," p. 441; folder 11, box 20, Cohen mss.; Carey McWilliams, "West Coast Letter," *The Chicago Jewish Forum*, 7 (Fall, 1948), 67.

54. Stern, "The Role of the Rabbi," p. 28; Bauman and Shankman, "The Rabbi as Ethnic Broker," p. 57; Berman, *Richmond's Jewry*, pp. 319, 321.

55. Folder 11, box 20, Oscar Cohen mss.; Lipson-Walker, " 'Shalom Y'All,' " pp. 134–136.

56. Harry L. Golden, "A Son of the South, and Some Daughters," *Commentary*, 12 (November, 1951), 380, 381.

57. Friedman, "An Episode," p. 171; Evans, *The Provincials*, p. 6; Harry Golden, "A Pulpit in the South," *Commentary*, 16 (December, 1953), 578; Stern, "The Role of the Rabbi," p. 30; Lebeau, "Profile of a Southern Jewish Community."

58. Evans, *The Provincials*, pp. 86, 141, 163, 329.

59. Joseph Wilfrid Vander Zanden, "The Southern White Resistance Movement to Integration" (unpublished Ph.D., Department of Sociology and Anthropology, University of North Carolina, 1957), pp. 310–311.

60. Nelson, *Terror in the Night*, p. 39; "Harvest of Violence," *The Economist*, 189 (October 25, 1958), 324; Stanley Meisler, "The Southern Segregationist and His Anti-Semitism," *The Chicago Jewish Forum*, 16 (Spring, 1958), 171, 172.

61. David Halberstam, "The White Citizens Councils," *Commentary*, 22 (October, 1956), 301.

62. Hero, "Southern Jews," pp. 221–222.

63. Quoted in Lewis M. Killian, *White Southerners* (revised edition; Amherst: University of Massachusetts Press, 1985; originally published, 1970), p. 80.

64. "Anti-Semitism in the South," Richmond *News-Leader*, July 7, 1958, p. 8.

65. "Hate Bombings, 1958," folder 3, box 4, JCRC mss., AJA; "Christians Condemn Bombings in South," CC, 75 (April 2, 1958), 398; Evans. *The Provincials*, pp. 283, 320; "A Prevalence of Scapegoats," *The Nation*, 206 (January 1, 1968), 5; Nelson, *Terror in the Night*, pp. 32, 35, 51, 69, 75–76; Nussbaum quote, p. 72.

66. Isaac Toubin, "Recklessness or Responsibility," *Southern Israelite*, February 27, 1959, p. 13.

67. Quoted in Lipson-Walker, " 'Shalom Y'All,' " p. 88.

68. Marvin Braiterman, "Mississippi Marranos," in Dinnerstein and Palsson, eds., *Jews in the South*, p. 353; Friedman, "An Episode," pp. 178, 180.

69. Quoted in Leonard Dinnerstein, *Uneasy at Home* (New York: Columbia University Press, 1987), p. 137; see also Marshall Field Stevenson, Jr., "Points of Departure, Acts of Resolve: Black-Jewish Relations in Detroit, 1937–1962"

(unpublished Ph.D., Department of History, University of Michigan, 1988), pp. 381–382.

70. Meisler, "The Southern Segregationist," p. 173; Minutes of the AJC Administrative Board, October 1, 1957, p. 7, November 6, 1958, p. 4; NCRAC Plenary Session, 1958, folder 3, box 4, JCRC mss.; Lebeau, "Profile of a Southern Jewish Community," pp. 434–435; T, September 26, 1959, p. 28; Friedman, "An Episode," p. 173; Evans, *The Provincials*, p. 317.

71. NCRAC Plenary Session, 1958, folder 3, box 4, JCRC mss.; William Bradford Huie, "The Untold Story of the Mississippi Murders," *Saturday Evening Post*, 237 (September 5, 1964), 11–15.

72. Huie, "The Untold Story," p. 12.; see also Jonathan Kaufman, *Broken Alliance: The Turbulent Times Between Blacks and Jews in America* (New York: Charles Scribner's Sons, 1988), pp. 70–71.

73. Joseph Lelyveld, "A Stranger in Philadelphia, Mississippi," *The New York Times Magazine*, December 27, 1964, p. 36.

74. "The Untold Story,", pp. 13–14.

75. Nelson, *Terror in the Night*, p. 64.

76. "The Untold Story," p. 12.

77. Quoted in Braiterman, "Mississippi Marranos," p. 355.

78. Quoted in ibid., p. 356.

79. Dinnerstein, *Uneasy at Home*, p. 140ff.

80. Quoted in ibid., p. 142.

81. Ibid., p. 140ff.

82. Quoted in Perry Nussbaum, "Christian Love for the Jews," *Jewish Digest*, 15 (June, 1970), 62.

83. Evans, *The Provincials*, p. 331; Minutes of the AJC Board of Governors, June 7, 1972, p. 4; folder 11, box 20, JCRC mss.

84. Lipson-Walker, " 'Shalom Y'All,' " p. 136.

85. Evans, *The Provincials*, p. 311.

86. Lipson-Walker, " 'Shalom Y'All,' " p. 116.

87. Lipson-Walker, " 'Shalom Y'All," pp. 210, 214; Wanda Katz Fishman and Richard L. Zeigenhaft, "Jews and the New Orleans Economic and Social Elites," JSS, 44 (Summer–Fall, 1982), 295; T (national edition), December 7, 1991, p. 7.

88. Quoted in Robert T. Handy, *A Christian America* (2nd edition; New York: Oxford University Press, 1984), p. 207.

89. Maas, "The Jews of Houston," p. 201; Lipson-Walker, " 'Shalom Y'All,' " pp. 58, 109.

Chapter 10

1. In 1967 Harold Cruse wrote that solidarity between blacks and Jews "was never a real fact . . . and it is misleading nonsense to claim that it ever was." The relationships between Jews and blacks, he continued, had always been "rather ambiguous." Harold Cruse, *The Crisis of the Negro Intellectual* (New York: William Morrow and Co., 1967), pp. 169–170. Then in the 1980s others also questioned whether an alliance between the two groups had actually existed. For discussions of the topic, see Marshall Field Stevenson, Jr., "Points of Departure, Acts of Resolve: Black-Jewish Relations in Detroit, 1937–1962"

(unpublished Ph. D., Department of History, University of Michigan, 1988) and Jonathan Kaufman, *Broken Alliance: The Turbulent Times Between Blacks and Jews in America* (New York: Charles Scribner's Sons, 1988). Stevenson writes, "The idea of an alliance is somewhat ambiguous given the tensions and differences we have described. Thus the major theme that emerges is that the concept of a black-Jewish alliance must be qualified by geographic, economic, and ideological considerations. . . . [T]he past notion of a single, smoothly operating black-Jewish alliance since the 1920s has been exaggerated," p. 483. Kaufman, who was inspired to write his tract because he believed that a black-Jewish alliance had at one time existed, wrote, "The alliance between blacks and Jews was never as strong as it appeared. It was rooted as much in the hard currency of politics and self-interest as in love and idealism. Even at those times when the alliance seemed strongest . . . the symbols of cooperation covered a cauldron of ambivalent feelings and conflicting emotions," p. 267.

2. L. D. Reddick, "Anti-Semitism Among Negroes," *The Negro Quarterly*, Summer, 1942, p. 114; Leo Laufer, "Anti-Semitism Among Negroes," *The Reconstructionist*, 14 (October 29, 1948), 7; John B. Sheerin, "The Myth of Black Anti-semitism," *Catholic World*, 209 (May, 1969), 50; Nathan Hurvitz, "Blacks and Jews in American Folklore," *Western Folklore*, 33 (October, 1974), 302; Arnold Rose, *The Negro's Morale: Group Identification and Protest, Minneapolis* (University of Minnesota Press, 1949), pp. 128–129; Henry Lewis Gates, Jr., "Black Demagogues and Pseudo-Scholars," T (national edition), July 20, 1992, p. A13; Naomi W. Cohen, "Antisemitism in the Gilded Age: The Jewish View," JSS, 41 (Summer/Fall, 1979), 190; Will Maslow, "Negro-Jewish Relations in America: a Symposium," *Midstream*, 12 (December, 1966), 74; Abraham G. Duker, "On Negro-Jewish Relations—A Contribution to a Discussion," JSS, 27 (January, 1965), 23; Lunabelle Wedlock, *The Reaction of Negro Publications and Organizations to German Anti-Semitism* (Washington, D.C.: the Howard University Studies in the Social Sciences, volume 3, #2, 1942; foreword by Ralph Bunche), pp. 29, 207; Joseph P. Weinberg, "Black-Jewish Tensions: Their Genesis," *CCAR Journal*, 21 (Spring, 1974), 31.

3. Richard Wright, *Black Boy* (New York: Harper & Bros., 1945; this edition: a Perennial Classic, 182), p. 70.

4. Joseph A. Johnson, Jr., *The Soul of the Black Preacher* (Philadelphia: United Church Press, 1971), p. 143.

5. Quoted in Joseph P. Weinberg, "Black-Jewish Tensions: Their Genesis," *CCAR Journal*, 21 (Spring, 1974), 34.

6. Weinberg, "Black-Jewish Tensions," p. 34; Thomas Wentworth Higginson, "Negro Spirituals," *Atlantic Monthly*, 19 (1867), 688; Arnold Shankman, *Ambivalent Friends: Afro-Americans View the Immigrant* (Westport, Conn.: Greenwood Press, 1982), p. 134; E. A. McIlhenny, *Befo' De War Spirituals* (Boston: The Christopher Publishing House, 1933), p. 39. See also, Hasia Diner, *A Time for Gathering* (Baltimore: Johns Hopkins University Press, 1992), p. 176.

7. Robert G. Weisbord and Arthur Stein, *Bittersweet Encounter: The Afro-American and the American Jew* (Westport, Conn.: Negro Universities Press, 1970; foreword by C. Eric Lincoln), p. 71, n. 15.

8. Weinberg, "Black-Jewish Tensions," p. 33; Ben A. Richardson, "This Is Our Common Destiny: Negroes Can Understand Jewish Position on Christ," *People's Voice*, September 11, 1943, p. 21.

9. *Harper's Magazine*, 19 (1859), 859–860.

10. Horace Mann Bond, "Negro Attitudes Toward Jews," JSS, 27 (January, 1965), 3–4.

11. James Baldwin, "The Harlem Ghetto: 1948," *Commentary*, 5 (February, 1948), 168.

12. Steven Bloom, "Interactions Between Blacks and Jews in New York City, 1900–1930, as Reflected in the Black Press" (unpublished Ph.D., Department of History, New York University, 1973), p. 102; Shankman, *Ambivalent Friends*, p. 122; David J. Hellwig, "Black Images of Jews: From Reconstruction to Depression," *Societas*, VIII (Summer, 1978), 207, 212; Seth Scheiner, *Negro Mecca* (New York: New York University Press, 1965), p. 131.

13. "Letter from Washington, D.C.," *Weekly Louisianan* (New Orleans), October 4, 1879, p. 1.

14. Scheiner, *Negro Mecca*, p. 131.

15. "Persecution of the Jews," *Washington Bee*, August 18, 1899.

16. "Charity Begins at Home," *Colored American* (Washington, D.C.), June 10, 1903, p. 8.

17. Jesse Fortune, "Among the Children of the East Side Jews," *New York Age*, January 15, 1905, p. 2.

18. "Why the Jew Prevails Everywhere," *New York Age*, February 8, 1905, p. 4; see also Hellwig, "Black Images," p. 216.

19. Louis R. Harlan, "Booker T. Washington's Discovery of Jews," in J. Morgan Kousser and James M. McPherson, eds., *Religion, Race, and Reconstruction* (New York: Oxford University Press, 1982), pp. 269, 270, 275, 276; Louis R. Harlan, *Booker T. Washington: The Wizard of Tuskegee, 1901–1905* (New York: Oxford University Press, 1983), p. 260; Louis R. Harlan, et al., eds., *The Booker T. Washington Papers* (11 volumes; University of Illinois Press, 1972–), III (1889–1895), 412.

20. Francis L. Broderick, *W. E. B. DuBois—Negro Leader in a Time of Crisis* (Stanford University Press, 1959), note on pp. 26–27.

21. W. E. Burghardt DuBois, "The Relation of the Negroes to the Whites in the South," *Annals of the American Academy of Political and Social Science*, 18 (1901), 126.

22. W. E. B. DuBois, *The Souls of Black Folk* (Chicago: A. C. McClurg and Co., 1903), p. 170.

23. Ibid., p. 169.

24. Ibid., p. 204.

25. W. E. B. DuBois, *The Souls of Black Folk* (Millwood, N. Y.: Kraus-Thomson Organization, Ltd., 1973), pp. 42–43; Herbert Aptheker, *"The Souls of Black Folk:* A Comparison of the 1903 and 1952 Editions," *Negro History Bulletin*, 34 (1971), 16.

26. Aptheker, *"Souls of Black Folk,"* p. 16.

27. DuBois, *Souls of Black Folk* (KTO, 1973), p. 41. There were over 1,000 east European Jews in Georgia during the early years of the twentieth century but whether DuBois ever met any is impossible for me to say.

28. "The Hebrew Race in America," *The Voice of the Negro* (Atlanta), 3 (January, 1906), 20; Shankman, *Ambivalent Friends*, pp. 15, 116, 117, 118, 119, 127, 130; Hellwig, "Black Images," p. 126; Isabel Boiko Price, "Black Responses to Anti-Semitism: Negroes and Jews in New York, 1880 to World War II" (unpublished Ph.D., Department of History, University of New Mexico, 1973), p. 139; "A Lesson from the Jews," *Colored American*, July 5, 1902, p. 8; Booker

T. Washington, *The Future of the American Negro* (New York: Haskell House, 1968; originally published 1900), pp. 182–183.

29. Hellwig, "Black Images," p. 207.

30. Hellwig, "Black Images," p. 208; Bloom, "Interaction Between Blacks and Jews," p. 94; see also *The Afro-American Ledger* (Baltimore), November 2, 1907, editorial; and *The Chicago Defender*, February 6, 1915, p. 8.

31. James Weldon Johnson, "The Negro and the Jew," *New York Age*, February 2, 1918, p. 4.

32. "The Jew Shows Us How in Many Ways," *Norfolk Journal and Guide*, May 22, 1926, p. 14; see also Hellwig, "Black Images," p. 207; Bloom, "Interactions Between Blacks and Jews," p. 148; Price, "Black Responses," p. 139; Harold L. Sheppard, "The Negro Merchant: A Study of Negro Anti-Semitism," *American Journal of Sociology*, 53 (September, 1947), 97; "Why The Jews Win," *The Chicago Defender*, April 14, 1928, II, 2.

33. Quoted in Seymour S. Weisman, "Black-Jewish Relations in the USA—II," *Patterns of Prejudice*, 15 (January, 1981), 49.

34. Lawrence W. Levine, *Black Culture and Black Consciousness* (New York: Oxford University Press, 1977), p. 385; Shankman, *Ambivalent Friends,* p. 141, fn. 29; Daryl Cumber Dance, *Shuckin' and Jivin': Folklore from Contemporary Black Americans* (Bloomington: Indiana University Press, 1978), p. 151 ff.; Nathan Hurvitz, "Blacks and Jews in American Folklore," *Western Folklore Quarterly*, 33 (October, 1974), 324–325; Richard M. Dorson, "Jewish-American Dialect Stories on Tape," in Raphael Patai, ed., *Studies in Biblical and Jewish Folklore* (Bloomington: Indiana University Press, 1960), p. 116; Richard M. Dorson, ed., *Negro Folktales in Michigan* (Cambridge: Harvard University Press, 1956), pp. 75, 76–77; Jack Nusan Porter, "John Henry and Mr. Goldberg: The Relationship Between Blacks and Jews," *Journal of Ethnic Studies* 7 (Fall 1960), 75–76; Lucy S. Davidowicz, "Can Anti-Semitism Be Measured?" *Commentary*, 50 (July 1970), 42; Richard M. Dorson, "More Jewish Dialect Stories," *Midwest Folklore*, 10 (Fall 1960), 138.

35. Price, "Black Responses," p. 133; Alan Spear, *Black Chicago: The Making of a Negro Ghetto: 1890–1920* (Chicago: University of Chicago Press, 1967), p. 194; Thomas Lee Philpott, *The Slum and the Ghetto* (New York: Oxford University Press, 1978), p. 198; Shankman, *Ambivalent Friends*, p. 123.

36. Donald S. Strong, *Organized Anti-Semitism in America* (Washington, D.C.: American Council on Public Affairs, 1941), p. 146; Minutes, VI (1932) 178.

37. "Prejudice," *The Philadelphia Tribune*, August 1, 1925, p. 4.

38. Wedlock, *Reaction of Negro Publications*, p. 91.

39. Price, "Black Responses," pp. 177, 180.

40. "As the Crow Flies," *The Crisis*, 40 (September, 1933), 97.

41. Quoted in Wedlock, *The Reaction of Negro Publications*, p. 116.

42. Quoted in Price, "Black Responses," p. 230.

43. Quoted in Robert G. Weisbord and Arthur Stein, "Negro Perceptions of Jews Between the World Wars," *Judaism*, 18 (1969), 443.

44. Wedlock, *The Response of Negro Publications*, p. 83; see also Price, "Black Responses," pp. 184, 228, 297; Edward L. Israel, "Jew Hatred Among Negroes," *The Crisis*, 43 (February, 1936), 39.

45. "Charity Begins at Home," *The Crisis*, 45 (April, 1938), 113; "Racism at Home," *The Colored Harvest* (Baltimore), October–November, 1938, p. 2.

46. "Jews in Germany vs. Negroes in America," *The Philadelphia Tribune*, October 12, 1933, p. 4.

47. "Germany vs. America," *The Philadelphia Tribune*, July 5, 1934, p. 4.

48. "Rabbis Pledge Friendship to Offset Anti-Semitism," *Chicago Defender*, July 16, 1938, p. 1; Wedlock, *The Reaction of Negro Publications*, passim.

49. A. Clayton Powell, Jr., "Harlem Negroes' View on Problems," *The New York Post*, March 27, 1935, p. 4; Allon Schoener, ed., *Harlem on My Mind* (New York: Random House, 1968), p. 171; Wil Haygood, *King of the Cats: The Life and Times of Adam Clayton Powell, Jr.* (Boston: Houghton Mifflin, 1993), p. 74.

50. Marie Syrkin, "Anti-Semitic Drive in Harlem," *Congress Weekly*, 8 (October 31, 1941), 7; see also Roi Ottley, "New World A-Coming," (Boston: Houghton Mifflin, 1943; reprinted, New York: Arno Press, 1968), p. 115.

51. "Harlem Jews vs. Negroes," *The Crusader*, July 4, 1942.

52. Winston McDowell, "Race and Ethnicity During the Harlem Jobs Campaign," *Journal of Negro History*, 69 (1984), 137–138; St. Clair Drake and Horace R. Cayton, *Black Metropolis: A Study of Negro Life in a Northern City* (New York: Harcourt, Brace and Co., 1945), pp. 441, 444.

53. Kelly Miller, "The Negro and the Jew in Business," *Richmond Planet*, April 13, 1935, p. 12.

54. Wedlock, *The Reaction of Negro Publications*, pp. 24, 116, 159; "Letters from Readers: Jew Hatred Among Negroes," *The Crisis*, 43 (April, 1936), 122; Drake and Cayton, *Black Metropolis*, p. 249; Israel, "Jew Hatred," p. 39; Harold Orlansky, "A Note on Anti-Semitism Among Negroes," *Politics*, 2 (August, 1945), 251.

55. Advertisement in *Cincinnati Enquirer*, September 21, 1941, clipping in *Cincinnati Enquirer*—Prejudice in Advertising, 1941–1955, folder 10, box 34, JCRC mss.

56. Drake and Cayton, *Black Metropolis* (New York: Harcourt, Brace & Co., 1945), p. 244n.

57. Ella Baker and Marvel Cooke, "The Bronx Slave Market," *The Crisis*, 42 (November, 1935), 330.

58. "Report of Committee on 'Street Corner Markets,'" pp. 5, 6 in folder, "Negro-Jewish Rel.—New York City, 1935–1968," Blaustein Library.

59. Baker and Cooke, "The Bronx Slave Market," p. 330.

60. Folder, "Negro-Jewish Re.—New York City," Blaustein Library; Weisbord and Stein, *Bittersweet Encounter*, p. 41; Wedlock, *Response of Negro Publications*, p. 169; Price, "Black Responses," pp. 197, 201; Nathan Zuckerman, ed., *The Wine of Violence* (New York: Association Press, 1947), p. 322.

61. "Anti-Semitism Among Negroes," *The Crisis*, June, 1938, p. 177.

62. "The Blacks and the Jews: a Falling Out of Allies," *Time*, 93 (January 31, 1969), 57.

63. Quoted in Israel, "Jew Hatred Among Negroes," p. 50.

64. Price, "Black Responses," p. 339; see also Wedlock, *Response of Negro Publications*, p. 182.

65. Mary Testa, "Anti-Semitism Among Italian-Americans," *Equality*, I (July, 1939), 27; "Rabbis Deplore Anti-Semitic Propaganda Among Negroes," *Boston Chronicle*, July 16, 1938, pp. 1, 8; same article with same title appears in *Chicago Defender* on the same day; George J. Schuyler, "Views and Reviews," *Pittsburgh Courier*, November 26, 1938, p. 10; Drake and Cayton, *Black Metropolis*, p. 198; William L. Patterson to Ira Latimer, July 11, 1938, William L. Pat-

terson to Walter White, July 11, 1938, A. Ovrum Tapper to Walter White, November 30, 1938, A. Ovrum Tapper to Charles F. Houston, December 2, 1938, Walter White to F. H. LaGuardia, September 18, 1939, folder,"Anti-Semitism, 1938," box C-208, folder, "Discrimination: Jews, 1939, July–September," box C-277, NAACP mss., LC; Adam C. Powell, "Soap Box: Negroes and Jews," *New York Amsterdam News*, April 16, 1938, p. 13; Weisbord and Stein, "Negro Perceptions of Jews," p. 444; Stevenson, "Points of Departure," pp. 93, 94, 127, and 141.

66. Weisbord and Stein, "Negro Perceptions of Jews," p. 444.

67. Quoted in Price, "Black Responses," pp. 266–267, 318.

68. Quoted in Zuckerman, *The Wine of Violence*, p. 315.

69. Cohen, "The Negro and Anti-Semitism," *Congress Weekly*, December 22, 1944, p. 5.

70. Price, "Black Responses," p. 318.

71. Ralph Bunche, "Foreword," in Wedlock, *The Reaction of Negro Publications*, p. 8.

72. Arthur Huff Fauset, "Anti-Semitism Among Us," *The Philadelphia Tribune*, August 19, 1944, p. 4.

73. "The Jew and The Negro," *National Baptist Voice* (Nashville, Tenn.), 38 (May 15, 1945), 1.

74. Leo Laufer, "Anti-Semitism Among Negroes," *The Reconstructionist*, 14 (October 29, 1948), 10; Shlomo Katz, ed., *Negro and Jew: An Encounter in America; A Symposium Compiled by Midstream Magazine* (New York: Macmillan Co., 1967), p. 42; see also "Dr. Kenneth Clark Believes Practical Action Will Improve Negro-Jewish Relations in U. S.," *Pittsburgh Courier*, March 23, 1946, p. 13.

75. Hasia R. Diner, *In the Almost Promised Land: American Jews and Blacks, 1915–1935* (Westport, Conn.: Greenwood Press, 1977), pp. xvii, 66–67; Edward S. Shapiro, *A Time for Healing: American Jewry Since World War II* (Baltimore: The Johns Hopkins University Press, 1992), p. 223; Gates, "Black Demagogues."

76. Quote from Peter I. Rose, "Blacks and Jews: The Strained Alliance," in *The Annals of the American Academy of Political and Social Science*, 454 (March, 1981), 55, see also p. 61; on Jewish involvement in the civil rights movement see David Rogers, *110 Livingston Street: Politics and Bureaucracy in the New York City Schools* (New York: Random House, 1968), pp. 147, 162; Robert G. Weisbord and Richard Kazarian, Jr., *Israel in the American Black Perspective* (Westport, Conn.: Greenwood Press, 1985), p. 33; Weisbord and Stein, *Bittersweet Encounter*, pp. 134–135; Haygood, *King of the Cats*, p. 75; Roy Wilkins, "Jewish-Negro Relations," *American Judaism*, 13 (Spring, 1963), 4ff; Natalie Gittelson, "AS: It's Still Around," *Harper's Bazaar*, 105 (February, 1972), 134; Claybourne Carson, Jr., "Blacks and Jews in the Civil Rights Movement," in Joseph R. Washington, Jr., ed., *Jews in Black Perspectives: A Dialogue* (Rutherford, NJ: Fairleigh Dickinson University Press, 1984), pp. 115–116, 131 fn. 4; Ellen Hume, "Falling Out: Blacks and Jews Find Confrontation Rising Over Jesse Jackson," *The Wall Street Journal*, May 29, 1984, p. 18; Harold Cruse, *Plural But Equal: A Critical Study of Blacks and Minorities and America's Plural Society* (New York: William Morrow and Co., Inc., 1987), p. 151; "When Blacks and Jews Fall Out," *The Economist*, 292 (July 7, 1984), 19; Joel Ziff, "Black Man and Jew," *Response*, Winter, 1968, p. 21; Bill Kovach, "Racist and Anti-Semitic Charges Strain Old Negro-Jewish Ties," T, October 23, 1968, p. 32; Eddie Ellis, "Semitism in the Ghetto," *Liberator*,

January, 1966, p. 7; August Meier and Elliot Rudwick, *CORE: A Study in the Civil Rights Movement* (Urbana: University of Illinois Press, 1975), pp. 225, 336; Richard Cummings, *The Pied Piper: Allard K. Lowenstein and the Liberal Dream* (New York: Grove Press, Inc., 1985), p. 236; Nathan Glazer, "Negroes and Jews: The New Challenge to Pluralism," *Commentary*, 38 (December, 1964), 33; Michael Kramer, "Blacks and Jews," *New York*, 18 (February 4, 1985), 28; James A. Jahannes, *Blacks and Jews: A New Dialogue* (Savannah: Savannah State College Press, 1983), p. 24; Shapiro, *A Time for Healing*, p. 223.

77. Memo to John Slawson from Harry Fleischman, Re: Notes for Meeting with Negro Leaders, April 13, [1960], in folder, "Race Rel Negroes 52–60," box 258, AJC mss., YIVO.

78. Quoted in Price, "Black Responses," p. 3.

79. Murray Friedman to Leonard Dinnerstein, February 17, 1992, in possession of the author.

80. James Q. Wilson, *Negro Politics* (Glencoe, Ill.: The Free Press, 1960), pp. 150 ff, 156, 160–161, 164; "Negro-Jewish Tensions," p. 26, AJC mimeographed report, 1958, Blaustein Library; Folder, "Jewish-Negro Relations, Tensions, AJC, Study Comments, Race Rel Negroes 58–59," letter from B. M. Joffee of Jewish Community Council of Detroit to S. Andhil Fineberg, January 22, 1959," box 258, AJC mss., YIVO; Tom Brooks, "Negro Militants, Jewish Liberals, and the Unions," *Commentary*, 32 (September, 1961), 214–215; Rogers, *110 Livingston Street*, pp. 144–145.

81. Louis Martin, "Dope and Data," *Chicago Defender*, April 5, 1958, p. 10; "Let Us Fight This Beast," *Pittsburgh Courier*, January 16, 1960, p. 12.

82. Peter Goldman, *Report from Black America* (New York: Simon and Schuster, 1970), p. 255; Jahannes, *Blacks and Jews*, p. 3; Katz, *Negro and Jew*, p. vii.

83. Charles Silberman, "Jesse and the Jews," *The New Republic*, 181 (December 29, 1979), 12; Gary T. Marx, *Protest and Prejudice: A Study of Belief in the Black Community* (New York: Harper and Row, 1967), pp. 70, 131, 133; Hume, "Falling Out," p. 1; Geraldine Rosenfield, "The Polls: Attitudes Toward American Jews," *Public Opinion Quarterly*, 46 (Fall, 1982), 432, 433; Gertrude J. Selznick and Stephen Steinberg, *The Tenacity of Prejudice* (New York: Harper & Row, 1969), pp. 121, 122, 123, 125; Harold E. Quinley and Charles Y. Glock, *Anti-Semitism in America* (New Brunswick: Transaction Books, 1983), pp. xxi, 55; Ronnie Tadao Tsukashima, "The Social and Psychological Correlates of Anti-Semitism in the Black Community" (unpublished Ph.D., Department of Sociology, University of California, Los Angeles, 1973), p. 189; Ronald Tado Tsukashima, "The Contact Hypothesis: Social and Economic Contact and Generational Changes in the Study of Black Anti-Semitism," *Social Forces*, 55 (September, 1976), 161; Gates, "Black Demagogues." A 1992 survey by the ADL also found that 37% of blacks and 17% of whites held strong antisemitic positions. This survey also classified 27% of college-educated blacks and 46% of blacks without a college education as "most antisemitic." ADL, "Highlights from an Anti-Defamation League Survey on Anti-Semitism and Prejudice in America," November 16, 1992, pp. 30, 33.

84. Ann G. Wolfe, "Negro Anti-Semitism: A Survey," March 15, 1966, unpublished paper in AJC, Blaustein Library.

85. T, August 15, 1963, p. 14.

86. Quoted in Peter Goldman, *The Death and Life of Malcolm X* (New York: Harper and Row, 1973), p. 15.

87. Lenora E. Berson, *The Negroes and the Jews* (New York: Random House, 1971), pp. 8–9.

88. Shad Polier to Roy Wilkins, June 10, 1963, Stephen Gill Spottswood to Cecil Moore, June 18, 1963, Bernard S. Frank to Kivie Kaplan, February 25, 1964, John A. Morsell to Thomas Atkins, October 29, 1964, John A. Morsell to Albert Sprague Coolidge, November 4, 1964, Albert Sprague Coolidge to Roy Wilkins, October 28, 1964, John A. Morsell to Frances L. Apt, December 18, 1964, in folder 5, "Jews, 1963–1965," box A175, 1956–1965, General Office File, NAACP Administration mss., LC; Katz, *Negro and Jew*, p. vii; Will Maslow, "Negro-Jewish Relations in the North," in folder, "Jewish-Negro Relations, Tensions, Race Rel, Negroes 38–60," "Memo from Harry Fleischman to Murray Friedman Re: Negro-Jewish Relations and *Pittsburgh Courier*," June 13, 1960, in folder, "Race Rel Negroes 52–60," box 258,AJC mss., YIVO; Brooks, "Negro Militants," pp. 209, 215.

89. Otis S. Johnson, "A Black Perspective on African American/Jewish American Relations," in Jahannes, *Blacks and Jews*, p. 4; Glazer, "Negroes and Jews," p. 32.

90. Kenneth B. Clark, "Black Power and Basic Power," *Congress Bi-Weekly*, 35 (January 8, 1968), 6; "Black-Jewish Relations [Andrew Young] 1978–1979," folder 1, box 77, JCRC mss.; Minutes of the AJC Board of Governors, October 3, 1967, p. 3; "Black Anti-Semitism and the Jewish Response," *Ideas*, 1 (Winter, 1968–1969), 9; Clayborne Carson, *In Struggle: SNCC and the Black Awakening of the 1960s* (Cambridge: Harvard University Press, 1981), p. 269; Carson, "Blacks and Jews," in Washington, ed., *Jews in Black Perspectives*, pp. 114, 119; Jerome Bakst, "Negro Radicalism Turns Anti-Semitic—SNCC's Volte Face," *Wiener Library Bulletin*, Winter, 1967–1968, p. 20; Kovach, "Racist And Anti-Semitic Charges," p. 32; "When Blacks and Jews Fall Out," p. 19; Alvin F. Poussaint, "Blacks and Jews: An Appeal for Unity," *Ebony*, 39 (July, 1974), 127.

91. Minutes of the AJC Board of Governors, March 1, 1955, p. 6, October 3, 1967, p. 5; Nat Hentoff, ed., *Black Anti-Semitism and Jewish Racism* (New York: Richard W. Baron, 1969), p. 133.

92. "Negro-Jewish Tensions," *American Mercury*, 103 (Spring, 1967), 17; Weisbord and Stein, *Bittersweet Encounter*, p. 139; Max Geltman, *The Confrontation: Black Power, Anti-Semitism, and the Myth of Integration* (Englewood Cliffs, N.J.: Prentice-Hall, Inc., 1970), p. 181; Jackie Robinson, "We Also Have Black Bigots," New York *Amsterdam News*, February 19, 1966, p. 15.

93. Quoted in Wolfe, "Negro Anti-Semitism," p. 17.

94. Wolfe, "Negro Anti-Semitism," p. 27; Meier and Rudwick, *CORE*, p. 411. Between 50% and 75% of all financial donations to the NAACP, SCLC, SNCC, and CORE came from Jewish donors, Kaufman, *Broken Alliance*, pp. 12, 98; Innis quote from Geltman, *The Confrontation*, p. 4.

95. Diner, *In The Almost Promised Land*, p. 38.

96. James Baldwin, "Negroes Are Antisemitic Because They're Anti-white," and Robert Gordis, "Negroes Are Antisemitic Because They Want a Scapegoat," in Leonard Dinnerstein, ed., *Antisemitism in the United States* (New York: Holt, Rinehart & Winston, 1971), pp. 116–131.

97. Quoted in Gerald S. Strober, *American Jews: Community in Crisis* (Garden City: Doubleday & Co., 1974), pp. 120–121.

98. Clifton F. Brown, "Black Religion—1968," in Patricia W. Romero, ed., *In Black America: 1968, The Year of Awakening* (Washington, D.C.: United Publishing Corp., 1969), p. 351; Hentoff, *Black Antisemitism*, pp. 56–57; folder, "Negro-Jewish Rel.—New York City, 1935–1968," Blaustein Library; Fred M. Hechinger, "Racism and Anti-Semitism in the School Crisis," T, September 16, 1968, p. 46; Marvin Weitz, "Black Attitudes to Jews in the United States from World War II to 1976" (unpublished Ph.D., Yeshiva University, 1977), p. 244ff.; AJYB 70 (1969) 81; Kaufman, *Broken Alliance*, pp. 148–149, 154–155.

99. Quoted in David Polish, "The Jewish-Negro Confrontation," *The American Zionist*, 54 (April, 1969), 19.

100. For a sympathetic discussion to all parties concerned in the Ocean Hill–Brownsville controversy, see Kaufman, *Broken Alliance*, p. 144ff.

101. Kovach, "Racist and Anti-Semitic Charges," p. 32; AJYB, 70 (1969), 76–78.

102. The full poem is printed in Kaufman, *Broken Alliance*, pp. 159–160.

103. Talk by Julius Lester, "Blacks, Jews and the Media," presented at conference on Black-Jewish Relations in the United States (Washington, D.C.), November 19, 1985; *Time*, January 31, 1969, p. 56; "The WBAI Incident," *Columbia Journalism Review*, 8 (Fall, 1969), 28; "New York: How Free the Air?" *Newsweek*, 73 (February 10, 1969), 25; T, January 16, 1969.

104. T, January 12, 1969, II, 25; *Time*, January 31, 1969, p. 56; Strober, *American Jews*, p. 127; Rose, "Blacks and Jews," p. 67.

105. Quoted in "Negro-Jewish Tensions," p. 16; see also Paul B. Levenson, "The Image of the Jew in the Negro Community," *Jewish Advocate*, May 31, 1962, p. 2.

106. Henry Siegman, "Negro Anti-Semitism," American Federation of Jews from Central Europe, *Conference on Anti-Semitism*, March 23, 1969, Marvin E. Aspen to Manuel Silver, June 6, 1969, both in folder, "Negro-Jewish Relations, 1969–1975," Blaustein.

107. Minutes of the AJC Board of Governors, November 8, 1967, p. 4.

108. Quoted in "Case Study of a Riot: The Philadelphia Story," in Minutes of the AJC Board of Governors, October 1, 1966, p. 4.

109. Jonathan D. Sarna and Nancy H. Klein, *The Jews of Cincinnati* (Cincinnati: Center for the Study of the American Jewish Experience on the Campus of the Hebrew Union College—Jewish Institute of Religion, 1989), p. 174.

110. Hillel Levine and Lawrence Harmon, *The Death of an American Jewish Community* (New York: The Free Press, 1992), p. 149.

111. Naomi W. Cohen, *Not Free To Desist: The American Jewish Committee, 1906–1946* (Philadelphia: JPS, 1972), pp. 405, 406.

112. Vernon E. Jordan, Jr., "Together!" *The Crisis*, 81 (October, 1974), 282.

113. Jordan, "Together," p. 282; Vernon E. Jordan, Jr., "Black and Jewish Communities: An Address to the Atlanta Chapter of the American Jewish Committee," *Vital Speeches*, 40 (August 1, 1974), 630; T, February 27, 1974, p. 15.

114. T, June 29, 1978, p. 1.

115. Louis Harris and Associates, Inc., *A Study of Attitudes Toward Racial and Religious Minorities and Toward Women* (New York: National Conference of Christians and Jews, 1978), p. 46.

116. Jimmy Carter, *Keeping Faith: Memoirs of a President* (New York: Bantam Books, 1982), p. 130; Zbigniew Brzezinski, *Power and Principle: Memoirs of*

the National Security Adviser, 1977–1981 (New York: Farrar, Straus, Giroux, 1983), p. 439; Hume, "Falling Out" p. 18.

117. Weisbord and Kazarian, *Israel in the American Black Perspective*, pp. 130, 138.

118. Thomas H. Landess and Richard M. Quinn, *Jesse Jackson and the Politics of Race* (Ottawa, IL: James Books, 1985), p. 154; Charles Silberman, "Jesse and the Jews," *The New Republic*, 181 (December 29, 1979), 13; William M. Phillips, Jr., *An Unillustrious Alliance: The African American and Jewish Communities* (New York: Greenwood Press, 1991), p. 115.

119. Quoted in *Afro-American*, August 28, 1979, p. 1 in folder "Black-Jewish Relations [Andrew Young] 1978–1979," folder 1, box 77, JCRC mss.

120. Quoted in William Safire, "Of Blacks and Jews," T, September 27, 1979, p. A19.

121. All quotes are from Milton Elerin, "The Young Resignation and Black Anti-Semitism," October 23, 1979, p. 1 in folder 1a, box 77, JCRC mss.

122. Weisbord and Kazarian, *Israel in the American Black Perspective*, p. 132; Hume, "Falling Out," p. 18.

123. Maulara Karenga, "Jesse Jackson and the Presidential Campaign," *The Black Scholar*, 15 (September/October 1984), 65.

124. Weisbord and Kazarian, *Israel in the American Black Perspective*, p. 137.

125. Leonard Fein, "Jesse Jackson and the Jews," *Moment*, 9 (April, 1984), 13.

126. Quoted in "Jackson and the Jews," *The New Republic*, 190 (March 19, 1984), 9.

127. Quoted in Weisbord and Kazarian, *Israel in the American Black Perspective*, p. 180.

128. Weisbord and Kazarian, *Israel in the American Black Perspective*, p. 176; Landess and Quinn, *Jesse Jackson and the Politics of Race*, p. 151; Julius Lester, "You Can't Go Home Again: Critical Thoughts about Jesse Jackson," *Dissent*, 32 (January, 1985), 22.

129. "Jews and Negroes, 1975– ," folder #1, Schomberg Clipping File. Schomburg Library, New York City.

130. Ellis Cose, "For Blacks, There Is No 'One' Voice," *Detroit Free Press*, October 18, 1979, p. 10-A; Elerin, "The Young Resignation and Black Anti-Semitism," p. 8.

131. Both quotes from Hume, "Falling Out," p. 18.

132. Quoted in Weisbord and Kazarian, *Israel in the American Black Perspective*, p. 179.

133. Folder, "Black Muslim Movement, 1963–1984," Blaustein.

134. Quoted in Hume, "Falling Out," p. 18.

135. Quoted in Ron Chepesiuk, "Spokesman for Civil Rights," *Modern Maturity*, 29 (April–May, 1986), 63.

136. Kitty O. Cohen, "Black-Jewish Relations in 1984: A Survey of Black US Congressmen," *Patterns of Prejudice*, 19 (April, 1985), 10–11.

137. Quoted in Hume, "Falling Out," p. 1.

138. Quoted in Fred Barnes, "Farrakhan Frenzy," *The New Republic*, 193 (October 28, 1985), 14; folder, "Black Muslim Movement, 1963–1984," Blaustein.

139. Quoted in Wolf Blitzer, *Between Washington and Jerusalem: A Reporter's Notebook*, (New York: Oxford University Press, 1985), p. 186.

140. AJYB, 87 (1987), 120, 122; Kenneth Bandler, "Farrakhan on the Cam-

pus: Challenges Facing Jewish Students," *Israel Horizons*, 34 (January–February, 1986), 15.

141. Julius Lester, "The Time Has Come," *The New Republic*, 193 (October 28, 1985), 11.

142. Michael Kramer, "Loud and Clear," *New York*, 18 (October 21, 1985), 22.

143. Quoted in ibid., p. 23.

144. Quoted in Roger Rosenblatt, "The Demagogue in the Crowd," *Time*, 126 (October 21, 1985), 102.

145. Quoted in Lester, "The Time Has Come," p. 14.

146. Quoted in Kramer, "Loud and Clear," p. 22.

147. Quoted in Lester, "The Time Has Come," p. 12.

148. Ibid.

149. Quoted in Rosenblatt, "The Demagogue in the Crowd," p. 102.

150. Lester, "The Times Has Come," p. 11.

151. Rosenblatt, "The Demagogue in the Crowd," p. 102.

152. Quoted in T, October 8, 1985, p. 14.

153. Quoted in *Chicago Tribune* (national edition), May 5, 1988, I, 7.

154. *Chicago Tribune* (national edition), May 4, 1988, p. 6. The Cokely story and charges appeared on the first page of the *Chicago Tribune* every day except May 5 during the first week in May 1988.

155. Quoted in Dirk Johnson, "Black-Jewish Hostility Rouses Leaders in Chicago to Action," T, July 29, 1988, p. 4.

156. Ron Dorfman, "Black Racism: Media Reality vs. Real Reality," *Quill*, 76 (September, 1988), 10–11; *Chicago Tribune* (national edition), May 5, 1988, I, 7; Johnson, "Black-Jewish Hostility," pp. 1, 4.

157. Florence Hamlish Levinsohn, "The Cokely Question," *Chicago*, 37 (August, 1988), 139.

158. Gates, "Black Demagogues," p. A13. For a scholarly response to Jeffries' statements see Seymour Drescher, "The Role of Jews in the Transatlantic Slave Trade," *Immigrants and Minorities*, 12 (July 1993), p. 117 ff.

159. Quoted in Eric Pooley, "Doctor J.," *New York*, September 2, 1991, p. 33, see also p. 34.

160. Taylor Branch, "The Uncivil War," *Esquire*, 111 (May, 1989), 112.

161. Quoted in T, August 11, 1991, IV, 6.

162. Branch, "The Uncivil War," p. 112.

163. Cornel West, "Black Anti-Semitism and the Rhetoric of Resentment," *Tikkun*, 7 (January–February, 1992), 16.

164. Kenneth B. Clark, "Candor About Negro-Jewish Relations," p. 121 and Gordis, "Negroes Are Antisemitic Because They Want a Scapegoat," p. 136 in Dinnerstein, ed., *Anti-Semitism in the United States*; Orlansky, "A Note on Anti-Semitism Among Negroes," p. 251; Scheiner, *Negro Mecca*, p. 134; B. Z. Sobel and May L. Sobel, "Negroes and Jews: American Minority Groups in Conflict," *Judaism*, 15 (Winter, 1966), 5, 10; Roland B. Gittlesohn, "Negro-Jewish Relations in America: A Symposium," *Midstream*, 12 (December, 1966), 30; Celia Stopnicka Heller and Alphonso Pinkney, "The Attitudes of Negroes Toward Jews," *Social Forces*, 43 (March, 1965), 366; Strober, *American Jews*, p. 133; David Riesman, "The Politics of Persecution," *Public Opinion Quarterly*, 6 (Spring, 1942), 49; Nathan Glazer and Daniel P. Moynihan, *Beyond the Melting Pot: The Negroes, Puerto Ricans, Jews, Italians, and Irish of New York City*

(Cambridge: Massachusetts Institute of Technology and Harvard University Presses, 1963), p. 77; Rose, *The Negro's Morale*, p. 139; "Introduction," Allon Schoener, ed., *Harlem on My Mind* (New York: Random House, 1968), n. p.

165. Quoted in Catherine Thorpe, "Varsity Athlete's Speech Offends Many," *The Spectator* (Columbia University student newspaper), May 16, 1990, p. 13.

166. Clark, "Black Power and Basic Power," p. 8.

167. Elly Bulkin, Minnie Bruce Pratt, Barbara Smith, *Yours In Struggle: Three Feminist Perspectives on Anti-Semitism and Racism* (Ithaca, N.Y.: Firebrand Books, 1984), p. 69.

168. Daniel Goleman, "Anti-Semitism: A Prejudice That Takes Many Guises," T, September 4, 1984; see also Eugene B. Brody and Robert L. Derbyshire, "Prejudice in American Negro College Students," *Archives of General Psychiatry*, 9 (December, 1963), 626.

169. Orlansky, "A Note on Anti-Semitism Among Negroes," p. 251; Sobel and Sobel, "Negroes and Jews," p. 10.

170. Katz, ed., *Negro and Jew*, p. 126; Lawrence W. Levine, *Black Culture and Black Consciousness* (New York: Oxford University Press, 1977), p. 305.

171. Richard L. Simpson, "Negro-Jewish Prejudice: Authoritarianism and Some Social Variables As Correlates," *Social Problems*, 7 (1959), 145.

172. Gates, "Black Demagogues."

173. Ibid.

174. Ibid.

175. Ibid; Jeffrey C. Alexander and Chaim Seidler-Feller, "False Distinctions and Double Standards: The Anatomy of Double Standards," in *Tikkun*, 7 (January–February, 1992), 12.

176. Quoted in Tom Tugend, "UCLA Blacks Slapped for Maligning Jews," *Arizona Jewish Post*, June 7, 1991, p. 7.

177. Quoted in ibid.

178. Quoted in T (national edition), February 3, 1993, p. A16.

179. Gates, "Black Demagogues."

180. Quoted in Leonard Dinnerstein, *Uneasy at Home: Antisemitism and the American Jewish Experience* (New York: Columbia University Press, 1987), p. 242.

181. Gordis, "Negroes Are Antisemitic Because They Want a Scapegoat," p. 137.

182. Gates, "Black Demagogues"; T, April 19, 1992, p. B3.

Chapter 11

1. Edward S. Shapiro, *A Time for Healing: American Jewry Since World War II* (Baltimore: The Johns Hopkins University Press, 1992), p. 229; Earl Raab, "Anti-Semitism in the 1980s," *Midstream*, February, 1983, p. 17; Peter Levine, *Ellis Island to Ebbets Field: Sport and the American Jewish Experience* (New York: Oxford University Press, 1992), p. 261; Charles S. Liebman and Steven M. Cohen, *Two Worlds of Judaism: The Israeli and American Experiences* (New Haven: Yale University Press, 1990), p. 41ff.; Lawrence H. Fuchs, *The American Kaleidoscope:Race, Ethnicity and the Civil Culture* (Hanover, N.H.: Wesleyan University Press, 1990), p. 560, fn. 35; Jeffrey C. Alexander and Chaim Seidler-Feller, "False Distinctions and Double Standards: The Anatomy of Double Standards,"

Tikkun, 7 (January–February, 1992), 12; William B. Helmreich, "Less Clout Means Fewer Enemies," *Los Angeles Times*, February 21, 1992, p. B7; Jerome A. Chanes, "Antisemitism in the United States," *Midstream*, January, 1990—I worked from a manuscript copy that Chanes sent to me; interview with Milton Himmelfarb, May 23, 1978, p. 18, in "Interviews, B-h," folder 13, box 20, Cohen mss.; David Singer and Renae Cohen, *Perceptions of Israel and American Jews: Findings of the May 1990 Roper Poll* (New York: AJC, 1990), table 8; Henry L. Feingold, *A Time for Searching: Entering the Mainstream, 1920–1945* (Baltimore: The Johns Hopkins University Press, 1992), p. 265; Tom W. Smith, *What Do Americans Think About Jews?* (New York: AJC, 1991), p. 40. Smith also noted that a National Opinion Research Center poll in 1991 found that 19 percent of those interviewed thought that Jews had "too much influence in American life and politics," p. 42, while a 1992 analysis conducted for the ADL found that 31% answered affirmatively to the question about whether Jews had too much power in the United States, "Highlights from an Anti-Defamation League Survey on Anti-Semitism and Prejudice In America" (New York: ADL, November 16), p. 19. Another 1992 study conducted for the American Jewish Committee in New York City during the summer of 1992 concluded that 47% of the city's residents thought that Jews had "too much influence in New York City life and politics." Whether this is just a local phenomenon or a national one is difficult to say at this point. Also, the wording of questions varies slightly, which affects answers. Moreover, it is difficult at this writing to assess whether the increased number of Americans who believe that Jews have "too much influence" is a blip in the charts, the beginning of a new wave of antisemitism (which I doubt), or whether the answer to that question no longer suggests increased hostility toward Jews or a strengthening of antisemitism in the United States but a realistic acknowledgment that Jews do exercise quite a bit of power in this country. See, for example, J. J. Goldberg, "Scaring The Jews," *The New Republic*, 208 (May 17, 1993), 22.

The AJC study covered only New York City. More than half of the city's residents are members of minority groups (African Americans, Hispanics, Asians) and therefore are statistically unrepresentative of the national population. Also, the question is somewhat differently worded than the one that asks "Do Jews have too much power?" A more revealing question in the study is the one that asks respondents' perceptions of how much discrimination exists toward Jews in New York. Broken down according to race and religion, the study shows that 13% of the whites, 14% of the Blacks, 19% of the Asians, 12% of the Hispanics, 8% of the Irish, and 18% of the Italians in New York think that there is "a lot" of discrimination toward Jews. Seventeen percent of the people in the 18–34-year-old bracket see "a lot" of antisemitism compared with 10% in the 35–54-year-old bracket, and 13% in the over-55 age bracket. Similarly, 15% of those who earn less than $20,000 a year and who have not graduated from high school perceive more antisemitism in the city than do those who earn more and have more years of schooling. Only 14% of the Jews, on the other hand, think that there is "a lot" of discrimination against them in the city. "1992 New York City Intergroup Relations Survey," (unpublished paper: survey conducted for the AJC by the Roper Organization, July 27–August 10, 1992), pp. 3, 27, 40.

2. Earl Raab, "The Black Revolution and the Jewish Question," *Commentary*, 47 (January, 1969), 23.

3. Oscar Cohen to Sheldon Steinhauser, September 27, 1972, in folder, "Pluralism in America," folder 4, box 14, clipping, May 29, 1974, folder, "New Anti-Semitism (American context)," folder 7, box 13, interview with Milton Himmelfarb, May 23, 1978," pp. 13, 17, folder 13, box 20, Cohen mss.; Steven M. Cohen, "Undue Stress in American Anti-Semitism? *Sh'ma*, September 1, 1989, p. 13; Chanes, "Antisemitism in the United States," p. 3; Smith, *What Do Americans Think*, p. 26; Helmreich, "Less Clout," p. B7; see also Alexander and Seidler-Feller, "False Distinctions," p. 12.

4. "1992 New York City Intergroup Relations Survey," conducted for the AJC by the Roper Organization, July 27–August 10, 1992, copy in possession of the author; Suzanne Garment, "Discrimination Puts on a New Face in Palm Beach," *The Wall Street Journal*, February 10, 1984, editorial page; *The Wall Street Journal*, April 12, 1990, p. A9; T national edition, December 1, 1990, p. 30, December 7, 1991; "Anti-Semitism: The 'Invisible' Form of Bigotry," *Sojourner*, 10 (November, 1982), 1, 2, 3; Letty Cottin Pogrebin, *Deborah, Golda, and Me: Being Female and Jewish in America* (New York: Crown Publishers, Inc., 1991), chapter 11; Susannah Heschel, "Anti-Judaism in Christian Feminist Theology," *Tikkun*, 5 (May–June, 1990), 26; Rachel Josefowitz Siegel, "Antisemitism and Sexism in Stereotypes of Jewish Women," *Women and Therapy*, 5 (Summer/Fall, 1986), 249, 257; Don Harrison, "Jews and the Cultural Revival of Philadelphia," in Murray Friedman, ed., *Philadelphia Jewish Life, 1940–1985* (Ardmore, Pa.: Seth Press, 1986), passim; Paul Bass, "When Pennies Hurt," *The Fairfield County Advocate* (Connecticut), February 25, 1993, p. 12; Norman Podhoretz, "The Hate That Dare Not Speak Its Name," *Commentary*, 82 (November, 1986), 21, 23; Stephen S. Rosenfeld, "Dateline Washington: Anti-Semitism and U. S. Foreign Policy," *Foreign Policy*, 47 (Summer, 1982), 174; Cara Gerdel Ryan, "Reviving the Politics of Anti-Semitism," *The Nation*, 233 (December 5, 1981), 634; "Again Anti-Semitism," *Newsweek*, 97 (February 16, 1981), 41; interviews with David and Judy Leonard, and Marilyn Heins; Ira Berkow, "Marge Schott: Baseball's Big Red Headache," T (national edition), November 29, 1992, pp. A21 and A24.

5. "The Virulent Disease of Anti-Semitism," CC, 91 (April 24, 1974), 443.

6. Franklin H. Littell, "American Protestantism and Antisemitism," in Naomi W. Cohen, ed., *Essential Papers on Jewish-Christian Relations* (New York: NYU Press, 1990), pp. 180–181; Katharina von Kellenbach, "Anti-Judaism in Christian-rooted Feminist Writings: An Analysis of Major U. S. American and West German Feminist Theologians" (unpublished Ph.D., Department of Philosophy, Temple University, 1990), p. 203; Jack R. Fischel, "Trouble in Paradise," *Journal of the Lancaster County Historical Society*, 87 (1983), 4.

7. Eugene J. Fisher, "Research on Christian Teaching Concerning Jews and Judaism: Past Research and Present Needs," *Journal of Ecumenical Studies*, 21 (Summer, 1984), 426.

8. Quoted in Fuchs, *American Kaleidoscope*, p. 378.

9. "Highlights from an Anti-Defamation League Survey," p. 30.

10. Cohen, "Undue Stress," p. 113.

11. Wolf Blitzer, *Between Washington and Jerusalem: A Reporter's Notebook* (New York: Oxford University Press, 1985), p. 13; Henry L. Feingold, "Finding a Conceptual Framework for the Study of American Antisemitism," *Jewish Social Studies*, 47 (Summer/Fall 1985), 324.

12. Micah L. Sifry, "Anti-Semitism in America," *The Nation*, 256 (January 25, 1993), 95; Mary McGrory, "Book Reveals Nixon and Kissinger as Two of a

Kind," *The Arizona Daily Star,* September 29, 1992, p. A11; Stanley I. Kutler, *The Wars of Watergate: The Last Crisis of Richard Nixon* (New York: Alfred A. Knopf, 1990), pp. 219, 294, 325, 455; AJYB 90 (1990), 211.

13. Clipping of Fikri Abbaza, interview with Richard Nixon, *Al-Mussawar,* July 12, 1974, folder, "Jewish Matters, 1969–1974," box 5, Leonard Garment mss., LC.

14. Folder "Jewish Matters," box 5, Garment mss.

15. "The Virulent Disease of Anti-Semitism," p. 443; T, April 10, 1974, p. 40.

16. Quoted in "The General and the Jews," *Newsweek,* 84 (November 25, 1974), 39.

17. Meg Greenfield, "The General and the Jews," *Newsweek,* 84 (December 9, 1974), 43.

18. Clipping of "The General and the Jews," *The Progressive,* January, 1975, pp. 8–9, in folder, "George S. Brown," Blaustein Library; "Brown Cowed," *Economist,* 253 (November 23, 1974), 56.

19. *Washington Post,* November 21, 1974, p. Al; "Interviews B–H," folder 13, box 20, Cohen mss.

20. Clipping, JTA News Bulletin, October 22, 1976, Bella S. Abzug to President Gerald Ford, June 29, 1976, in folder, "George S. Brown," Blaustein Library.

21. Naomi Levine, "The Nazis of Skokie," *Patterns of Prejudice,* 12 (January–February, 1978), 19; David G. Dalin, "Jews, Nazis, and Civil Liberties," AJYB, 80 (1980), 4; "Skokie and the Nazis," *Newsweek,* 92 (July 31, 1978), 31; "Skokie as Symbol," CC, 95 (April 19, 1978), 411; "Anti-Semitic Incidents, 1976–1981," folder 1, box 76, JCRC mss.

22. Kenneth S. Stern, "Bigotry on Campus," AJC (1990), 1; Courtney Leatherman, "More Anti-Semitism Is Being Reported on Campuses but Educators Disagree on How to Respond to It," *The Chronicle of Higher Education,* 36 (February 7, 1990), Al; "College and Anti-Semitism," folder 7, box 5, Cohen mss.

23. "College and Anti-Semitism," folder 7, box 5, "Incidents," folder 2, box 10, Cohen mss.

24. "College Anti-Semitism," folder 7, box 5, "Incidents," folder 2, box 10, Cohen mss.; quote from Neal Weinberg, "Umass Fights Anti-Semitism," *The Sunday Republican* (Springfield, Mass.), December 11, 1983, p. B-1.

25. "Incidents," folder 2, box 10, Cohen mss.; Abraham K. Korman, *The Outsiders: Jews and Corporate America* (Lexington, Mass.: Lexington Books, 1988), pp. 177–178; "Combatting Bigotry on Campus," ADL (1989), pp. 6–7; anonymous source.

26. Riv-Ellen Prell, "Rage and Representation: Jewish Gender Stereotypes in American Culture," in Faye Ginsburg and Anna Lowenhaupt Tsing, eds., *Uncertain Terms: Negotiating Gender in American Culture* (Boston: Beacon Press, 1990), p. 253ff.; Gary Spencer, "JAP-Baiting on a College Campus: An Example of Gender and Ethnic Stereotyping" (unpublished paper, Department of Sociology, Syracuse University, October, 1987); AJYB 90 (1990), 216–217; "Combatting Bigotry on Campus," p. 9; Helen Teitlebaum, "There's No Excuse for JAP Slur," *The Diamondback* (University of Maryland student newspaper), October 27, 1988, p. 4; Pogrebin, *Deborah, Golda, and Me,* p. 231ff.

27. "ADL Conference on Campus Prejudice" (ADL, 1990), pp. 28–29.

28. *Weiss* v. *US*, 36 *Fair Employment Practice Cases* 1 (1984) @2 (Eastern District, Virginia).

29. Korman, *The Outsiders*, pp. 96–97.

30. Quoted in Jean Baer, *The Self-Chosen: "Our Crowd" Is Dead; Long Live Our Crowd* (New York: Arbor House, 1982) p. 82; see also Shapiro, *A Time for Healing*, p. 115.

31. Interview with John Coleman, pp. 7, 8, "Interviews, B-H," folder 13, box 20, Cohen mss.

32. Robert A. Bennett, "No Longer a WASP Preserve," T, June 29, 1986, III, 1, 8; Steven L. Slavin and Mary A. Pradt, "Anti-Semitism in Banking," *The Bankers Magazine*, 162 (July–August, 1979), 21.

33. Bennett, "No Longer a Wasp Preserve," p. 1.

34. Ibid., pp. 1, 8; Shapiro, *A Time for Healing*, p. 116.

35. Samuel Z. Klausner, "Succeeding in Corporate America: The Experience of Jewish M.B.A.s," (AJC, 1988), 35.

36. Marcia Graham Synott, *The Half-Opened Door: Discrimination and Admissions at Harvard, Yale, and Princeton, 1900–1970* (Westport, Conn.: Greenwood Press, 1979), pp. xix–xx, 209, 222; David Silverberg, "Jewish Studies on the American Campus," *Present Tense*, p. 53 and *Washington Post* editorial, May 5, 1977, in "College Anti-Semitism," folder 7, box 5, Cohen mss.

37. Shapiro, *A Time for Healing*, pp. 94–95, 98; Seymour Martin Lipset and Everett Carl Ladd, Jr., "Jewish Academics in the United States: Their Achievements, Culture and Politics," AJYB, 72 (1971), 92, fn. 10; Baer, *The Self-Chosen*, p. 243.

38. Arthur Hertzberg, "What Future for American Jews?" *The New York Review of Books*, November 23, 1989, p. 26; Leatherman, "More Anti-Semitism," p. A40; interview with Aram Epstein, New Rochelle, New York, May 30, 1992.

39. Robert Andrew Everett, "Judaism In Nineteenth-Century American Transcendentalist and Liberal Protestant Thought," *Journal of Ecumenical Studies*, 20 (Summer, 1983), 397; Carl D. Evans, "The Church's False Witness Against Jews," CC, 99 (May 5, 1982), 530, 531.

40. Rosemary Radford Ruether, "Anti-Semitism and Christian Theology," in Eva Fleischner, ed., *Auschwitz: Beginning of a New Era? Reflections on the Holocaust* (New York: KTAV Publishing House, Inc., 1977), p. 42; Walter Burghardt, S. J., "Response to Rosemary Ruether," in Fleischner, *Auschwitz*, pp. 93, 94.

41. Evans, "The Church's False Witness," p. 530.

42. Ibid., p. 532.

43. Everett, "Judaism in Nineteenth-Century American Transcendentalist and Liberal Protestant Thought," p. 414.

44. Fisher, "Research on Christian Teaching," p. 426; John T. Pawlikowski, "Judaism in Christian Education and Liturgy," in Fleischner, *Auschwitz*, pp. 161–162; Judith H. Banki, "The Image of Jews in Christian Teaching," *Journal of Ecumenical Studies*, 21 (Summer, 1984), 449; interview with Dr. David R. Hunter, January 23, 1978, "Interviews, B-H," folder 13, box 20, Cohen mss.; Chanes, "Antisemitism in the United States."

45. Anthony J. Bevilacqua, "Catholic-Jewish Relations in the USA," *Christian Jewish Relations*, 17 (September, 1984), 51.

46. Ibid., p. 54; "Jewish-Catholic Relations, 1972–1973," folder 5, box 85, JCRC mss.; AJYB, 88 (1988), 158; Robert Wuthnow, *The Restructuring of American Religion: Society and Faith Since World War II* (Princeton University Press, 1988),

p. 96; Doron P. Levin, "Bitter Memories of Bigotry Live on in Father Coughlin's Parish," T, May 25, 1992, p. 7.

47. James F. Moore to Leonard Dinnerstein, September 4, 1991.

48. Shapiro, *A Time for Healing*, pp. 43, 231ff.

49. T, national edition, June 7, 1991, p. A8; see also Debra Nussbaum Cohen, "Comprehensive New Survey Finds Most American Jews Intermarry," *Arizona Jewish Post*, June 21, 1991, pp. 1, 2.

50. T, national edition, June 7, 1991, p. A8.

51. Amongst Jews, however, anxiety over intermarriage and its potentially debilitating effects on the future of the Jewish people in the United States has replaced the fear of antisemitism as a staple in the Jewish press and with rabbis nationally. See Shapiro, *A Time for Healing* pp. 229, 231.

52. Melvin Tumin, "Anti-Semitism and Status Anxiety: A Hypothesis," JSS, 33 (October, 1971), 315.

53. Nathan Glazer, "Exposed American Jew," *Commentary*, 59 (June, 1975), 5.

54. *Commentary*, 47 (March, 1969), 33–42, and 57 (May, 1974), 53–55; "Combatting Anti-Semitism," folder 10, box 5, Cohen mss.; Neil C. Sanberg, *Jewish Life in Los Angeles* (New York: University Press of America, 1986), p. 127; Jeffrey M. Silberman, "Anti-Semitism in the Suburbs," *Humanistic Judaism*, 13 (Spring, 1985), 15; Cohen, "Undue Stress," p. 114; Nathan Glazer, "New Perspectives in American Jewish Sociology," AJYB, 87 (1987), p. 7n; Shapiro, *A Time for Healing*, pp. 48–49, 56–57; T, March 2, 1992, p. A6.

55. Craig Horowitz, "The New Anti-Semitism," *New York*, 26 (January 11, 1993), 23.

56. Gates, "Black Demagogues"; Smith, "What Do Americans Think About Jews?" p. 27; Wuthnow, *The Restructuring of American Religion*, p. 222; Chanes, "Antisemitism in the United States"; Shmuel Almog, "Antisemitism as a Dynamic Phenomenon: The 'Jewish Question' in England at the End of the First World War," *Patterns of Prejudice*, 21 (Winter, 1987), 3; Cohen, "Undue Stress," pp. 113–114.

57. Liebman and Cohen, *Two Worlds of Judaism*, p. 43.

58. "The Virulent Disease of Anti-Semitism," p. 443.

59. "Interviews, B-H," folder 13, box 20, Cohen mss.

Conclusion

1. At the Republican governors' conference in November 1992 Mississippi Governor Kirk Fordice stated, "The United States of America is a Christian nation." South Carolina Governor Carroll A. Campbell suggested that "the value base of this country comes from the Judeo-Christian heritage," thereby adding the "Judeo" aspect. Governor Fordice responded, however, "If I wanted to do that I would have done it." T (national edition), November 21, 1992, p. 15.

2. "The Talk of the Town," *The New Yorker*, 66 (November 26, 1990), 37.

3. Paul Berman, "Gentlemen's Disagreement," *The New Yorker*, October 12, 1992, p. 114.

4. "Troubled Racial and Ethnic Relations In New York City Revealed in New American Jewish Committee/Roper Survey," AJC Press Release, October 15, 1992.

Bibliography

Manuscript Collections

Abzug, Bella, COHC
Borah, William E., LC
Brooklyn Jewish Community Council, AJA
Celler, Emanuel, LC
Clapper, Raymond, LC
Cohen, Benjamin V., LC
Cohen, Oscar, AJA
Fineberg, Solomon A., AJA
Frankfurter, Felix. LC
Garment, Leonard, LC
Hobson, Laura Z., COHC
Hofstadter, Richard, COHC
Ickes, Harold, LC. Diaries of
JCRC, AJA
NAACP, LC
Randall, James G. LC
Rapp, Michael G., AJA
Roosevelt, Franklin D., Roosevelt Library, Hyde Park, N.Y.
Russel, Charles E., LC
Sabath, A. J., AJA
Saltonstall, Leverett, Massachusetts Historical Society

Newspapers

Afro-American Ledger (Baltimore). Also known as *The Ledger*.
Amerian Sentinel (New York City)
American Hebrew (AH) (New York)
American Israelite (AI) (Cincinnati)
American Jewish World (Minneapolis)
The Arizona Daily Star (Tucson)
The Arizona Jewish Post (Tucson)
Boston Daily Evening Transcript
Chicago Defender
Chicago Tribune
Colored American (Washington, D.C.)
Daily Alta California
Detroit Free Press
The Evening News (Detroit)
Freedman's Journal (New York City)
The Freeman (Indianapolis)
Israelite; see *American Israelite*
Jerusalem Post
Jewish Messenger (New York City)
Los Angeles Times
Louisville Defender
Milwaukee Daily Sentinel
Minneapolis Journal
The Modern View (St. Louis)
New York Age
New York *Amsterdam News*
New York Freeman
New York *Post*
New York *Sun*
New York Times
New York *Tribune*
New York *World*
Norfolk Journal and Guide (Virginia)
Philadelphia Tribune
Pittsburgh Courier
PM (New York)
Richmond News-Leader (Virginia)
Richmond Planet (Virginia)
Schomburg Clipping File (Schomburg Library, New York)
Spectator (London)
St. Louis Argus
Times-Dispatch (Richmond, Virginia)
Village Voice (New York City)
Voice of the Negro (Atlanta)
Wall Street Journal
Washington Bee
Washington Post

Books

Abelow, Samuel P. *History of Brooklyn Jewry*. Brooklyn: Scheba Publishing Co., 1937

Abernathy, Arthur T. *The Jew and Negro*. Moravia Falls, N.C.: Dixie Publishing Co., 1910.

Adorno, T. W. Else Frenkel-Brunswick, Daniel J. Levinson, R. Nevitt Sanford. *The Authoritarian Personality*. New York: Harper & Bros., 1950.

Ahlstrom, Sidney. *A Religious History of the American People*. New Haven: Yale University Press, 1972.

Alexander, Charles C. *The Ku Klux Klan in the Southwest*. Lexington: University of Kentucky Press, 1965.

Ashkenazi, Elliott. *The Business of Jews in Louisiana, 1840–1875*. Tuscaloosa: University of Alabama Press, 1988.

Auerbach, Jerold S. *Unequal Justice: Lawyers and Social Change in Modern America*. New York: Oxford University Press, 1976.

Baltzell, E. Digby. *The Protestant Establishment*. New York: Random House, 1964.

Bayor, Ronald H. *Neighbors in Conflict: The Irish, Germans, Jews, and Italians in New York City, 1929–1941*. 2nd edition. University of Illinois Press, 1988.

Berlin, George L. *Defending the Faith: Nineteenth-Century American Jewish Writings on Christianity and Jesus*. Albany: State University of New York Press, 1989.

Berman, Myron. *Richmond's Jewry*. Charlottesville: University Press of Virginia, 1979.

Berson, Lenora E. *The Negroes and the Jews*. New York: Random House, 1971.

Bettleheim, Bruno and Morris Janowitz. *Social Change and Prejudice; Including Dynamics of Prejudice*. New York: The Free Press of Glencoe, 1964.

Bishop, Claire Huchet. *How Catholics Look at Jews*. New York: Paulist Press, 1974.

Blau, Joseph L. and Salo W. Baron, eds. *The Jews of the United States, 1790–1840: A Documentary History*. New York: Columbia University Press, 1963. 3 volumes.

Blumberg, Janice Rothschild. *One Voice*. Macon, Ga.: Mercer University Press, 1985.

Borden, Morton. *Jews, Turks, and Infidels*. Chapel Hill: University of North Carolina Press, 1984.

Brinkley, Alan. *Voices of Protest: Huey Long, Father Coughlin and the Great Depression*. New York: Vintage, 1983.

Broun, Heywood and George Britt. *Christians Only*. New York: The Vanguard Press, 1931.

Capeci, Dominic J., Jr. *Race Relations in Wartime Detroit: The Sojourner Truth Housing Controversy of 1942*. Philadelphia: Temple University Press, 1984.

Capeci, Dominici J. Jr. *The Harlem Riot of 1943*. Philadelphia: Temple University Press, 1977.

Carlson, John Roy. *Under Cover*. New York: E. P. Dutton & Co., 1943.

Carson, Clayborne. *In Struggle: SNCC and the Black Awakening of the 1960s*. Cambridge: Harvard University Press, 1981.

Carter, Dan T. *Scottsboro*. Baton Rouge: Louisiana State University Press, 1969.

Cash, W. J. *The Mind of the South*. New York: Alfred A. Knopf, 1941.

Chalmers, David. *Hooded Americans: The First Century of the Ku Klux Klan, 1865–1956*. New York: Doubleday & Co., Inc., 1965.

Cohen, Naomi W. *A Dual Heritage: The Public Career of Omar S. Strauss*. Philadelphia: JPS, 1969.

Cohen, Naomi W., *Encounter with Emancipation: The German Jews in the United States, 1830–1914*. Philadelphia: JPS, 1984.

Cohen, Naomi W., ed. *Essential Papers on Jewish-Christian Relations: Imagery and Reality*. New York: NYU Press, 1990.

Cohen, Naomi W. *Jews in Christian America: The Purusit of Religious Equality*. New York: Oxford University Press, 1992.

Cohen, Naomi W. *Not Free To Desist: The American Jewish Committee, 1906–1966*. Philadelphia: JPS, 1972.

Cole, Wayne S. *America First: The Battle Against Intervention, 1940–1941*. Madison: University of Wisconsin Press, 1953.

Cole, Wayne S. *Senator Gerald P. Nye and American Foreign Relations*. Minneapolis: University of Minnesota Press, 1962.

Cruse, Harold. *The Crisis of the Negro Intellectual*. New York: William Morrow & Co., 1967.

Dance, Daryl Cumber. *Schuckin' and Jivin': Folklore from Contemporary Black Americans*. Bloomington: Indiana University Press, 1978.

Davies, Rosemary Rieves. *The Rosenbluth Case: Federal Justice on Trial*. Ames: Iowa State University Press, 1970.

Diner, Hasia R. *A Time for Gathering: The Second Migration, 1820–1880*. Johns Hopkins University Press, 1992.

Diner, Hasia R. *In the Almost Promised Land: American Jews and Blacks, 1915–1935*. Westport, Conn.: Greenwood Press, 1977.

Dinnerstein, Leonard, ed. *Antisemitism in the United States*. New York: Holt, Rinehart & Winston, 1971.

Dinnerstein, Leonard. *America and the Survivors of the Holocaust*. New York: Columbia University Press, 1982.

Dinnerstein, Leonard. *The Leo Frank Case*. New York: Columbia University Press, 1968.

Dinnerstein, Leonard. *Uneasy at Home: Antisemitism and the American Jewish Experience*. New York: Columbia University Press, 1987.

Dinnerstein, Leonard and Mary Dale Palsson, eds. *Jews in the South*. Baton Rouge: Louisiana State University Press, 1973.

Dobkowski, Michael. *The Tarnished Dream*. Westport, Conn.: Greenwood Press, 1979.

DuBois, W. E. B. *The Souls of Black Folk*. Millwood, N.Y.: Kraus-Thomson Organization, Ltd., 1973.

DuBois, W. E. B. *The Souls of Black Folk*. Chicago: A.C. McClurg and Co., 1903.

Dundes, Alan ed., *The Blood Libel Legend: A Casebook in Anti-Semitic Folklore*. University of Wisconsin Press, 1991.

Elson, Ruth Miller. *Guardians of Tradition: American Schoolbooks of the Nineteenth Century*. Lincoln: University of Nebraska Press, 1964.

Endelman, Judith E. *The Jewish Community of Indianapolis: 1849 to the Present*. Bloomington: Indiana University Press, 1984.

Epstein, Benjamin R. and Arnold Forster. *"Some of My Best Friends . . . "* New York: Farrar, Straus and Cudahy, 1962.

Essays in American Jewish History. Cincinnati: American Jewish Archives, 1958.

Evans, Eli N. *The Provincials: A Personal History of Jews in the South.* New York: Atheneum, 1973.

Faber, Eli. *A Time for Planting: The First Migration, 1654–1820.* Baltimore: Johns Hopkins University Press, 1992.

Fein, Isaac M. *The Making of an American Jewish Community: The History of Baltimore Jewry from 1773 to 1920.* Philadelphia: JPS, 1971.

Feingold, Henry. *A Time for Searching: Entering the Mainstream, 1920–1945.* Baltimore: Johns Hopkins University Press, 1992.

Feldman, Egal. *Dual Destinies: The Jewish Encounter with Protestant America.* Urbana: University of Illinois Press, 1990.

Flannery, Edward H. *The Anguish of the Jews: Twenty-Three Centuries of Antisemitism.* Revised and Updated. New York: Paulist Press, 1985.

Fleischner, Eva, ed. *Aushwitz: Beginning of a New Era? Reflections on the Holocaust.* New York: KTAV Publishing House, Inc., 1977.

Forster, Arnold and Bejamin R. Epstein. *The Trouble-Makers.* Garden City, N.Y.: Doubleday, 1952.

Forster, Arnold. *Square One.* New York: Donald I. Fine, 1988.

Friedman, Saul S. *The Incident at Massena.* New York: Stein & Day, 1978.

Freund, Miriam K. *Jewish Merchants in Colonial America.* New York: Behrman's Jewish Book House, 1939.

Gabler, Neal. *An Empire of Their Own: How the Jews Invented Hollywood.* New York: Crown Publishers, Inc., 1988.

Gal, Allon. *Brandeis of Boston.* Cambridge: Harvard University Press, 1980.

Geltman, Max. *The Confrontation: Black Power, Anti-Semitism, and the Myth of Integration.* Englewood Cliffs, N.J.: Prentice-Hall, Inc., 1970.

Gerber, David A., ed. *Anti-Semitism in American History.* Urbana: University of Illinois Press, 1986.

Glanz, Rudoof. *The Jew in the Old American Folklore.* New York: n. p., 1961.

Glock, Charles Y. and Rodney Stark. *Christian Beliefs and Anti-Semitism.* New York: Harper & Row, 1966.

Goodman, Abram Vossen. *American Overture.* Philadelphia: JPS, 1947.

Gossett, Thomas F. *Race: The History of an Idea.* New York: Schochen Books, 1965.

Graeber, Isacque and Stuart Henderson Britt, eds. *Jews in a Gentile World.* New York: The Macmillan Co., 1942.

Grinstein, Hyman B. *The Rise of the Jewish Community of New York, 1654–1860.* Philadelphia: JPS, 1945.

Handy, Robert T. *A Christian America.* 2nd edition. New York: Oxford University Press, 1984.

Harlan, Louis R. *Booker T. Washington: The Wizard of Tuskegee, 1901–1915.* New York: Oxford University Press, 1983.

Harlan, Louis R. et al., eds. *The Booker T. Washington Papers.* Urbana: University of Illinois Press, 1974.

Haygood, Wil. *King of the Cats: The Life and Times of Adam Clayton Powell, Jr.* Boston: Houghton Mifflin, 1993.

Hendrick, Burton. *Jews in America.* New York: Doubleday, Page & Co., 1923.

Hentoff, Nat, ed. *Black Anti-Semitism and Jewish Racism.* New York: Richard W. Baron, 1969.

Hertzberg, Steven. *Stranger Within the Gate City: The Jews of Atlanta, 1845–1915.* Philadelphia: JPS, 1978.

Higham, John. *Strangers in the Land.* New Brunswick: Rutgers University Press, 1955.

Higham, John. *Send These to Me: Jews and Other Immigrants in Urban America.* Revised Edition. Baltimore: Johns Hopkins Press, 1984.

Howe, Irving. *World of Our Fathers: The Journey of the East European Jews to America and the Life They Found and Made.* New York: Harcourt Brace Jovanovich, 1976.

Hsia, R. Po-Chia. *The Myth of Ritual Murder: Jews and Magic in Reformation Germany.* New Haven: Yale University Press, 1988.

Isaac, Jules. *The Teaching of Contempt: Christian Roots of Anti-Semitism.* New York: Holt, Rinehart and Winston, 1964.

Janowsky, Oscar I. ed. *The American Jew.* New York: Harper & Bros., 1942, 183–204.

Jeansonne, Glenn. *Gerald L. K. Smith: Minister of Hate.* New Haven: Yale University Press, 1988.

Jick, Leon A. *The Americanization of the Synagogue, 1820–1870.* Hanover, N.H.: Brandeis University Press, 1976.

Johnson, Joseph A. Jr. *The Soul of the Black Preacher.* Philadelphia: United Church Press, 1971.

Jones, Charles C. *A Catechism of Scripture Doctrine and Practice for Families and Sabbath-schools Designed also for the Oral Instruction of Coloured Persons.* Philadelphia: Presbyterian Board of Education, 1852.

Jones, J. B. *A Rebel War Clerk's Diary.* 2 volumes. Philadelphia: J.B. Lippincott & Co., 1866.

Kaganoff, Nathan M. and Melvin I. Urofsky, eds. *"Turn to the South": Essays on Southern Jewry.* Charlottesville: University Press of Virginia, 1979.

Kantowicz, Edward R. *Polish-American Politics in Chicago, 1880–1940.* Chicago: University of Chicago Press.

Katz, Irving. *August Belmont.* New York: Columbia University Press, 1968.

Katz, Jacob . *From Prejudice to Destruction.* Cambridge: Harvard University Press, 1980.

Katz, Jacob. *Out of the Ghetto.* Cambrdige: Harvard University Press, 1973.

Kaufman, Jonathan. *Broken Alliance: The Turbulent Times Between Blacks and Jews in America.* New York: Charles Scribner's Sons, 1988.

Klingenstein, Susanne. *Jews In The American Academy, 1900–1940.* New Haven: Yale University Press, 1991.

Korman, Abraham K. *The Outsiders: Jews and Corporate America.* Lexington, Mass.: Lexington Books, 1988.

Korn, Bertram Wallace. *American Jewry and the Civil War.* Philadelphia: JPS, 1951.

Korn, Bertram Wallace. *Eventful Years and Experiences.* Cincinnati: The American Jewish Archives, 1954.

Korn, Bertram Wallace. *The American Reaction to the Mortara Case: 1858–1859.* Cincinnati: The American Jewish Archives, 1957.

Lebeson, Anita Libman. *Pilgrim People: A History of Jews in America from 1492 to 1974.* Revised Edition. New York: Minerva Press, 1975.

Lee, Albert. *Henry Ford and the Jews.* New York: Stein and Day, 1980.

Levine, Lawrence W. *Black Culture and Black Consciousness.* New York: Oxford University Press, 1977.

Levine, Peter. *Ellis Island to Ebbets Field: Sport and the American Jewish Experience.* New York: Oxford University Press, 1992.

Lewisohn, Ludwig. *Up Stream.* New York: Boni and Liveright, 1922.

Liebman, Arthur. *Jews and the Left.* New York: John Wiley, 1979.

Liebman, Charles S. and Steven M. Cohen. *Two Worlds of Judaism: The Israeli and American Experiences.* New Haven: Yale University Press, 1990.

Lipstadt, Deborah E. *Beyond Belief: The American Press and the Coming of the Holocaust, 1933–1945.* New York: The Free Press, 1986.

Littell, Franklin H. *The Crucifixion of the Jews.* New York: Harper & Row, 1975.

Marcus, Jacob R. *The Colonial American Jew, 1492–1776.* 3 volumes. Detroit: Wayne State University Press, 1970.

Marcus, Jacob Rader. *United States Jewry 1776–1985.* Volume 2: The Germanic Period. Detroit: Wayne State University Press, 1991.

Marcus, Sheldon. *Father Coughlin: The Tumultuous Life of the Priest of the Little Flower.* Boston: LIttle, Brown, & Co., 1973.

Marty, Martin E. *Righteous Empire: The Protestant Experience in America.* New York: Dial Press, 1970.

Marx, Gary T. *Protest and Prejudice: A Study of Belief in the Black Community.* New York: Harper and Row, 1967.

Mayo, Louise A. *The Ambivalent Image: Nineteenth-Century America's Perception of the Jew.* Rutherford, N.J.: Fairleigh Dickinson University Press, 1988.

McIlhenny, E.A. *Befo' De War Spirituals.* Boston: The Christopher Publishing House, 1933.

McWilliams, Carey. *A Mask for Privilege: Anti-Semitism in America.* New York: Little, Brown & Company, 1948.

Merwick, Donna. *Boston's Priests, 1848–1910.* Cambridge, Mass.: Harvard University Press, 1973.

Moore, Deborah Dash. *At Home in America: Second-Generation New York Jews.* New York: Columbia University Press, 1981.

Murray, Robert K. *Red Scare: A Study in National Hysteria 1919–1920.* Minneapolis: University of Minnesota Press, 1955.

Nelson, Jack. *Terror in the Night: The Klan's Campaign Against the Jews.* New York: Simon and Schuster, 1993.

Nicholas, H.G., ed. *Washington Dispatches, 1941–1945.* London: Weidenfeld and Nicolson, 1981.

Novick, Peter. *That Noble Dream.* Cambridge: Cambridge University Press, 1988.

Nugent, Walter T. K. *The Tolerant Populists.* Chicago: University of Chicago Press, 1963.

Olson, Bernhard E. *Faith and Prejudice: Intergroup Problems in Protestant America.* New Haven: Yale University Press, 1963.

Oren, Dan A. *Joining the Club: A History of Jews and Yale.* New Haven: Yale University Press, 1985.

Ottley, Roi *"New World A-Coming."* Boston: Houghton Mifflin Co., 1943.

Parsons, Talcott and Kenneth B. Clark, eds. *The Negro American.* Boston: Houghton Mifflin Co., 1966.

Pettigrew, Thomas F. *A Profile of the Negro American.* Princeton, N.J.: D. Van Nostrand Co., 1964.

Pogrobin, Letty Cottin. *Deborah, Golda, and Me: Being Female and Jewish in America*. New York: Crown Publishers, Inc., 1991.

Poliakov, Leon. *The History of Antisemitism*. New York: The Vanguard Press, 1965.

Quinley, Harold E. and Charles Y. Glock. *Anti-Semitism in America*. New York: Free Press, 1979.

Reed, John Shelton. *The Enduring South*. Chapel Hill: University of North Carolina Press, 1972.

Reznikoff, Charles *The Jews of Charleston*. Philadelphia: JPS, 1950.

Ribuffo, Leo P. *The Old Christian Right*. Philadelphia: Temple University Press, 1983.

Roberts, Kenneth L. *Why Europe Leaves Home*. Indianapolis: The Bobbs-Merrill Co., 1922.

Rockaway, Robert A. *The Jews of Detroit: From the Beginning, 1762–1914*. Detroit: Wayne State University Press, 1986.

Rogaw, Arnold A., ed. *The Jew in a Gentile World*. New York: The Macmillan Co., 1961.

Rosenstock, Morton. *Louis Marshall, Defender of Jewish Rights*. Detroit: Wayne State University Press, 1965.

Rosovsky, Nitza. *The Jewish Experience at Harvard and Radcliffe*. Cambridge: Harvard University Press, 1986.

Ross, Edward Alsworth. *The Old World in the New*. New York: The Century Co., 1914.

Ross, Robert W. *So It Was True: The American Protestant Press and the Nazi Perseuction of the Jews*. Minneapolis: University of Minnesota Press, 1980.

Rothschild, Janice O. *As But a Day*. Atlanta, Ga.: Hebrew Benevolent Congregation, 1967.

Ruether, Rosemary Radford. *Faith and Fratricide: The Theological Roots of Anti-Semitism*. New York: The Seabury Press, 1974.

Sachar, Howard M. *A History of the Jews in America*. New York: Alfred A. Knopf, 1992.

Saperstein, Marc. *Moments of Crisis in Jewish-Christian Relations*. London: SCM Press, 1989.

Sarna, Jonathan D. *Jacksonian Jew: The Two Worlds of Mordecai Noah*. New York: Holmes and Meier, 1981.

Sarna, Jonathan D. and Nancy H. Klein, *The Jews of Cincinnati*. Cincinnati: Center for Study of the American Jewish Experience on the Campus of the Hebrew Union College—Jewish Institute of Religion, 1989.

Schappes, Morris U. *A Documentary History of the Jews in the United States, 1654–1875*. 3rd edition. New York: Schocken Books, 1971.

Schiff, Ellen. *From Stereotype to Metaphor: The Jew in Contemporary Drama*. Albany: State University of New York Press, 1982.

Schmier, Louis, ed. *Reflections of Southern Jewry: The Letters of Charles Wessolowsky, 1878–1879*. Macon, Ga.: Mercer University Press, 1982.

Selznick, Gertrude J. and Stephen Steinberg. *The Tenacity of Prejudice*. New York: Harper and Row, 1969.

Shankman, Arnold. *Ambivalent Friends: Afro-Americans View the Immigrant*. Westport, Conn.: Greenwood Press, 1982.

Shapiro, Edward S. *A Time for Healing: American Jewry Since World War II*. Baltimore: Johns Hopkins University Press, 1992.

Simmel, Ernst, ed. *Anti-Semitism: A Social Disease*. New York: International Universities Press, 1946.

Singerman, Robert. *Antisemitism: An Annotated Bibliography and Research Guide*. New York: Garland Publishing Co., 1982.

Solomon, Barbara Muller. *Ancestors and Immigrants*. New York: John Wiley & Sons, 1965.

Sorin, Gerald. *A Time for Building: The Third Migration, 1880–1920*. Baltimore: Johns Hopkins University Press, 1992.

Stack, John F. Jr. *International Conflict in an American City: Boston's Irish, Italians, and Jews, 1935–1944*. Westport, Conn.: Greenwood Press, 1979.

Stark, Rodney, Bruce D. Foster, Charles Y. Glock, and Harold E. Quinley. *Wayward Shepherds: Prejudice and the Protestant Clergy*. New York: Harper & Row, 1976.

Stember, Charles Herbert et al. *Jews in the Mind of America*. New York: Basic Books, 1966.

Strong, Donald S. *Organized Anti-Semitism in America: The Rise of Group Prejudice During the Decade 1930–1940*. Washington, D.C.: American Council on Public Affairs, 1941.

Synott, Marcia Graham. *The Half-Opened Door: Discrimination and Admissions at Harvard, Yale, and Princeton, 1900–1970*. Westport, Conn.: Greenwood Press, 1979.

Trachtenberg, Joshua. *The Devil and the Jews*. New Haven: Yale University Press, 1943.

Tsukoshima, Ronald T. *The Social and Psychological Correlates of Black Anti-Semitism*. San Francisco: R + E Research Associates, 1978.

Tull, Charles J. *Father Coughlin and the New Deal*. Syracuse University Press, 1965.

Tumin, Melvin Marvin. *An Inventory and Appraisal of Research on American Anti-Semitism*. New York: Freedom Books, 1961.

Unger, Irwin. *The Greenback Era: A Social and Political History of American Finance, 1865–1876*. Princeton: Princeton University Press, 1964.

Vance, Zebulon. *The Scattered Nation*. New York: Marcus Schnitzer, 1916.

Vorspan, Max and Lloyd P. Gartner. *History of the Jews of Los Angeles*. Philadelphia: JPS, 1970.

Washington, Joseph R. Jr., ed. *Jews in Black Perspectives: A Dialogue*. Rutherford, N.J.: Fairleigh Dickinson University Press, 1984.

Wechsler, Harold S. *The Qualified Student*. New York: John Wiley and Sons, 1977.

Wedlock, Lunabelle. *The Reaction of Negro Publications and Organizations to German Anti-Semitism*. Washington, D.C.: The Howard University Studies in the Social Sciences III, No. 2, 1942.

Weisbord, Robert G. and Richard Kazarian, Jr. *Israel in the American Black Perspective*. Westport, Conn.: Greenwood Press, 1985.

Weisbord, Robert G. and Arthur Stein. *Bittersweet Encounter: The Afro-American and the American Jew*. Westport, Conn.: Negro Universities Press, 1970.

Whiteman, Maxwell. *Copper for America: The Hendricks Family and a National Industry, 1755–1939*. New Brunswick: Rutgers University Press, 1971.

Williamson, Clark. *Has God Rejected His People? Anti-Judaism in the Christian Church*. Nashville: Abingdon, 1982.

Wise, Isaac M. *Reminiscences*. Edited by David Philipson. New York: Arno Press, 1973.

Wistrich, Robert S. *Antisemitism: The Longest Hatred*. London: Metheun London, 1991.

Wolf, Edwin II and Maxwell Whiteman. *The History of the Jews in Philadelphia from Colonial Times to the Age of Jackson*. Philadelphia: JPS, 1957.

Wolfskill, George & John A. Hudson. *All But the People*. New York: Macmillan Co., 1969.

Worthington, H. P. C. *Hell for the Jews*. New York: Printed for the author, 1879.

Wright, Richard. *Black Boy*. New York: Harper and Bros., 1945.

Wyman, David S. *Paper Walls: America and the Refugee Crisis, 1938–1941*. University of Massachusetts Press, 1968.

Wyman, David S. *The Abandoment of the Jews*. New York: Pantheon Books, 1984.

Zuckerman, Nathan. *The Wine of Violence*. New York: Association Press, 1947.

Articles

Adler, Selig. "Zebulon B. Vance and the Scattered Nation," *Journal of Southern History*, 7 (1941), 357–77.

Alexander, Charles C. "Prophet of American Racism: Madison Grant and the Nordic Myth," *Phylon*, 23 (Spring, 1962), 73–90.

Alexander, Jeffrey C. and Chaim Seidler-Feller, "False Distinctions and Double Standards: The Anatomy of Double Standards," *Tikkun*, 7 (January–February, 1992), 12–14.

Allport, Gordon W. and Bernard M. Kramer. "Some Roots of Prejudice," *Journal of Psychology*, 22 (1946), 9–39.

Anscar, Hans. "Catholics and Anti-Semitism," *Catholic World*, 150 (November, 1939), 173–78.

Anthony, Alfred Williams. "Explaining the Jew," CC, 50 (August 16, 1933), 1034–36.

"Anti-Jewish Prejudice at Camps," AH, 102 (November 23, 1917), 69.

"Anti-Jewish Sentiment in California—1855," AJA 12 (April, 1960), 15–33.

"Anti-Semitic Vandalism," CC, 77 (February 17, 1960), 198, 200.

"Anti-Semitism Among Negroes," *Crisis*, 45, (June, 1938), 177.

"Anti-Semitism Is Here," *The Nation*, 147 (August 20, 1938), 167–168.

"Anti-Semitism Masked as Sociology," AH, 95 (September 18, 1914), 611–12.

"Anti-Semitism: The 'Invisible' Form of Bigotry," *Sojourner*, 10 (November, 1982), 1–4.

"Anti-Semitism: The Banality of Evil," *The Economist*, 286 (January 15, 1983), 91.

"Anti-Semitism," *Life*, 23 (December 1, 1947) 44.

Appel, John and Selma Appel, "Anti-Semitism in American Caricature," *Society*, 24 (November–December, 1986), 78–83.

Aptheker, Herbert. "The Souls of Black Folk: A Comparison of the 1903 and 1952 Editions," *Negro History Bulletin*, 34 (1971), 15–17.

"Are You a Jewess?" AJA 15 (April, 1963), 59.

Ash, Steven V. "Civil War Exodus: The Jews and Grant's General Orders No. 11," *The Historian*, 44 (August, 1982), 505–523.

Attwood, William. "The Position of the Jews in America Today," *Look*, 19 (November 29, 1955), 27–35.

Auerbach, Jerold S. "From Rags to Robes: The Legal Profession, Social Mobility, and the American Jewish Experience," AJHQ, 66 (December, 1966), 249–284.

Auerbach, Jerold S. "Fighting Anti-semitism at Wellesley," *Sh'ma*, 15 (November 16, 1984), 1–2.

Baker, Ella and Mavel Cooke. "The Bronx Slave Market," *Crisis*, 42 (November, 1935), 330–31, 340.

Bakst, Jerome. "Negro Radicalism Turns Anti-Semitic: SNCC's Volte Face," *Weiner Library Bulletin*, (Winter, 1967–68), 20–22.

Baldwin, James. "Negroes Are Antisemitic Because They're Antiwhite," in Leonard Dinnerstein, ed, *Antisemitism in the United States*. New York: Holt, Rinehart and Winston, 1971.

Baldwin, James. "The Harlem Ghetto, 1948," *Commentary*, 5 (February, 1948), 165–170.

Banki, Judith H. "The Image of Jews in Christian Teaching," *Journal of Educmenical Studies*, 21 (Summer, 1984), 437–451.

Basso, Hamilton. "The Riot in Harlem," *New Republic*, 82 (April 3, 1935), 209–210.

Bauman, Mark K. and Arnold Shankman. "The Rabbi as Ethnic Broker: The Case of David Marx," *Journal of American Ethnic History*, 2 (Spring, 1983), 51–68.

Bayton, James A. "The Racial Stereotypes of Negro College Students," *Journal of Abnormal and Social Psychology*, 36 (January, 1941), 97–102.

Beckwith, Martha W. "The Jew's Garden . . . ," *Journal of American Folklore*, 64 (1951), 224–25.

Bender, Eugene I. "Reflections on Negro-Jewish Relationships: The Historical Dimension," *Phylon*, 30 (January, 1969), 56–65.

Benson, Arnold "The Catholic Church and the Jews," *American Jewish Chronicle*, 1 (February 15, 1940), 5–7.

Berdyaev, Nicholas A. "The Crime of Anti-Semitism," *Commonweal*, 29 (April 21, 1939), 706–09.

Berman, Hyman. "Political Antisemitism in Minnesota During the Great Depression," JSS, 38 (Summer/Fall, 1976), 247–264.

Berman, Myron. "Rabbi Edward Newton Calisch and the Debate over Zionism in Richmond, Virginia," AJHQ, 62 (March, 1973), 295–305.

Berman, Paul. "Gentlemen's Disagreement," *The New Yorker*, 68 (October 12, 1992), 114–118.

Bernheimer, Charles S. "Prejudice Against Jews in the United States," *The Independent*, 65 (1908), 1105–08.

Bettelheim, Bruno. "The Dynamism of Anti-Semitism in Gentile and Jew," *Journal of Abnormal and Social Psychology*, 42 (1947), 153–68.

Bishop, Claire Huchet. "Learning Bigotry," *Commonweal*, 80 (May 22, 1964), 264–66.

Blandford, B. W. "Commodore Uriah P. Levy." AH, 116 (April 17, 1925), 763, 770–71.

Bloore, Stephen. "The Jew in American Democratic Literature, 1794–1930," PAJHS, 40 (June, 1951), 345–360.

Blum, Barbara Sandra and John H. Mann. "The Effect of Religious Membership on Religious Prejuidce," *The Journal of Social Psychology*, 52 (1960), 97–101.

Boas, Ralph Philip. "Jew-Baiting in America," *Atlantic Monthly*, CXXVII (May, 1921), 658–65.

Boas, Ralph Philip. "Who Shall Go to College?" *Atlantic Monthly*, CXXX (October, 1922), 44–48.

Bond, Horace Mann. "Negro Attitudes Toward Jews," JSS, 27 (January, 1965), 3–9.

Boxerman, Burton A. "The St. Louis Jewish Coordinating Council: Its Formative Years," *Missouri Historical Review*, 65 (October, 1970), 51–71.

Boxerman, Burton Alan. "Rise of Anti-Semitism in St. Louis, 1933–1945," *Yivo Annual of Jewish Social Science*, 14 (1969), 251–69.

Boyd, Ernest. "As a Gentile Sees It," *Scribner's Magazine*, 94 (October, 1933), 242–43.

Braiterman, Marvin. "Negro Anti-Semitism: Fact or Fiction," *Dimensions in American Judaism*, 2 (Summer, 1968), 40–41.

Britt, George. "Poison in the Melting Pot," *The Nation*, 148 (April 1, 1939), 374–76.

Brody, David "American Jewry, The Refugees and Immigration Restriction (1932–1942)," PAJHS, 45 (June, 1956), 219–47.

Brody, Eugene B. and Robert L. Derbyshire. "Prejudice in American Negro College Students," *Archives of General Psychiatry*, 9 (December, 1963), 619–628.

Broom, Leonard, Helen P. Beem, Virginia Harris. "Characteristics of 1,107 Petitioners for Change of Name," *American Journal of Sociology*, 20 (February, 1955), 33–99.

Brown, James. "Christian Teaching and Anti-Semitism," *Commentary*, 24 (December, 1957), 494–501.

Brown, Lewis P. "The Jew Is Not a Slacker," *North American Review*, CCVII (June, 1918), 857–62.

Browne, Lewis. "What Can the Jews Do?" *Virginia Quarterly Review*, 15 (Spring, 1939), 216–26.

Browne, Lewis. "Why Are Jews Like That?" *American Magazine*, 107 (January, 1929), 7–9, 104–06.

Burnham, Kenneth E., John F. Connors III, and Richard C. Leonard. "Religious Affiliation, Church Attendance, Religious Education, and Student Attitudes Toward Race," *Sociological Analysis*, 30 (Winter, 1969), 235–44.

Burt, Struthers. "Why Hate the Jews?" *The Forum*, 101 (May and June, 1939), 243–47, 291–95.

Cahnman, Werner J. "Theories of Anti-Semitism," *Chicago Jewish Forum*, 8 (Fall, 1949), 50–53.

Capeci, Dominic J. Jr. "Black-Jewish Relations in Wartime Detroit: The Marsh, Loving, Wolf Surveys and the Race Riot of 1943," JSS, 47 (Summer–Fall, 1985), 221–42.

Carson, Claybourne, Jr. "Blacks and Jews in the Civil Rights Movement," in Joseph R. Washington, Jr., ed., *Jews in Black Perspectives: A Dialogue*. Rutherford, N.J.: Fairleigh Dickinson University Press, 1984.

Chanes, Jerome A. "Antisemitism in the United States," *Midstream*, 36 (January, 1990), 26–30.

Childs, Marquis W. "They Still Hate Roosevelt," *New Republic*, 96 (September 14, 1938), 147–49.

"Christians Condemn Bombings in South," CC, 75 (April 2, 1958), 397–98.

Chworowsky, Karl M.C. "Are Jews Like That?," *Christian Register*, 124 (May, 1945), 191–92.

Clark, Kenneth B. "Candor About Negro-Jewish Relations," Leonard Dinnerstein, ed., *Antisemitism in the United States*. New York: Holt, Rinehart and Winston, 1971, 116–24.

Clinchy, Everett R. "Anti-Semitism Must Be Uprooted!" CC, 55 (November 23, 1938), 1461–62.

Clinchy, Everett R. "The Borderland of Prejudice," CC, 47 (May 14, 1930), 623–27.

Cohen, Israel. "The Reign of Anti-Semitism," *New Statesman and Nation*, 4 (July 23, 1932), 96–97.

Cohen, Kitty O. "Black-Jewish Relations in 1984: A Survey of Black U.S. Congressmen," *Patterns of Prejudice*, 19 (April, 1985), 3–18.

Cohen, Naomi W. "Antisemitic Imagery: The Nineteenth Century Background," JSS, 47 (Summer/Fall 1985), 307–11.

Cohen, Naomi W. "Antisemitism in the Gilded Age: The Jewish View," JSS, 41 (Summer/Fall, 1979), 187–201.

Cohen, Naomi W. "Pioneers of American Jewish Defense," *AJA*, 29 (November, 1977), 116–150.

Cohen, Samuel I. "Jewish-American Defense Agencies: A Study," *Chicago Jewish Forum*, 20 (Fall, 1961), 2–8.

Cohen, Steven M. "Undue Stress on American Anti-Semitism?" *Sh'ma*, September 1, 1989, 113–15.

Cohn, David L. "What Can the Jew Do?" *Saturday Review of Literature*, 28 (January 27, 1945), 7–9.

Cooper, Charles I. "The Jews of Minneapolis and Their Christian Neighbors," JSS, 8 (1946), 31–38.

Cooper, Eunice and Marie Jahoda. "Evasion of Propaganda: How Prejudiced People Respond to Anti-Prejudice Propaganda," *Journal of Psychology*, 23 (1947), 15–25.

Cox, Oliver C. "Race Prejudice and Intolerance—A Distinction," *Social Forces*, 24 (December, 1945), 216–19.

Cray, Ed. "The Rabbi Trickster," *Journal of American Folklore*, 77 (October, 1964), 331–45.

Danby, Herbert. "Have Jews Always Hated Christians?" CC, 44 (May 12, 1927), 588–90.

Davidowicz, Lucy S. "Can Anti-Semitism Be Measured?" *Commentary*, 50 (July, 1970), 36–43.

DeGlequier, A. "The Antisemitic Movement in Europe," *Catholic World*, 48 (March, 1889), 741–51.

Dobkowski, Michael N. "American Anti-Semitism: A Reinterpretation," *American Quarterly*, 29 (Summer, 1977), 166–81.

Dobkowski, Michael N. "American Antisemitism and American Historians: A Critique," *Patterns of Prejudice*, 14 (April, 1980), 33–43.

Drachman, Bernard. "Anti-Jewish Prejudice in America," *The Forum*, 52 (July, 1914), 31–40.

Drescher, Seymour, "The Role of Jews in the Transatlantic Slave Trade," *Immigrants and Minorities* (London), 12 (July 1993), 113–25.

Duker, Abraham G. "Negroes Versus Jews I: Anti-Semitism Is Asserted," *Patterns of Prejudice,* (London) 3, No. 2 (March-April, 1969), 9–12.

Duker, Abraham G. "On Negro-Jewish Relations—A Contribution to a Discussion," JSS, 27 (January, 1965), 18–29.

Duker, Abraham G. "Twentieth-Century Blood Libels in the United States," in Leo Landman, ed., *Rabbi Joseph H. Lookstein Memorial Volume.* New York: KTAV Publishing House, Inc., 1980, pp. 85–109.

Dundes, Alan. "Study of Ethnic Slurs: The Jew and the Polack in the United States," *Journal of American Folklore,* 84 (April, 1971), 186–203.

Dundes, Alan and Thomas Hauschild. "Auschwitz Jokes," *Western Folklore,* 42 (October, 1983), 249–60.

Dwork, Deborah. "Health Conditions of Immgirant Jews on the Lower East Side of New York," *Medical History,* 25 (1981), 1–40.

Ehrenberg, Sigrid. "Anti-Semitism: The 'Invisible' Form of Bigotry," *Sojourner,* 10 (November, 1982), 1–4.

Einstein, Albert. "Why Do They Hate the Jews?" *Collier's,* 102 (November 26, 1938), 9–10, 38.

Eitches, Edward. "Maryland's 'Jew Bill'," AJHQ, 60 (March, 1971), 258–79.

Ellman, Yisrael. "Intermarriage in the United States: A Comparative Study of Jews and Other Ethnic and Religious Groups," JSS, 49 (Winter, 1987), 1–26.

Engle, Paul. " 'Those Damn Jews . . . ,' " *American Heritage,* 30 (December, 1978), 72–79.

Erskine, Hazel Gaudet. "The Polls: Religious Prejudice, Part 2: Anti-Semitism," *Public Opinion Quarterly,* 29 (Winter, 1965–66), 649–64.

Evans, Carl D. "The Church's False Witness Against Jews," CC, 99 (May 5, 1982), 530–33.

Fein, Isaac M., "Niles' Weekly Register on the Jews," PAJHS, L (September, 1960), 3–22.

Fein, Leonard "Jesse Jackson and the Jews," *Moment,* 9 (April, 1984), 13–14.

Everett, Robert. Andrew. "Judaism in Nineteenth-Century American Transcendentalist and Liberal Protestant Thought." *Journal of Ecumenical Studies,* 20 (Summer, 1983) 396–414.

Feingold, Henry L. "Finding a Conceptual Framework for the Study of American Antisemitism." JSS, 47 (Summer/Fall, 1985), 313–26.

Ferretti, Fred. "New York's Black Anti-Semitism Scare," *Columbia Journalism Review,* 8 (Fall, 1969), 18–29.

Feuer, Lewis S. "America's First Jewish Professor: James Joseph Sylvester at the University of Virginia," *AJA,* 36 (November 1984), 152–201.

Feuer, Lewis S. "The Stages of Social History of Jewish Professors in American Colleges and Universities," AJH, 71 (June, 1982), 432–65.

Fineberg, Solomon Andhil. "Can Anti-Semitism Be Outlawed?" *Contemporary Jewish Record,* VI (December, 1943), 619–31.

Fink, Reuben. "Visas, Immigration, and Official Anti-Semitism," *The Nation,* 112 (June 22, 1921), 870–72.

Fishman, Walda Katz and Richard L. Zweigenhaft. "Jews and the New Orleans Economic and Social Elites," JSS, 44 (Summer–Fall, 1982), 291–98.

Fleck, G. Peter. "Jesus in the Post-Holocaust Jewish-Christian Dialogue," CC, 100 (October 12, 1983), 904–06.

Forster, Arnold and Benjamin R. Epstein. "The New Antisemitism," *The National Jewish Monthly*, 88 (April, 1974) 26–29.

Frank, Jerome. "Red, White, and Blue," *Saturday Evening Post*, 214 (December 6, 1941), 9–11, 83–86.

Freedman, Morris. "The Jewish College Student: 1951 Model," *Commentary*, 12 (October, 1951), 305–13.

Freedman, Morris. "The Knickerbocker Case," *Commentary*, 8 (August, 1949), 118–28.

Frenkel-Brunswick, Else & R. Nevitt Sanford. "Some Personality Factors in Anti-Semitism," *Journal of Psychology*, 20 (1945), 271–91.

Friedman, Murray. "Black Anti-Semitism on the Rise," *Commentary*, 68 (October, 1979), 31–35.

Friedman, Murray. "One Episode in Southern Jewry's Response to Desegregation: A Historical Memoir," *AJA*, 33 (November, 1981), 170–83.

Gannett, Lewis S. "Is America Anti-Semitic?" *The Nation*, 116 (March 21, 1923), 330–32.

Gans, Herbert J. "Negro-Jewish Conflict in New York City," *Midstream*, 15 (March, 1969), 3–15.

Gans, Herbert J. "The Future of American Jewry," *Commentary*, 21 (June, 1956), 555–63.

Gerber, David A. "Cutting Out Shylock: Elite Anti-Semitism and the Quest for Moral Order in the Mid-Nineteenth Century American Market Place," *Journal of American History*, 69 (December, 1982), 615–37.

Gershman, Carl. "The Andrew Young Affair," *Commentary*, 68 (November, 1979), 25–33.

Gettelsohn, Roland B. "Negro-Jewish Relations in America: A Symposium," *Midstream*, 12 (December, 1966), 29–34.

Ginsberg, M. "Anti-Semitism," *Sociological Review*, 35 (January–April, 1943), 1–11.

Gittelson, Natalie. "Anti-Semitism: It's Still Around," *Harper's Bazaar*, 105 (February, 1972), 76–77, 134.

Glazer, Nathan. "Negroes and Jews: The New Challenge to Pluralism," *Commentary*, 38 (December, 1964), 29–34.

Goldberg, J. J. "Scaring the Jews," *The New Republic*, 208 (May 17, 1993), 22, 24.

Golden, Harry L. "A Pulpit in the South," *Commentary*, 16 (December, 1953), 574–79.

Golden, Harry L. "A Son of the South, and Some Daughters," *Commentary*, 12 (November, 1951), 379–81.

Golden, Harry L. "Jew and Gentile in the New South," *Commentary*, 20 (November, 1955), 403–12.

Golden, Harry L. "Jews in the South," *Congress Weekly*, 18 (December 31, 1951), 7–11.

Golden, Harry L. "The Jewish People of North Carolina," *North Caroilina Historical Review*, 32 (1955) 194–216.

Gordis, Robert. "Negroes are Antisemitic Because They Want a Scapegoat," in Leonard Dinnerstein, ed., *Antisemitism in the United States*. New York: Holt, Rinehart and Winston, 1971.

Gottheil, Gustav. "The Position of the Jews in America," *North American Review*, 126 (1878), 293–308.

Greenberg, Leonard A. "Some American Anti-Semitic Publications of the Late 19th Century," PAJHS, 37 (1947), 421–25.

Greenberg, Leonard A. and Harold J. Jonas. "An American Anti-Semite in the Nineteenth Century," in Joseph L. Blau, ed., *Essays on American Jewish Life and Thought*. New York: Columbia University Press, 1959, pp. 265–83.

Greenberg, Michael and Seymour Zenchelsky. "Private Bias and Public Responsibility: Anti-Semitism at Rutgers in the 1920s and 1930s," *History of Education Quarterly*, 33 (Fall, 1993), 295–319.

Gresham, Foster B. "The Jew's Daughter: An Example of Ballad Variation," *Journal of American Folklore*, 47 (1934), 358–61.

Hadden, Jeffrey K. "Churchly Particularism and the Jews," CC, 83 (August 10, 1966), 987–88.

Handlin, Oscar. "The American Jewish Committee," *Commentary*, 23 (January, 1957), 1–10.

Handy, Robert T. "The Protestant Quest for a Christian America, 1830–1930," *Church History*, 22 (March, 1953), 8–20.

Harlan, Louis R. "Booker T. Washington's Discovery of Jews," in J. Morgan Kousser and James M. McPherson, eds., *Region, Race, and Reconstruction*. New York: Oxford University Press, 1982.

Harper, Jimmy. "Alabama Baptists and the Rise of Hitler and Fascim, 1930–1938," *Journal of Reform Judaism*, 32 (Spring, 1985), 1–11.

"Have the Jews a Jewish Citizenship?" CC, 62 (January 3, 1945), 4–5.

Heller, Celia Stopnicka and Alphonso Pinkney. "The Attitudes of Negroes Toward Jews," *Social Forces*, 43 (March, 1965), 364–69.

Hellwig, David J. "Black Images of Jews: From Reconstruction to Depression," *Societas*, VII (Summer, 1978), 205–23.

Hellwig, David J. "Black Leaders and United States Immigration Policy, 1917–1929," *Journal of Negro History*, 66 (Summer, 1981), 110–27.

Hendrick, Burton J. "The Great Jewish Invasion," *McClure's Magazine*, 28 (January, 1907), 307–21.

Hendrick, Burton J. "Jewish Invasion of America," *McClure's Magazine*, 40 (March, 1913), 125–65.

Hendrick, Burton J. "The Jews in America," *World's Work*, 45 (1922–23), 266–86, 366–77, 594–601.

Herberg, Will. "Anti-Semitism Today," *Commonweal*, 60 (July 16, 1954), 359–63.

Heschel, Susannah. "Anti-Judaism in Christian Feminist Theology," *Tikkun*, 5 (May–June, 1990), 25–28, 95–97.

Higginson, Thomas Wentworth. "Negro Spirituals," *Atlantic Monthly*, 19 (1867), 685–95.

High, Stanley. "Star-Spangled Fascists," *Saturday Evening Post*, 211 (May 27, 1939), 5–7, 70–73.

Higham, John. "Anti-Semitism in the Gilded Age: A Reinterpretation," *Mississippi Valley Historical Review*, 43 (March, 1957), 559–78.

Higham, John. "Social Discrimination Against Jews in America, 1830–1930," PAJHS, 47 (September, 1957), 1–33.

Hoge, Dean R. and Jackson W. Carroll, "Christian Beliefs, Nonreligious Factors, and Anti-Semitism," *Social Forces*, 53 (June, 1975), 581–94.

Holmes, William F. "Whitecapping: Anti-Semitism in the Populist Era," AJHQ, 63 (March, 1974), 244–61.

Hook, Sidney, "Anti-Semitism in the Academy: Some Pages of the Past," *Midstream*, 25 (January, 1979), 49–54.

Hook, Sidney. "Reflections on the Jewish Question," *Partisan Review*, 16 (May, 1949), 463–82.

Horowitz, Craig, "The New Anti-Semitism," *New York*, 26 (January 11, 1993), 21–27.

Hourwich, Isaac A. "Is There Anti-Semitism in America?" AH, 93 (October 17, 1913), 683–84.

Houseman, Gerald L. "Antisemitism in City Politics: The Separation Clause and the Indianapolis Nativity Scene Controversy, 1976–1977," JSS, 42 (Winter, 1980), 21–36.

Huie, William Bradford. "The Untold Story of the Mississippi Murders," *Saturday Evening Post*, 237 (September 5, 1964), 11–15.

Hurvitz, Nathan. "Blacks and Jews in American Folklore," *Western Folklore*, 33 (October, 1974), 301–25.

Hurvitz, Nathan. "Jews and Jewishness In the Street Rhymes of American Children," JSS, 16 (1954), 135–50.

"I Married a Gentile," *Atlantic*, 163 (March, 1939), 321–26.

"I Married a Jew," *Atlantic*, 163 (January, 1939), 38–46.

"I Was a Jew," *Forum*, 103 (March, 1940), 8–11.

"Intolerance in California," *The Occident and American Jewish Advocate*, 13 (1855), 123–32.

Irwin, Theodore. "Inside the 'Christian Front,'" *Forum*, 103 (March, 1940), 102–8.

Jacobs, Samuel J. "The Blood Libel Case at Massena—A Reminiscence and a Review," *Judaism*, 28 (1979), 465–74.

"Jews and College Life," *Harper's Weekly*, LXXII (January 15, 1916), 53–55.

"Jews and Democracy," CC, 54 (June 9, 1937), 734–36.

Johnson, Alvin. "The Jewish Problem in America," *Social Research*, 14 (December, 1947), 399–412.

Johnson, Alvin. "The Rising Tide of Anti-Semitism," *Survey Graphic*, 28 (February, 1939), 113–16.

Jones, Nathaniel R. "The Future of Black-Jewish Relations," *Crisis*, 82 (January, 1975), 24–27.

"Judaism and Christian Education: The Strober Report," CC, 86 (November 5, 1969), 1410.

Kallen, Horace. "The Roots of Anti-Semitism," *The Nation*, 116 (February 28, 1923), 241.

Kanof, Abram. "Uriah Phillips Levy: The Story of a Pugnacious Commodore," PAJHS, 39 (September 1949), 1–66.

Kernan, William C. "Coughlin, the Jews, and Communism," *The Nation*, 147 (December 17, 1938), 655–58.

King, Dennis. "The Farrakhan Phenomenon: Ideology, Support, Potential," *Patterns of Prejudice*, 20 (January, 1986), 11–22.

Kingdom, Frank. "Discrimination in Medical Colleges," *The American Mercury*, 61 (October, 1945), 391–99.

Korff, S.A. "The Great Jewish Conspiracy," *Outlook*, CXXVII (February 2, 1921), 180–82.

Korn, Bertram W. "The Jews of the Confederacy," *AJA*, 13 (April, 1961), 3–90.

Krause, P. Allen. "Rabbis and Negro Rights in the South, 1954–1967," *AJA*, 21 (April, 1969), 20–47.

Lamb, Blaine. "Jews in Early Phoenix, 1870–1920," *Journal of Arizona History*, 18 (1977), 299–318.

Laser, Louise. "The Only Jewish Family in Town," *Commentary*, 28 (December, 1959), 489–96.

Lasker, Bruno. "Jewish Handicaps in the Employment Market," *Jewish Social Service Quarterly*, II (March, 1926), 170–77.

Laski, Harold J. "A Note on Anti-Semitism," *New Statesman and Nation*, 25 (February 13, 1943), 107–08.

Laufer, Leo. "Anti-Semitism Among Negroes," *The Reconstructionist*, 14 (October 29, 1948), 10–17.

Lebeau, James. "Profile of a Southern Jewish Community: Waycross, Georgia," AJHQ, 58 (June, 1969), 429–42.

Lefkovitz, Maurice. "Minneapolis Jewry—An Appraisal," *American Jewish World*, 12 (September 7, 1923), 21, 71.

Leo, John. "Black Anti-Semitism," *Commonweal*, 89 (February 14, 1969), 618–20.

Levine, Naomi. "The Nazis of Skokie," *Patterns of Prejudice*, 12 (January–February, 1978), 19–20.

Levinsohn, Florence Hamlish. "The Cokely Question," *Chicago*, 37 (August, 1988), 112–15, 138–40.

Levitan, Albert. "Leave the Jewish Problem Alone!" CC, 51 (April 25, 1934), 555–57.

Lewin, Kurt. "Bringing Up the Child," *The Menorah Journal*, 28 (Winter, 1940), 29–45.

Lewin, Kurt. "Self-Hatred Among Jews," *Contemporary Jewish Record*, IV (June, 1941), 219–32.

Lewis, David Levering. "Henry Ford's Anti-Semitism and its Repercussions," *Michigan Jewish History*, 24 (1984), 3–10.

Lindeman, E.C. "Sapiro, The Spectacular," *The New Republic*, 50 (April 13, 1927), 216–18.

Lippmann, Walter. "Public Opinion and the American Jew," AH, 110 (April 14, 1922), 575.

Luce, W. Ray. "The Cohen Brothers of Baltimore: From Lotteries to Banking," *Maryland Historical Magazine*, 68 (Fall, 1973), 288–308.

Maranell, Gary M. "An Examination of Some Religious and Political Attitude Corrolates of Bigotry," *Social Forces*, 45 (March, 1967), 356–62.

Maritain, Jacques. "On Anti-Semitism," *Commonweal*, 36 (September 25, 1942), 534–37.

Mazrui, A. "Negritude, the Talmudic Tradition, and the Intellectual Performance of Blacks and Jews," *Ethnic and Racial Studies*, 1 (January, 1978), 19–36.

McDonald, Rufus Cyrene. "The Jews of the North End of Boston," *Unitarian Review*, 35 (May, 1891), 362–69.

McDonogh, Edward C. "Status Level of American Jews," *Sociology and Social Research*, 32 (July, 1948), 944–53.

McWilliams, Carey. "Minneapolis: The Curious Twin," *Common Ground,* 7 (Autumn, 1946), 61–65.

Meisler, Stanley. "The Southern Segregationist and his Anti-Semitism," *Chicago Jewish Forum,* 16 (Spring, 1958), 171–73.

"Michigan: Grosse Pointe's Gross Points," *Time,* 75 (April 25, 1960), 25.

Miller, Kelly. "The Harvest of Race Prejudice," *Survey Graphic,* 53 (March 1, 1925), 682–83, 711–12.

Modras, Ronald. "Father Coughlin and Anti-Semitism: Fifty Years Later," *Journal of Church and State,* 31 (Spring, 1989), 231–47.

Montagu, M.F. Ashley. "Anti-Semitism in the Academic World," *Chicago Jewish Forum,* 4 (Summer, 1946), 219–24.

Montagu, M. F. Ashley. "Some Psychological Factors in Race Prejudice," *The Journal of Social Psychology,* 30 (November, 1949), 175–88.

Morais, Nina. "Jewish Ostracism in America," *North American Review,* 133 (September, 1881), 256–75.

Morgan, David T. "Judaism in Eighteenth-Century Georgia," *Georgia Historical Quarterly,* 58 (Spring, 1974), 41–55.

Mosely, Charlton. "Latent Klanism in Georgia, 1890–1915," *Georgia Historical Quarterly,* 56 (Fall, 1972), 365–87.

Mostov, Stephen G. "Dun and Bradstreet Reports as a Source of Jewish Economic History: Cincinnati, 1840–1875," *AJH,* 72 (March, 1983), 333–53.

Mostow, Stephen G. "A Sociological Portrait of German Jewish Immigrants in Boston: 1815–1861," *AJS Review,* 3 (1978), 121–52.

Murphy, Paul L. "Sources and Nature of Intolerance in the 1920s," *Journal of American History,* 51 (June, 1964), 60–76.

Nock, Albert Jay. "The Jewish Problem in America," *Atlantic Monthly,* 167 (June, 1941), 699–706.

Nock, Albert Jay. "The Jewish Problem in America, II," *Atlantic Monthly,* 168 (July, 1941), 68–76

O'Gara, James. "Christian Anti-Semitism," *Commonweal,* 80 (May 22, 1964), 252.

Ogles, Robert M. and Herbert H. Howard. "Father Coughlin in the Periodical Press, 1931–1942," *Journalism Quarterly,* 61 (Summer, 1984), 280–86.

O'Grady, Joseph P. "Politics and Diplomacy: The Appointment of Anthony M. Keiley to Rome in 1885," *Virginia Magazine of History and Biography,* 76 (April, 1968), 191–209.

Orlansky, Harold. "A Note on Anti-Semitism Among Negroes," *Politics,* 2 (August, 1945), 250–52.

Peck, Abraham J. "That Other 'Peculiar Institution': Jews and Judaism in the Nineteenth Century South," *Modern Judaism,* 7 (February, 1987), 99–114.

Peskin, Allan. "The Origins of Southern Anti-Semitism," *The Chicago Jewish Forum,* 14 (Winter, 1955–1956), 83–88.

Pickard, Kate E.R. "The Kidnapped and the Ransomed," *AJA,* 9 (April, 1957), 3–31.

Pitt, Leonard. "The Beginnings of Nativism in California," *Pacific Historical Review,* 30 (February, 1961), 23–38.

Podhoretz, Norman. "My Negro Problem—And Ours," *Commentary* (February, 1963), 376–87.

Podhoretz, Norman. "Is It Good for the Jews?" *Commentary,* 53 (February, 1972), 7–14.

Pogrebin, Letty Cottin. "Anti-Semitism in the Women's Movement," *MS*, 10 (June, 1982), 45–48, 62, 64–65, 69–70, 73–74.

Polos, Nicholos C. "Black Anti-Semitism in Twentieth-Century America: Historical Myth or Reality?" *AJA*, 27 (1975), 8–31.

Porter, Jack Nusan. "John Henry and Mr. Goldberg: The Relationship Between Blacks and Jews," *Journal of Ethnic Studies*, 7 (Fall, 1979), 73–86.

Poussaint, Alvin F. "Blacks and Jews: An Appeal for Unity," *Ebony*, 39 (July, 1974), 120–22, 124–26, 128.

"Prejudice Subsides," CC, 67 (April 19, 1950), 483.

Prell, Riv-Ellen. "Rage and Representation: Jewish Gender Stereotypes in American Culture," in Faye Ginsburg and Anna Lowenhaupt Tsing, eds. *Uncertain TERMS: Negotiating Gender in American Culture.* Boston: Beacon Press, 1990.

"Present State of the Jews," *North American Review*, (October, 1856), 352–81.

"Protestantism and Tolerance," CC, 62 (February 14, 1945), 198–200.

Prothro, E. Terry and John A. Jensen. "Group Differences in Ethnic Attitudes of Louisiana College Students," *Sociology and Social Research*, 34 (March, 1950), 252–58.

Raab, Earl. "Is there a New Anti-Semitism?" *Commentary*, 57 (May, 1974), 53–55.

Raab, Earl "The Black Revolution & the Jewish Question," *Commentary*, 47 (January, 1969), 23–33.

Raab, Earl. "Anti-Semitism in the 1980s," *Midstream*, 29 (February, 1983), 11–18.

Raab, Earl. "In Promised Dixieland," *Commentary*, 5 (May, 1948), 460–64.

"Race Prejudice Against Jews," *Independent*, LXV (December 17, 1908), 1451–56.

"Racism at Home," *The Colored Harvest* (Baltimore), October–November, 1938, p. 2.

Rand, Jerry, "The Ordeal of Uriah Levy," *American Mercury*, 56 (June 1943), 728–34.

Ravage, Marcus Eli. "A Real Case Against the Jews," *Century*, 115 (January, 1928), 346–50.

"Rector Resigns Under Pressure," CC, 80 (February 6, 1963), 165–66.

Reddick, L. D. "Anti-Semitism Among Negroes," *The Negro Quarterly*, (Summer, 1942), 112–22.

Reed, John Shelton. "Ethnicity in the South: Some Observations on the Acculturation of Southern Jews," *Ethnicity*, 6 (1979), 97–106.

Reid, Sydney, "Because You're a Jew," *Independent*, LXV (November 26, 1908), 1212–17.

"Religious Liberty," *Niles' Weekly Register*, 16 (March 29, 1819), 226–33.

Rhine, Alice Hyneman. "Race Prejudice at Summer Resorts," *Forum*, III (1887), 523–31.

Ribufo, Leo P. "Henry Ford and the International Jew," AJH, 69 (June, 1980), 437–77.

Rivkin, Ellis. "A Decisive Pattern in American Jewish History," *Essays in American Jewish History*. Cincinnati: American Jewish Archives, 1958, pp. 23–62.

Robinson, Duane and Sylvia Rohde. "Two Experiments with an Anti-Semitism Poll," *Journal of Abnormal and Social Psychology*, 41 (April, 1946), 136–44.

Rockaway, Robert A. "Anti-Semitism in an American City: Detroit, 1850–1914," AJHQ, 64 (September, 1974), 42–54.

Rockaway, Robert A. "Louis Brandeis on Detroit," *Michigan Jewish History*, 17 (July, 1977), 17–19.

Rockaway, Robert A. "The Detroit Jewish Ghetto Before World War I," *Michigan History*, 52 (Spring, 1968), 28–36.

Rockaway, Robert A. "Ethnic Conflict in an Urban Environment: The German and Russian Jew in Detroit, 1881–1914," AJHQ, 60 (December, 1970), 133–50.

"Roosevelt Censures Annapolis Editor," AH, 111 (June 16, 1922), 162.

Rose, Peter I. "Blacks and Jews: The Strained Alliance," *The Annals of the American Academy of Political and Social Science*, 454 (March, 1981), 55–69.

Rosenfield, Geraldine. "The Polls: Attitudes Toward American Jews," *Public Opinion Quarterly*, 46 (Fall, 1982), 431–43.

Ross, Edward Alsworth. "The Hebrews of Eastern Europe in America," *The Century*, 88 (Septermber, 1914), 785–792.

Ruchames, Louis. "Danger Signals in the South," *Congress Weekly*, 11 (April 21, 1944), 8–91.

Ruchames, Louis, "The Abolitionists and the Jews: Some Further Thoughts," Bertram A. Korn, ed., *A Bicentennial Festschrift for Jacob Rader Marcus*, Waltham, Mass.: American Jewish Historical Society, 1976.

Ryan, John A. "Anti-Semitism in the Air," *Commonweal*, 29 (December 30, 1938), 260–62.

Sager, Gordon. "Swastika over Philadelphia," *Equality*, I (July, 1939), 3–8.

Sarna, Jonathan D. "Anti-Semitism and American History," *Commentary*, 71 (March, 1981), 42–47.

Sarna, Jonathan D. "The American Jewish Response to Nineteenth-Century Christian Missions," *Journal of American History*, 68, (June, 1981), 35–51.

Sarna, Jonathan D. "The Pork on the Fork: A Nineteenth-Century Anti-Jewish Ditty," JSS, 44 (Spring, 1982), 169–72.

Saveth, Edward N. "Discrimination in the Colleges Dies Hard," *Commentary*, 9 (February, 1950), 115–21.

Schappes, Morris U. "Anti-Semtism and Reaction, 1795–1800," PAJHS, 38 (December, 1948), 109–37.

Schmier, Louis. "The First Jews of Valdosta," *Georgia Historical Quarterly*, 62 (Spring, 1978), 32–50.

Schwartz, Henry. "The Uneasy Alliance: Jewish-Anglo Relations in San Diego, 1850–1860," *Journal of San Diego History*, 20 (Summer, 1974), 52–60.

"Sees American Judaism in Rapid Decline," CC, 63 (March 13, 1946), 323–24.

Seldes, George. "Father Coughlin: Anti-Semite," *New Republic*, 96 (November 2, 1938), 353–54.

Seldes, George. "Lindbergh's Nazi Pattern," *New Republic*, 105 (September 22, 1941), 360–61.

Seldes, George and Michael Straight, "The Anti-Semitic Conspiracy," *New Republic*, 105 (September 22, 1941), 362–63.

Seller, Maxine S. "Isaac Leeser: A Jewish-Christian Dialogue in Antebellum Philadelphia," *Pennsylvania History*, 35 (July 1968), 231–42.

Sheerin, John P. C.S.P. "The Myth of Black Anti-Semitism," *The Catholic World*, 209 (May, 1969), 50–51.

Short, K. R. M. "Hollywood Fights Anti-Semitism, 1940–1945," in K. R. M.

Short, ed., *Film and Radio Propaganda in World War II*. Knoxville: University of Tennessee Press, 1983.

Shuster, Georege. "The Conflict Among Catholics," *American Scholar*, 10 (Winter, 1940–1941), 5–16.

Sifry, Micah L. "Anti-Semitism in America," *The Nation*, 256 (January 25, 1993), 92–96.

Simpson, Richard L. "Negro-Jewish Prejudice: Authoritarianism and Some Social Variables as Correlates," *Social Problems*, 7 (1959), 138–46.

"Skokie as Symbol," CC, 95 (April 19, 1978), 411–12.

Slavin, Steven L. and Mary A. Pradt, "Anti-Semitism in Banking," *The Banking Magazine*, 162 (July–August, 1979), 19–21.

Smertenko, Johan J. "Hitlerism Comes to America," *Harper's Magazine*, 167 (November, 1933), 660–70.

Smertenko, Johan J. "The Jew: A Problem For America," *Outlook*, 152 (August 7, 1929), 574–77, 599–600.

Smith, Alson J. "The Christian Terror," CC, 56 (August 23, 1939), 1107–19.

Smith, Goldwyn, D.C.L. "New Light on the Jewish Question," *North American Review*, 153 (August, 1891), 129–53.

Sobel, B.Z. and May L. Sobel. "Negroes and Jews: American Minority Groups in Conflict," *Judaism*, 15 (Winter, 1966), 3–22.

Stark, Rodney and Stephen Steinberg. "Jews and Christians in Suburbia," *Harper's Magazine*, 235 (August, 1962), 73–78.

Stegner, Wallace. "Who Persecutes Boston?" *Atlantic Monthly*, 174 (July, 1944), 45–52.

Stockley, W.F.P. "Popes and Jewish 'Ritual Murders,'" *Catholic World*, 139 (1934), 450–60.

Straight, Michael. "The Anti-Semitic Conspiracy," *The New Republic*, 105 (September 22, 1941), 362–63.

Streiker, Lowell D. "Christian Education and the Jewish People," CC, 84 (February 8, 1967), 168–71.

Strum, "Louis Marshall and Anti-Semitism at Syracuse University," *AJA*, 35 (April, 1983), 1–12.

Supple, Barry E. "A Business Elite: German Jewish Finances in Nineteenth-Century New York," *Business History Review*, 31 (Summer, 1957), 143–78.

Sutterland, John F. "Rabbi Joseph Krauskopf of Philadelphia: The Urban Reformer Returns to the Land," *AJHQ*, 67 (June, 1978), 342–62.

Syrkin, Marie. "Anti-Semitic Drive in Harlem," *Congress Weekly*, 8 (October 31, 1941), 6–8.

Tarshish, Allan. "Jew and Christian in a New Society: Some Aspects of Jewish-Christian Relationships in the United States, 1848–1881," in Bertram Wallace Korn, ed., *A Bicentennial Festschrift for Jacob Rader Marcus*. Waltham, Mass.: American Jewish Historical Society, 1976.

"The Gentile Beast," CC, 84 (March 8, 1967), 299–300.

"The Jew and the Club," *Atlantic Monthly*, CXXXIV (October, 1924), 450–56.

"The Jew and the Negro," *National Baptist Voice* (Nashville, Tenn.), 38 (May 15, 1945), 1.

"The Jewish Law Student and New York Jobs: Discriminatory Effects in Law Firm Hiring Practices," *Yale Law Journal*, 73 (March, 1964), 625–60.

"The Jewish Problem," CC, 51 (Febraury 28, 1934), 279–81.

"The Jews and the Underwriters," *Banking and Insurance Chronicle*, 2 (May, 1867), 138.

"The Nazis Are Here," *The Nation*, 148 (March 4, 1939), 253.

"The Position of the Jews in America." *North American Review*, 126 (1878), 293–308.

"The Tragedy Is the Romance," CC, 50 (July 19, 1933), 924–25.

"The WBAI Incident," *Columbia Journalism Review*, 8 (Fall, 1969), 28–29.

Toby, Jackson. "Bombing in Nashville," *Commentary*, 25 (May, 1958), 385–89.

"Tolerance Is Not Enough!" CC, 53 (July 1, 1936), 926–28.

"Too Many Jews in Government?" CC, 55 (December 28, 1938), 1614–15.

Travis, James, "The Secret of Antisemitism," *Catholic World*, 156 (January, 1943), 420–26.

Trilling, Diana. "Lionell Trilling, a Jew at Columbia," *Commentary*, 67 (March, 1979), 40–46.

Tsukashima, Ronald Tadao and Darrel Montero. "The Contact Hypothesis: Social and Economic Contact and Generational Changes in the Study of Black Anti-Semitism," *Social Forces*, 55 (September, 1976), 149–65.

Tumin, Melvin. "Anti-Semitism and Status Anxiety: A Hypothesis," JSS, 33 (October, 1971), 307–16.

Twain, Mark. "Concerning the Jews," *Harper's Magazine*, 99 (1899), 527–35.

Vance, Zebulon. "The Scattered Nation," in *Modern Eloquence*, edited by Thomas B. Reed et al., 17 vols. Philadelphia: John D. Morris & Co., 1900. VI, 1115–42.

Wall, James M. "The Virulent Disease of Anti-Semitism," CC, 91 (April 24, 1974), 443–44.

Ward, Charles D. "Anti-Semitism at College: Changes Since Vatican II," *Journal for the Scientific Study of Religion*, 12 (March, 1973), 85–88.

"Washington Notes," *New Republic*, 77 (January 10, 1934), 250–51.

Weber, Laura E. " 'Gentiles Preferred': Minneapolis Jews and Employment, 1920–1950," *Minnesota History*, 52 (Spring, 1991), 166–82.

Wechsler, Israel S. "The Psychology of Anti-Semitism," *The Menorah Journal*, 11 (April, 1925), 159–66.

Wechsler, James. "The Coughlin Terror," *The Nation*, 149 (July 22, 1939), 92–97.

Weinberg, Joseph P. "Black-Jewish Tensions: Their Genesis," *CCAR Journal*, 21 (Spring, 1974), 31–37.

Weisbord, Robert G. and Arthur Stein, "Negro Perceptions of Jews Between the World Wars," *Judaism*, 18 (1969), 428–47.

Weisman, Seymour S. "Black-Jewish Relations in the U.S.—II," *Patterns of Prejudice*, 15 (January, 1981), 45–52.

West, Cornel. "Black Anti-Semitism and the Rhetoric of Resentment," *Tikkun*, 7 (January–February, 1992), 15–16.

"When Blacks and Jews Fall Out," *The Economist*, 292 (July 7, 1984), 19–20.

"When Christians Taught Jews to Hate," CC, 44 (May 26, 1927), 650–52.

"When Minor Is Major," CC, 91 (December 18, 1974), 1187–88.

"Why Is Anti-Semitism?" CC, 54 (July 7, 1937), 862–64.

Whyte, William Foote. "Race Conflicts in the North End of Boston," *The New England Quarterly*, 12 (December, 1939), 623–42.

Williams, Beverly S. "Anti-Semitism and Shreveport, Louisiana: The Situation in the 1920s," *Louisiana History*, 21 (Fall, 1980), 387–98.

Williams, Oscar R. Jr. "Historical Impressions of Black-Jewish Relations Prior to World War II," *Negro History Bulletin*, 40 (July–August, 1977), 728–31.

X, "The Jew and the Club," *Atlantic*, 134 (October, 1924), 450–56.

Yale, William. "The Non-Assimilation of Israel," *Atlantic Monthly*, 130 (1922), 276–78.

Unpublished Essays, Theses, and Dissertations

Adland, Jon. "American Jewish Reaction to the Damascus Affair, 1840." Essay, Hebrew Union College, 1978, Small collection, AJA.

Badger, Chuck. "The Response of Christian Fundamentalists to the Holocaust, 1933–1945." Seminar paper, Department of History, University of Arizona, December, 1987.

Bloom, Steven. "Interactions Between Blacks and Jews in New York City, 1900–1930, as Reflected in the Black Press." Ph.D. dissertation, History Department, New York University, 1973.

Boxerman, Burton Alan. "Reaction of the St. Louis Jewish Community to Anti-Semitism, 1933–1945." Ph.D. dissertation, St. Louis University, 1967.

Brackman, Harold David. "The Ebb and Flow of Conflict: A History of Black-Jewish Relations Through 1900." Ph.D. dissertation, Department of History, University of California at Los Angeles, 1977.

Chamblee, Jr., Roy Zebulon. "The Sabbath Crusade: 1810–1920." Ph.D. dissertation, George Washington University, 1968. Ann Arbor, Mich.: University Microfilms, Inc., 1968.

Clark, Jack Freshman. "Beyond Pontius Pilate and Judge Lynch: The Pardoning Power in Theory and Practice as Illustrated in the Leo Frank Case." Senior thesis, Departments of History and Government, Harvard College, March, 1986.

Cooper, Elizabeth K. "Attitudes of Children and Teachers Toward Mexican, Negro, and Jewish Minorities." M.A. thesis in Education, University California at Los Angeles, 1945.

Davidowicz, Lucy S. "Louis Marshall's Yiddish Newspaper, *The Jewish World*." M.A. thesis, Department of History, Columbia University, 1961.

Flower, Edward. "Anti-Semitism in the Free Silver and Populist Movements and the Election of 1896." M.A. thesis, Columbia University, 1952.

Frommer, Morris. "The American Jewish Congress: A History, 1914–1950." Ph.D. dissertation, History Department, Ohio State University, 1978.

Halpern, Rose A. "American Reaction to the Dreyfus Case." M.A. thesis, Department of History, Columbia University, 1941.

Kaplan, Helga Eugenie. "Century of Adjustment: A History of the Akron Jewish Community." Ph.D. dissertation, History Department, Kent State University, 1978.

Kellenbach, Katharina von. "Anti-Judaism in Christian-Rooted Feminist Writings: An Analysis of Major U. S. American and West German Feminist Theologians." Ph.D. dissertation, Department of Philosophy, Temple University, 1990.

Lamb, Blaine Peterson. "Jewish Pioneers in Arizona, 1850–1920." Ph.D. dissertation, Department of History, Arizona State University, 1982.

Lipson-Walker, Carolyn. "Shalom Y'All: The Folklore and Culture of Southern

Jews." Ph.D. dissertation, Department of Folklore in the Program of American Studies, Indiana University, 1986.

Maas, Elaine H. "The Jews of Houston: An Ethnographic Study." Ph.D. dissertation, Sociology Department, Rice University, 1973.

McCarthy, Edward C. "The Christian Front Movement in New York City, 1938–1940." M.A. thesis, Columbia, 1965.

Meltzer, Lois J. "Anti-Semitism in the United States Army During World War II." M.A. thesis, Baltimore Hebrew College, 1977.

Mostov, Stephen G., "A Jerusalem on the Ohio: The Social and Economic History of Cincinnati's Jewish Community, 1840–1875." Ph.D. dissertation, Department of Near Eastern and Judaic Studies, Brandeis University, 1981.

Neuringer, Sheldon Morris. "American Jewry and United States Immigration Policy, 1881–1953." Ph.D. dissertation, Department of History, University of Wisconsin, 1969.

Pettibone, Dennis Lynn. "Caesar's Sabbath: The Sunday-Law Controversy in the United States, 1879–1892." Ph.D. dissertation, History Department, University of California at Riverside, 1979.

Price, Isabel Boiko. "Black Responses to Anti-Semitism: Negroes and Jews in New York, 1880 to World War II." Ph.D. dissertation, Department of History, University of New Mexico, 1973.

Rahden, Till van. "Beyond Ambivalence: Variations of Catholic Antisemitism in Turn of the Century Baltimore." M.A. thesis, Department of History, Johns Hopkins University, 1992.

Raphael, Mark Lee. "Intra-Jewish Conflict in the United States, 1869–1915." Ph.D. dissertation, Department of History, University of California at Los Agneles, 1973.

Rapp, Michael Gerald. "An Historical Overview of Anti-Semitism in Minnesota, 1920–1960—With Particular Emphasis on Minneapolis and St. Paul." Ph.D. dissertation, Department of History, University of Minnesota, 1977.

Rappaport, Joseph. "Jewish Immigrants and World War I: A Study of American Yiddish Press Reactions." Ph.D. dissertation, Department of History, Columbia University, 1951.

Reitman, Herbert Samuel. "Defense and Development: A History of Minneapolis Jewry, 1930–1950." Ph.D. dissertation, Department of History, University of Minnesota, 1970.

Sable, Jacob M. "Some American Jewish Organizational Efforts To Combat Anti-Semitism." Ph.D. dissertation, Yeshiva University, 1964.

Scholnick, Myron I. "The New Deal and Anti-Semitism in America." Ph.D. dissertation, Department of History, University of Maryland, 1971.

Spencer, Gary. "JAP-Baiting on a College Campus: An Example of Gender and Ethnic Stereotyping." Paper, Department of Sociology, Syracuse University, October, 1987.

Stafford, Bert Lanier. "The Emergence of Anti-Semitism in the America First Committee, 1940–1941." M.A. thesis, New School, 1948.

Stevenson, Marshall Field, Jr. "Points of Departure, Acts of Resolve: Black-Jewish Relations in Detroit, 1937–1962." Ph.D. dissertation, Department of History, University of Michigan, 1988.

Sutker, Solomon. "The Jews of Atlanta: Their Social Structure and Leadership

Patterns." Ph.D. dissertation, Department of Sociology, University of North Carolina, 1950.

Urquhart, Ronald Albert. "The American Reaction to the Dreyfus Affair: A Study of Anti-Semitism in the 1850s." Ph.D. dissertation, Department of History, Columbia, 1972.

Weingarten, Irving. "The Image of the Jew in the American Periodical Press, 1881–1921." Ph.D. dissertation, School of Education, Health, Nursing and Art Professions, New York University, 1979.

Weitz, Marvin. "Black Attitudes to Jews in the United States From World War II to 1976." Ph.D. dissertation, Yeshiva University, 1977.

Zanden, James Wilfrid Vander. "The Southern White Resistance Movement to Integration." Ph.D. dissertation, Department of Sociology and Anthropology, University of North Carolina, 1957.

Index